METHODS IN
MICROBIOLOGY

METHODS IN
MICROBIOLOGY

Volume 15

Edited by

T. BERGAN

Department of Microbiology, Institute of Pharmacy and
Department of Microbiology,
Aker Hospital, University of Oslo,
Oslo, Norway

1984

ACADEMIC PRESS

(*Harcourt Brace Jovanovich, Publishers*)
London   Orlando   San Diego   San Francisco   New York
Toronto   Montreal   Sydney   Tokyo   São Paulo

ACADEMIC PRESS INC. (LONDON) LTD.
24–28 Oval Road
London NW1 7DX

U.S. Edition published by
ACADEMIC PRESS INC.
(Harcourt Brace Jovanovich, Inc.)
Orlando, Florida 32887

British Library Cataloguing in Publication Data

ISBN   0–12–521515–0
LCCCN 68–57745
ISSN   0580–9517

Filmset in Monophoto Times New Roman by Latimer Trend & Company Ltd, Plymouth
Printed in Great Britain by St Edmundsbury Press,
Bury St Edmunds, Suffolk

CONTRIBUTORS

T. Bergan Department of Microbiology, Institute of Pharmacy, P.O. Box 1108, University of Oslo, Blindern, Oslo 3, Norway

Y. J. Erdman Department of Microbiology, Middlesex Hospital Medical School, London W1, UK

G. Kapperud Institute of Zoology and Department of Microbiology, Institute of Pharmacy, University of Oslo, Oslo, Norway

B. Lányi National Institute of Hygiene, Gyali ut 2–6, 1097 Budapest, Hungary

A. A. Lindberg Karolinska Institute, Department of Clinical Bacteriology, Huddinge University Hospital, 19186 Huddinge, Sweden

L. Le Minor Institut Pasteur, 28 rue du Docteur Roux, 75724 Paris Cedex 15, France

T. L. Pitt Division of Hospital Infection, Central Public Health Laboratory Service, 175 Colindale Avenue, London NW9 5HT, UK

R. Sakazaki Enterobacteriology Laboratories, National Institute of Health, 10–35 Kamiosaki 2-chome, Shinagawa-ku, Tokyo, Japan

M. B. Slade Department of Immunology, Royal Postgraduate Medical School, Hammersmith Hospital, DuCane Road, London W12 0HS, UK. (Present address: AMD Limited, 65 Dickson Avenue, Artarmon, New South Wales 2064, Australia)

K. Sørheim Department of Microbiology, Institute of Pharmacy, University of Oslo, Blindern, Oslo 3, Norway

A. I. Tiffen Department of Microbiology, University of Reading, London Road, Reading RG1 5AQ

PREFACE

This volume is the second in the series of volumes giving detailed descriptions of the serology of the *Enterobacteriaceae*.

Chapter 1 is devoted to the genus *Salmonella*, which represents one of the best examples of extensive and detailed characterization of the antigenic mosaic of a bacterial genus. The extent to which species names have been assigned to serotypes may be challenged, but some of the serologically characterized species names will undoubtedly remain as first described, and serology will have a primary place in the diagnosis of salmonellas.

Chapters 2 and 3 deal with species which have to be differentiated from *Salmonella*. The serology of serratias have yet to be characterized for species other than *Serratia marcescens*, which is the only species described here. Bacteria with relevance both to environmental phytopathology, and by association to other enterobacteria, to clinical bacteriology are included in Chapters 4 and 5. Chapter 6 describes the situation for the genus *Yersinia*. In this group of bacteria, it may be assumed that considerable changes of the serogrouping scheme are possible, since this group of bacteria is presently under considerable taxonomic re-evaluation.

Chapter 7 presents a technique, gas liquid chromatography, which is used in addition to the traditional bacteriological determinant methods and serology.

Oslo   T. Bergan
January 1984

CONTENTS

1

Serology of *Salmonella*

A. A. LINDBERG

Karolinska Institute, Department of Clinical Bacteriology,
Huddinge University Hospital, Huddinge, Sweden

and

L. LE MINOR

Institut Pasteur, 199 INSERM Paris, Cedex, France

I. Introduction

The genus *Salmonella*, a member of the family Enterobacteriaceae (Ørskov and Ørskov, 1978), is composed of phenotypically and genotypically related bacteria. The base composition of their DNA is 50–52% (G + C). *Salmonella* (including *Arizona*) strains are at least 70% related to each other using DNA/DNA hybridization (Crosa *et al.*, 1973). Accordingly, all the strains called either *Salmonella* or *Arizona* belong to one genospecies.*

The genus *Salmonella* has been subdivided by Kauffmann (1966) into four "subgenera": subgenus III corresponds to the taxon formerly called *Arizona* or *Salmonella arizonae*. This subdivision is based on biochemical characteristics (Table I).

Subgenera II and III are related to each other by their biochemical characteristics and by properties of their H-antigens. They are frequently

METHODS IN MICROBIOLOGY
VOLUME 15 ISBN 0–12–521515–0

*See Addendum on p. 141

TABLE I

Salmonella subgenera

Biochemical characteristics	Subgenera			
	I	II	III	IV
ONPG[a]	−	−/		−
Lactose	−	−	+/x	−
Malonate	−	+	+	−
Gelatin	−	+	+	+
KCN[b]	−	−	−	+
Dulcitol	+	+	−	−
D-Tartrate	+	−/x	−/x	−/x
Mucate	+	+	d	−
Salicin	−	−	−	+
Galacturonate	−	+	d[c]	+

x, late and irregularly positive; d, different types.
[a] ONPG, hydrolysis of orthonitrophenyl-β-D-galactopyranoside.
[b] KCN, culture in the presence of KCN.
[c] Diphasic strains are positive and monophasic strains are negative.

isolated from reptiles, and are isolated rarely from man. Strains belonging to the subgenus IV are rare, and isolated from the natural environment.

Subgenera are divided into serotypes, defined by a particular association of O-, H- and occasionally K-antigenic factors. About 2000 serotypes are known at present (WHO, 1980). Some serotypes have binomial Latin names as in a species (e.g. *Salmonella typhi*), whereas others are designated only by their antigenic formula. The reasons are historical. The early described serotypes had an important medical or veterinary role. The Latin names given reflected that role; the agent of human typhoid fever was called *Salmonella typhi*, the agent of abortion in sheep was called *S. abortus-ovis* and the agent of a typhoid-like disease of the mouse named *S. typhimurium*. Sometimes, *Salmonella* serotypes first thought to be associated with an animal species were later found to be ubiquitous, e.g. *Salmonella typhimurium* and *S. bovis-morbificans*. For this reason, new *Salmonella* serotypes were subsequently named after the geographical origin of the first strains isolated, e.g. *S. london*, *S. oslo*, *S. panama* and so on. This tradition is continued with *Salmonella* subgenus I serotypes. New serotypes of subgenera II, III and IV are designated only by their antigenic formula.

The antigenic structure of *Arizona* (i.e. *Salmonella* subgenus III) (mostly isolated from cold-blooded animals) had been studied independently of the antigenic structure of the *Salmonella* subgenus I (mostly isolated from man and warm-blood animals). Thus different symbols were used to designate related or similar O- and H-factors of *Salmonella* and *Arizona*. Now a connection between the *Arizona* formula and the formula of the other *Salmonella* has been established (Rohde, 1979) and it is possible to use the same system of symbols for all serotypes of the genus *Salmonella* (including *Arizona*).

The nomenclature used to describe the serotypes of *Salmonella* might be controversial but the taxonomy of this genus is probably the best known of all microbes.

II. The antigens

It is not practical to identify and study all the antigens that a micro-organism may possess. Using quantitative immunoelectrophoretic methods almost 100 immunoprecipitates were found in a system using a sonicated *S. typhi* antigen preparation and the corresponding rabbit antiserum (Espersen *et al.*, 1980). Only the main types of antigens now used to classify *Salmonella* are considered here.

A. Serological studies

1. *Methodology*

The serological typing of *Salmonella* is based on the bacterial agglutination test. Other serological methods can be used. But since agglutination is easy to perform with living or killed bacteria, and gives a clear and sufficiently rapid reaction without extraction of antigens of the bacterial cell, it is the method most commonly used.

Agglutinations can be done by different techniques: macro-technique in tubes, micro-technique using micro-agglutination combined with a mechanical system to dispense suspensions and antisera (quantitative methods), and slide agglutination (qualitative method) used to disclose the presence of antigenic factors of a bacteriological culture (called serological analysis).

Antisera for typing are obtained by immunizing rabbits. The preparation of bacterial suspensions is described in Section III.

(a) *O-antigen.* The O-antigens are heat-stable and alcohol-stable. In contrast, the H-antigens are heat-labile and denatured by alcohol. Certain factors (i.e. antigen determinants) of O-antigens are fully expressed only in smooth colonies.

Consequently, preparing an O-antigen suspension of motile *Salmonella* requires addition of an equal volume of alcohol to a heavy suspension in saline which maintains agglutinability of O-antigen and destroys the agglutinability of H-antigen. In preparing O-antigen-immune sera, rabbits are immunized either with non-flagellated bacteria, killed by chemicals, e.g. 0.5% formalin, or motile bacteria killed by heating for 2 h at 100°C.

When titrating O-antibodies in a serum, geometric dilutions of the serum are mixed with the suspension of bacteria, and incubated either for 18 h at 50°C, or 2 h at 37°C followed by an incubation for 18 h at room temperature. Reading is done with the naked eye. Using a concave mirror is helpful. The titres obtained with the various methods are similar. O-agglutinates are granular, and difficult to dissociate by shaking the tube as bacteria adhere to one another via bridging antibodies cell to cell. It is necessary when using classical methods to wait 18 h or more before reading because of the reduced probability of collision between the bacteria. The reaction can be enhanced by centrifugation immediately after the bacteria and antiserum have been mixed. Hereby bacteria come in close contact with one another in the pellet, and shaking the tube renders the agglutinates visible. In slide-agglutination, heavy suspensions of bacteria and high titres of antibodies are used. This increases the probability of contact between bacterial cells, and agglutination may be obtained within seconds.

Cross-absorption studies have identified many antigenic determinants, 67 of which are used for serological identification. O-antigen labelled with the same number are closely related, but not always antigenically identical. O-antigen factors can be classified as follows.

1. Major factors that identify the O-antigenic group, e.g. factor O:9 characterizes *Salmonella* O-antigen Group D. Major factors are determined by locus *rfb*, which is near *his* on the chromosomal map (Fig. 5, Mäkelä and Stocker, 1969).
2. Minor factors
   (a) that have less or no discriminating value because they are always associated with another factor, e.g. factor O:12 is always associated with O:2, O:4 or O:9 in Groups A, B and D.
   (b) that result from a chemical modification of major antigens. For example *Salmonella* O-antigen Group B that possesses antigenic factors O:4 and O:12, characteristic of this group, may possess, in addition, the following minor factors:
   — O:5 results from acetylation of abequose present in the repeating units of the polysaccharide responsible for specificity O:4,12.
   — O:1 results from insertion of a glucose residue linked α-1,6 to the galactose in the repeating unit due to conversion by phage P22 (Table II).
   — O:27 results from the conversion by another phage, called 27, which changes the linkage between the repeating units from α-1,2 to α-1,6 (Table II).

TABLE II

Structure of O-antigenic polysaccharide chain in various *Salmonella* serogroups and of various O-antigen immunodeterminants

Serogroup	Structure	Identified immunodominant structures	Reference
A = O:1, 2, 12	D-Glcp 1 α↓ 4/6 Parp 1 α↓ 3 →2 D-Manp 1→4 L-Rhap 1→3 D-Galp 1→ α α α	Parp 1→3 D-Manp = O:2 α D-Glcp 1→4 D-Galp = O:12_2 α D-Glcp 1→6 D-Galp = O:1 α	Hellerqvist *et al.* (1971a)
B = O:1, 4, [5], 12, 27	D-Glcp 1 α↓ 4/6 OAc-2Abep 1 α↓ 3 →2 D-Manp 1→4 L-Rhap 1→3 D-Galp 1→ α α α	Abep 1→3 D-Manp = O:4 α *OAc*-2Abep = O:5 D-Glcp 1→4 D-Galp = O:12_2 α D-Glcp 1→6 D-Galp = O:1 α D-Galp 1→6 D-Manp = O:27 α	Hellerqvist *et al.* (1968, 1969a, c)
C1 = O:6, 7, 14	D-Glcp 1 1↘ 3 →2 D-Manp I 1→2 D-Manp II 1→2 D-Manp III 1→2 D-Manp IV 1→ α/β α/β α/β β 3↖	D-Glcp 1→3 D-GlcpNac 1→ *see text for explanations* α/β β	S. B. Svensson *et al.*, (unpublished data)
C2 = O:6, 8	Abep OAc-2D-Glcp 1 1 α↓ 3 α↓ 3 →4 L-Rhap 1→2 D-Manp 1→2 D-Manp 1→3 D-Galp 1→ β α α ↓2 OAc	Abep 1→3 L-Rhap = O:8 α	Hellerqvist *et al.* (1970a, 1971b, 1972)

TABLE II—*continued*

Serogroup	Structure	Identified immunodominant structures	Reference
C3 = O:8, 20	Abep 1 α↓ 3 OAc-2Glc 1 α↓ 4 →4 L-Rhap 1→2 D-Manp 1→2 D-Manp 1→3 D-Galp 1→ (β)	D-Glcp 1→4 D-Galp=O:20 (α)	Hellerqvist et al. (1970a, 1971b, 1972)
D1 = O:1, 9, 12	Tyvp 1 α↓ 3 D-Glcp 1 α↓ 4/6 →2 D-Manp 1→4 L-Rhap 1→3 D-Galp 1→ (α, α)	Tyvp 1→3 D-Manp=O:9 (α) D-Glcp 1→4 D-Galp=O:12_2 (α) D-Glcp 1→6 D-Galp=O:1 (α)	Hellerqvist et al. (1969b, 1970b, 1971f)
D2 = O:9, 46	Tyvp 1 α↓ 3 D-Glcp 1 α↓ 4 →6 D-Manp 1→4 L-Rhap 1→3 D-Galp 1→ (β, α) OAc |6	Tyvp 1→3 D-Manp=O:9 (α) D-Galp 1→6 D-Manp 1→4 L-Rhap = O:46 (α, β) D-Glcp 1→Y D-Galp=O:12_2	Hellerqvist et al. (1970b, 1971f)
E1 = O:3, 10	→6 D-Manp 1→4 L-Rhap 1→3 D-Galp 1→ (β, α)	D-Manp 1→4 L-Rhap=O:3 (β) OAc-6 D-Galp=0:10	Hellerqvist et al. (1971c)
E2 = O:3, 15	D-Glcp 1 α↓ 3 →6 D-Manp 1→4 L-Rhap 1→3 D-Galp 1→ (β, β)	D-Galp 1→6 D-Manp 1→4 L-Rhap=O:15 (β, β)	Hellerqvist et al (1971e)
E3 = O:3, (15), 34	D-Glcp 1 α↓ 4 →6 D-Manp 1→4 L-Rhap 1→3 D-Galp 1→ (β, β)	D-Glcp 1→4 D-Galp=O:34 (α)	Robbins and Uchida (1962a)

Serogroup	Structure	Identified immunodominant structures	Reference

E4 = O:*1*, 3, 19	D-Glcp
	1
	α ↓
	6
	→6 D-Manp 1→4 L-Rhap 1→3 D-Galp 1→
	    β α
	                  D-Glcp
	                  1
	                  4
		D-Glcp 1→6 D-Galp = O:1
		           α
			Hellerqvist et al. (1971d)

G = O:13, 22	D-Galp 1→3 D-GalpNAc 1→3 D-GalpNAc 1→4 L-Fucp
	  β
	             D-GlcpNAc
	              ↓1
	              4
			Simmons et al. (1965a)

L = O:21	D-Galp 1→3 D-GalpNAc 1→3 D-GalpNAc 1→
	  β
	            D-Glcp
	            1
	           →
	           4
			Lüderitz et al. (1966)

N = O:30	D-Glcp 1→3 D-GalpNAc 1→4 L-Fucp β		Simmons et al. (1965b)
U = O:43	D-Galp 1→3 D-Galp 1→3 D-GalpNAc 1→3 D-GlcpNAc 1→4 L-Fucp α β		Lüderitz et al. (1965)

Without discriminating between the chromosomally determined O-antigen factors and the phage-determined O-antigen factors, illogical recognition of serotypes have been done. For example, in Group E, *S. anatum* (O:3,10; H:e,h:1,6), *S. newington* (O:3,15; H:e,h:1,6) and *S. minneapolis* (O:3,15,34; H:e,h:1,6) have been considered as three different serotypes. In fact, *S. newington* and *S. minneapolis* are variants of *S. anatum* converted by the phage ε15 (*S. newington*) or ε15 and ε34 (*S. minneapolis*). Agglutinability by serum anti-12_2 appears with factor O:34. It is probably the same for all the so-called serotypes of Groups E2 (O:3,*15*) and E3 (O:3,*15,34*). A similar situation exists in Group C: the serotypes quoted in Group C4 (O:6,7,*14*) are only variants of the corresponding serotypes of Group C1 (O:6,7) converted by phage 14.

Lysogenic conversion may cause other changes in O-antigen specificity: O:*20*-antigen of the Group C3, O:*1,37* antigens of Group G, O:*1*-antigens of Group R, T and 51 and other subfactors of minor importance.

Various mutations causing the absence of O-antigen-specific chains (Section II.B) allows transformation to the R-form, which has defective somatic antigen, produces rough colonies, and has lost its pathogenicity. In rare cases, when the specific O-antigen is completely lost, a new specificity called T (transient) by Kauffmann (1966) may be expressed. Two T-antigens, T:1 and T:2 have been described. T:1 is the best known. T:1-forms resemble the S-forms in colony morphology and pathogenicity. Evolution to the R-form is often rapid. Specificity of the T:1-forms is the same even if derived from S-forms with different O-antigen specificities.

Some antigenic determinants are subject to form or phase variation described by Kauffman. This means that different colonies from the same culture may show different expressions of particular O-antigen determinants, e.g. O:1 and O:12_2 (see below).

O-antigen formulas have been established on the basis of agglutination in sera of immunized rabbits before and after cross-absorptions. For example in the case of O-9,12; O-*1*,4,[5],12 and O-1,3,19, cells of the first two O-antigen combinations cross-react because both share the O:12, the last two groups cross-react because they have O:1 in common. O:1 is in italic in O:*1*,4,[5],12 because it is caused by phage conversion. The O:5 is put in brackets because it may be present in any given strain. This convention is adopted throughout this chapter. As it is a consequence of O:4 modification, it can exist only in strains possessing O:4. If a O:*1*,4,[5],12 strain is converted by phage 27, O:*27*, is added to the formula. O:9, O:4 and O:3 are major antigenic factors. All the serotypes possessing one of these factors are combined in a single O-group. At the beginning of the serological classification of *Salmonella*, letters of the alphabet were assigned to such groups: all the strains possessing O:4 were classified in the Group B, those possessing O:9 in the Group D and those possessing O:3 in the Group E. Later other O-antigen groups were discovered and the number of letters in the alphabet was insufficient. After Group Z, corresponding to O:50, the O-antigen groups are designated (more logically) by their major O-antigen factor: O:51 to O:67. Accordingly, the cross-reactions with strains of another group are shown in the O-antigenic formula which may comprise several O-antigen factors or only one factor (e.g. O:11 for Group F).

The great majority of *Salmonella* in S-form are lysed by bacteriophage Felix O1. This property is useful for routine diagnosis (Cherry *et al.*, 1954; Lindberg, 1973).

(b) *Vi-antigen.* Among the K-antigens of the Enterobacteriaceae, the Vi-antigen is the most important diagnostically in the genus *Salmonella.* The Vi-antigen occurs in only three serotypes: *S. typhi, S. paratyphi*-C and in rare strains of *S. dublin.* Vi-agglutinates are granular and difficult to dissociate by agitation.

Most freshly isolated strains of *S. typhi* are O-inagglutinable, i.e. their O-antigen in living bacteria cannot be agglutinated by antiserum O:9, but strongly by Vi-antiserum (V-forms). After heating the bacterial cells for 10 min at 100°C, such cultures become O-agglutinable and Vi-inagglutinable. S-cultures devoid of Vi-antigen are O-agglutinable (W-forms). Cultures containing a mixture of bacteria in the two forms are both O- and Vi-agglutinable.

Two genes, called *viaA* and *viaB*, located at distant sites of the chromosome, control the expression of the Vi-antigen (Fig. 5, Mäkelä and Stocker, 1969).

The Vi-antigen corresponds to the receptor of bacteriophages able to lyse only Vi-cultures. These phages are called Vi phages. One of them, phage ViII, was used by Craigie and Yen (1936), who established the phage-typing system of *S. typhi.* This typing system was later improved by Felix (1955) and Anderson (1962).

The Vi-antigen may exist in rare strains of *Citrobacter freundii* (formerly called "Ballerup"). Such strains, which possess O- and H-antigens never found in the genus *Salmonella*, are conveniently used to prepare immune sera for detecting Vi-antigen in *Salmonella* cultures.

(c) *M-antigen.* Cultures of *Salmonella* giving mucoid colonies are rare. They are interesting since they have a mucoid surface layer, called antigen M by Kauffmann. This is developed most abundantly during growth at low temperatures and at high salt concentrations. The M-antigen may cause bacterial agglutination, or be precipitated when reacted with an anti-M-serum. The M-antigen of *Salmonella* appears to be identical in all serotypes. The M-antigen is not interesting diagnostically. Since it inhibits the agglutinability of O-antigens and as mucoid cultures often segregate into mucoid and normal colonies, it is diagnostically simplest to study the antigenic formula of non-mucoid segregants. The evolution is in the sense of the loss of the M-character. Normal colonies do not revert to the M-form.

(d) *"Fim"-antigen.*   Rare cultures of *Salmonella* may possess fimbriae that can render the bacteria O-hypoagglutinable or even O-inagglutinable (Rohde *et al.*, 1975). Heating for 2.5 h at $100°C$ destroys Fim-antigens (proteins) and restores O-agglutinability. Fim I-antigens have been found only in certain serotypes of Groups B and D, in subgenus I (Rohde *et al.*, 1975). The Fim II-antigen is found in about one-third of the *Salmonella* subgenus III regardless of O-antigen group. It has also been found in a few subgenus II strains, but so far never in subgenus I strains (Rohde *et al.*, 1975).

(e) *Flagellar (H) antigen.*   Flagella are polymers of flagellin, a protein of 50 000 molecular weight. H-antibodies have two important properties:

1. H-agglutination of bacteria is flaky (bacteria are attached to each other by their flagella), quickly achieved (the probability of contact between flagella of different cells is high) and easy to dissociate by shaking (flagella break easily).
2. When H-antibodies are fixed on the flagella, bacteria become immobilized.

A few serotypes (e.g. *S. enteritidis*, *S. typhi*) produce only one kind of flagellar antigen. Such strains are called monophasic.

The great majority of serotypes can alternatively express two specificities (Andrewes, 1922, 1925) for their flagellar antigen (called diphasic). For example *S. typhimurium* can produce flagella of either specificities H:*i* or H:*1,2*. One isolated bacterial cell of these serotypes has flagella of one specificity, for example H:*i*. After several generations, one bacterial cell possessing flagella of the other specificity, H:*1,2*, appears in the population with a frequency of 10^{-4} (Stocker, 1949). A culture (i.e. billions of bacteria) of *S. typhimurium* is for this reason usually a mixture of bacterial cells, some possessing flagella of the H:*i*-specificity and others flagella of the H:*1,2*-specificity. If the proportion is balanced, such a culture will be agglutinated by antisera against both H:*i* and H:*1,2*. If a great majority of the bacterial population has flagella with H:*1,2*-specificity, the culture will not be agglutinated by anti-H:*i*-serum. As determination of the two phases of the H-antigen is essential for the identification of the serotype, it is essential to select the minority of the bacterial population possessing flagella of the H-phase other than H:*1,2*. This is easy using the immobilizing property of H-antibodies, on a semi-solid agar supplemented with serum anti-H:*1,2*, the bacterial cells possessing flagella of this specificity will be immobilized. The bacterial cells possessing flagella of another specificity will swarm on the medium, and the culture in this example will be strongly agglutinated by anti-H:*i* serum

Diphasic serotypes possess two non-allellic structural genes, *H:1* and *H:2* that code for flagellin of H-antigen phase 1 or phase 2. The switch from the expression of one gene to that of the other is controlled by a genetic element linked to the *H:2* gene (Lederberg and Iino, 1956; Iino and Lederberg, 1964; Iino, 1964) and called phase determinant (Kutsukate and Iino, 1980) (see below).

The symbols used to refer to H-antigenic factors are letters or numbers. At the beginning of the serological studies on *Salmonella*, only a few serotypes were known. Some antigenic factors appeared to be specific for one serotype and letters were used to refer to them, for instance *b* for *S. paratyphi B* and *c* for *S. paratyphi C*. The corresponding phase was called "specific phase". Other factors related to each other were found in the H-antigen of the other phase. Consequently, this phase was called "not specific" and the H-antigen factors designated by numbers, e.g. H:*1,2* and H:*1,5*. Later new factors were found and alphabetic letters were not sufficient; they were designated by *z* and a number: H:z_1 ... The last identified is H:z_{68}. Such H:*z*-antigen factors are no more related to each other than is the factor *a* to the factor *b*. Specificities found in the phase originally called specific (e.g. *c* and *z*) were later found to be associated one to the other in subsequently identified serotypes. For this reason, H-antigen factors are now classified in phase 1 or phase 2. Phage χ attacks motile *Salmonella* (Sertic and Boulgakov, 1936). The receptor site is on the flagella. Sensitivity to χ is determined, not only by the presence of flagella, but also by the specificity of the H-antigen. For example, *Salmonella* possessing H:d are susceptible, whereas *Salmonella* with H:g-antigen are resistant to phage χ (Lindberg, 1973).

2. *Designation of serotypes*

H-antigen factors referred to the same symbol are strongly related, but not always identical. Only those antigens that are of diagnostic importance are quoted in the Kauffmann–White scheme. "The Kauffmann–White scheme should be regarded as a diagnostic blueprint, making no pretence of listing all the existing antigens" (Kauffmann, 1966, 1978).

The most recent Kauffmann–White scheme was published by WHO in 1980. Each year a supplement to this scheme is published in *Annales de Microbiologie* (Pasteur Institute, Paris). The most recent developments have been included in the complete list of recognized *Salmonella* serotypes in Appendix I.

The antigenic formula of *S. typhi* is written 9,12,[Vi]:d:-, the O-antigen factors are 9 and 12, the Vi-antigen may be present or absent, the H-antigen phase 1 has the specificity called *d* and a second phase of the H-antigen does not exist.

The antigenic formula of *S. paratyphi* B is written *1*,4,[5],12:b:1,2. This means that the serotype possesses the O-antigen factors 4 and 12, the H-antigen phase 1:*b*, the antigen H-phase 2:*1,2*. Factor O:5 may be present or absent according to the strain and the presence of determinant O:1 is connected with phage conversion. It is present only if the culture is lysogenized by the corresponding converting phage (see below).

The most frequent *Salmonella* serotypes are listed in Table III.

TABLE III

The most frequently occurring serotypes of *Salmonella* (in decreasing order of frequency (Kelterborn, 1979)

S. typhimurium	*S. senftenberg*	*S. orion*
S. enteritidis	*S. bareilly*	*S. cholerae-suis* subsp. kun.
S. infantis	*S. cholerae-suis*	*S. paratyphi* A
S. heidelberg	*S. brandenburg*	*S. muenster*
S. newport	*S. tennessee*	*S. kentucky*
S. dublin	*S. agona*	*S. kottbus*
S. anatum	*S. braenderup*	*S. adelaide*
S. saint-paul	*S. gallinarum-pullorum*	*S. worthington*
S. paratyphi B	*S. meleagridis*	*S. havana*
S. typhi	*S. stanley*	*S. reading*
S. panama	*S. bovis-morbificans*	*S. cerro*
S. derby	*S. manhattan*	*S. livingstone*
S. montevideo	*S. newington*	*S. wien*
S. blockley	*S. chester*	*S. eimsbuettel*
S. thompson	*S. java*	*S. manchester*
S. oranienburg	*S. javiana*	*S. isangi*
S. muenchen	*S. schwarzengrund*	*S. sofia*
S. give	*S. cubana*	*S. emek*
S. bredeney	*S. san-diego*	*S. virchow*

B. Immunochemistry and genetics

1. *The O-antigens*

(a) *General structure of the lipopolysaccharide.* The O-antigenic specificities of *Salmonella* bacteria reside in the polysaccharide chain of the lipopolysaccharide (LPS) of the bacterial outer membrane. The LPS is not covalently linked to the membrane, but it is held there in complex with proteins and phospholipids (Nikaido and Nakae, 1979).

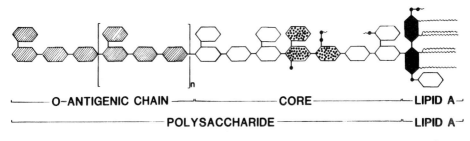

O-ANTIGENIC CHAIN ──────────────── CORE──────────── LIPID A

─────────────── POLYSACCHARIDE ───────────────── LIPID A

Fig. 1. Structure of *Salmonella* lipopolysaccharide: ⬭ , monosaccharide; ●, phosphate;~~~, ethanolamine; ⸒⸒⸒⸒ , long chain fatty acid.

The arbitrary designation of O-antigens with figures such as O:4, O:9 and so, which initially was done on the basis of a remaining antibody activity after absorptions has in the last two decades been explained in structural, biosynthetic and genetic studies (Lüderitz *et al.*, 1981; Osborn, 1979; Stocker and Mäkelä, 1971; Mäkelä and Stocker, 1981). Each LPS molecule (Figs 1–2) has an innermost part, called lipid A, which is made up of a disaccharide of β-1,6-linked-D-glucosamine residues.

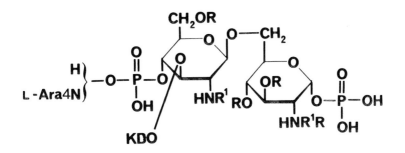

Fig. 2. Structure of lipid A in *Salmonella*: R, fatty acid; R^1, 3-hydroxy-D-myristic acid; L-Ara4N, 4-amino-4-deoxy-L-arabinose; KDO, 3-deoxy-D-*manno*-octulosonic acid; P, phosphorus.

The disaccharide carries phosphate residues in positions 1 and 4′. This hydrophilic lipid A backbone carries a number of substituents:

1. the ester-bound phosphoryl residue at 4′ of GlcNII is partially substituted by a 4-amino-4-deoxy-L-arabinosyl residue (4-Ara-N);
2. the Cl-phosphate group is partially substituted by a phosphoryl ethanolamine residue;

3. each of the two amino groups of the disaccharide backbone is acylated with a β-OH-myristic acid residue;
4 approximately five of the hydroxyl groups are esterified with fatty acids;
5. the hydroxyl group at the 3′ position of GlcNII represents the attachment site of the polysaccharide portion of the LPS (Gmeiner *et al.*, 1971).

The polysaccharide chain, a heteropolysaccharide, can according to composition and structure be subdivided into the core and the O-specific chain (Fig. 1). There are a vast number of O-polysaccharides, but apparently only one core structure common to all *Salmonella* bacteria (Fig. 3). Its innermost portion contains 2-keto-3-deoxyoctonate (KDO or dOc1A), which forms a ketosidic linkage to the lipid A backbone, and L-glycero-D-mannoheptose (L-D-Hep). Both form a branched trisaccharide. The distal part of the core is built by D-glucose, D-galactose and *N*-acetyl-D-glucosamine. The O-antigen-specific chain is linked to the subterminal D-GlcII residue of the core. The core is substituted with polar groups such as phosphate and ethanolamine. As is found for lipid A and the O-antigen polysaccharide chain, the core is heterogeneous. The substituents of the main chain such as HepIII, P, EtN,D-GalII and possibly D-Glc*N*Ac are not present in equimolar ratios (Hellerqvist and Lindberg, 1971).

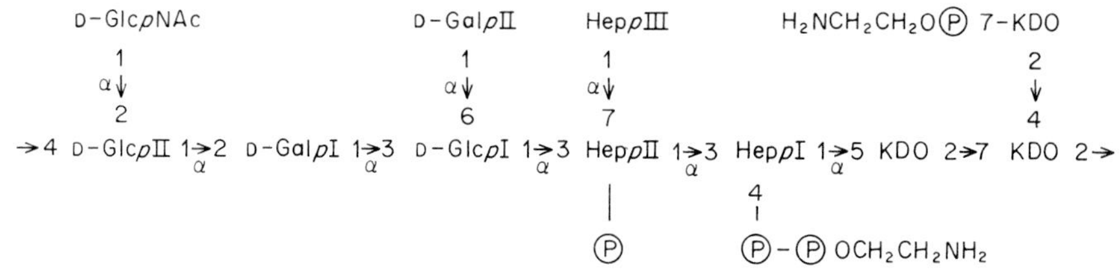

Fig. 3. Structure of core polysaccharide from *S. typhimurium*: Glc, D-glucose; Gal, D-galactose; GlcNAc, *N*-acetyl-D-glucosamine; Hep, L-*glucero*-D-*manno*-heptose; KDO, 3-deoxy-D-*manno*-octulosonic acid; P, phosphorus.

Acid-labile (Fumahara and Nikaido, 1980) and alkali-labile (Hellerqvist and Lindberg, 1971) substituents have occasionally been detected. The nature of these substituents is unknown.

The O-antigen polysaccharide chain is a linear polymer of an oligosaccharide repeat unit. The structure of the repeat unit determines the O-antigen specificity of the bacterium. Since the structure of the repeat unit differs widely between different serogroups, there is a wide difference in O-antigen specificity. But there are also similarities between serogroups, and these similarities are also explainable on the basis of their structures. Let us consider Groups A, B and D (Table II). The basic structure of the repeating unit in all three serogroups is the →2 Man α-1→4 Rha α-1→3 Gal α-1→ trisaccharide. To the D-mannose residue is α-1, 3-linked paratose in Group A, abequose in Group B and tyvelose in Group D. The nature of these dideoxyhexosyls gives the O-antigen specificity (serogroup specificity) to the bacteria: O:2 in Group A, O:4 in Group B and O:9 in Group D.

(b) *Antigen determinants and antibody specificity.*   The combining site of an antibody accommodates a saccharide structure equivalent to the hexa- or a heptasaccharide in size, as was shown for the dextran system by Kabat and co-workers (Kabat, 1961, 1966). This was also shown to hold true for *Salmonella* O-antigens in that an oligosaccharide structure larger than a tetrasaccharide but smaller, or equal to, an octasaccharide corresponded with rabbit antibodies with O:4-antigen specificity (Jörbeck *et al.*, 1979). Since the repeat units of the *Salmonella* O-antigenic polysaccharide chains consist of from three to five monosaccharides, this would suggest that the antigens contain more than one repeat unit. This is not so, however, since the largest binding contribution in the interaction between the antigen and the combining site is given by one monosaccharide (Kabat, 1966). This monosaccharide can be present in a terminal non-reducing position such as abequose in Group B bacteria (Table II), or in an internal chain position such as D-galactose in Group E2 bacteria (Table II). The name immunodominant sugars has been coined for these monosaccharides. It is, however, not only the monosaccharide, but also its type of linkage to the next monosaccharide which makes it immunodominant. The disaccharide is a still better representative of the antigenic determinant, and the complimentariness between the antigen and the antibody increases with the size of the saccharide. In the following section the structure of various O-antigen determinants are given. For some of them, such as the O:4-antigenic determinant, the structure is based on conclusive immunochemical studies, for others the structure is tentative and more research is needed.

The O-antigenic formulae in the Kauffmann–White scheme are based on the interaction of bacteria with rabbit antisera, most of them obtained after one or more absorptions with related bacteria to remove cross-reactive antibodies. The specificity of the absorbed antiserum preparation is then consequently dependent on the chemical relatedness of the absorbing bacterial strain(s). Therefore, the specificity of antisera can vary, and a given serum can, indeed, contain antibodies directed against more than one antigenic determinant. The Kauffmann–White scheme must, therefore, be considered as a simplified scheme. As predicted by Kauffmann (1961) most of the assigned O-antigens can be subdivided into two or more antigenic determinants. Consequently, more specific antisera than those existing today can be presumed for future work. This cannot be done through absorbtions, since the range of absorbing strains is limited. Two alternatives are available. One is to produce mono-clonal antibodies (Köhler and Milstein, 1975), an approach not yet exploited, but with great theoretical potential. It must be noted, however, that the hybridoma antibodies may not be identical to rabbit antibodies. Therefore, elaborate specificity testing must be performed before using such antibody preparations. Another approach is to synthesize antigenic determinants and as haptens link them covalently to an immunogenic carrier. Antibodies with superior specificity as compared to conventional antisera, usable without prior absorptions, have successfully been prepared against O:2-, 4-, 8- and 9-antigens, and tested clinically (Ekborg et al., 1977; Svenungsson and Lindberg, 1977, 1978a,b, 1979; Svenungsson et al , 1979).

(c) O:4-antigen.   The chemical nature of the Salmonella O-antigenic de-terminants has been most carefully studied for O:4-antigen in S. typhimurium (Jörbeck et al., 1979b). Precipitation–inhibition studies using a S. typhimurium O-antigenic polysaccharide free of lipid A as the antigen and a rabbit factor O:4-serum as the antiserum, and a series of saccharides representative of the S. typhimurium and related Salmonella O-polysaccharide chains as inhibitors revealed that (i) abequose-OMe was the only monosaccharide derivative that caused inhibition, (ii) for abequose-containing saccharides the inhibitory power increased with the molecular weight of the saccharide up to an octasaccharide from S. typhimurium which was 300-fold more effective than the abequose α-OMe derivative and (iii) no saccharides from S. paratyphi A (O:2,12) or S. enteritidis (O:9,12) caused inhibition even in con-centrations more than 1000-fold higher than that of effective inhibitors. These results confirmed earlier observations by Staub et al. (1959) that the dideoxyhexosyls are immunodominant sugars. However, attempts to elicit anti-dideoxyhexosyl-specific rabbit antibodies for diagnostic purposes by coupling the monosaccharides as haptens to ovalbumin as an immunogenic carrier failed (Lüderitz et al., 1966b). The reason may be that a monosaccharide is too small a hapten to elicit an immune response in the rabbit. For this purpose a structure the size of isomaltotrionic acid is required as a hapten (Arakatsu et al., 1966). Through synthesis of the disaccharide abequose α-1→3 mannose, which subsequently was covalently linked to bovine serum albumin, as carrier, an immunogenic conjugate was prepared which elicited in rabbits an antibody response specific for the disaccharide hapten (Jörbeck et al., 1979b). The antibodies were complementary to abequose-containing saccharides only, as shown in precipitation inhibition studies, and the disaccharide abequose α-1→3 mannose was the best inhibitor of all of a series of saccharides tested. The antibodies recognized only the disaccharide in their combining sites, the rest of the antibody–antigen interaction occurred between the linkage arm (joining the hapten and the carrier) and the ε-aminolysyl residue (the linkage point of the carrier molecule) and the antibody (Jörbeck *et al.*, 1980). This finding explains why the antibodies elicited against the synthetic abequose α-1→3 mannose disaccharide hapten were specific for the O:4 antigenic determinant in coagglutination and immunofluorescence studies (Svenungsson and Lindberg, 1977, 1978). Antibodies elicited by the conjugate are superior to conventionally prepared O:4-antiserum since high titred sera collected from rabbits can be used without prior absorption(s).

Besides the O:4-antigen determinant, the O:2-antigen paratose α-1→3 mannose, O:8-antigen abequose α-1→3 rhamnose and O:9-antigen tyvelose α-1→3 mannose, determinants have been synthesized and, as haptens, used to elicit specific antibody preparations (Svenungsson and Lindberg, 1977, 1978a, 1979).

The O:5-antigen determinant 2 *O*Ac-abequose, present in *Salmonella* Group B, has likewise been synthesized, and as a hapten linked to an immunogenic carrier it has elicited a specific O:5-antibody response (Stellner *et al.*, 1972).

(d) *O:1- and O:12-antigens.* The O-antigenic formula of *S. typhimurium* is usually given as O:*1*,4,[5],12. The chemical structures responsible for the O:4- and O:5-antigens are given above (Table II). O:12-antigen (which can be subdivided in partial antigens 12_1, 12_2 and 12_3) is subject to form variation which denotes quantitative changes in the O-antigens (Mäkelä and Mäkelä, 1966). When Kauffmann tested 20 colonies from an original culture by slide agglutination he found the following distribution: nine colonies of the $1 + +$, $12_2 \pm$ form; one colony of the $1 + +$, $12_2 + +$ form; five colonies of the $1 \pm$, $12_2 \pm$ form and five colonies of the $1 \pm$, $12_2 + +$ form. On subculturing the four different antigenic forms further dissociations were seen, although all four forms were not always present at the same time (Kauffmann, 1954).

The structural background for the O:1- and O:12_2-antigen determinants has been firmly established, and both are as a result of glucosylation of the D-galactose residue of the repeat unit (Table II; Stocker *et al.*, 1961b; Staub, 1961). O:1-antigen specificity is found when D-glucose is α-1, 6 linked to D-galactose, whereas O:12_2-antigen specificity is found when D-glucose is α-1,4 linked. The presence of O:1-antigen is a consequence of lysogenic conversion, i.e. the presence of a prophage in the bacterium, such as iota and P22 in groups A, B and D, which gives the bacterium the information required to form the antigen (Iseki and Kasiwagi, 1957; Zinder, 1957). When the bacteria are cured of the prophage they lose the ability to express the O:1-antigen. O:1-antigen specificity is also found in Groups E4 (O:1,3,19) (see below), G(O:1,13,23,37), R (O:1,40), T (O:1,42) and 51 (O:1,51). In all cases except Group E4 lysogenic conversion is responsible for the appearance of O:1-antigenic specificity (Le Minor *et al.*, 1963; Le Minor, 1963, 1966). Although not proven by structural analyses, the addition of α-1,6-linked D-glucose residues to D-galactose is most likely responsible for the appearance of O:1-antigen specificity in these groups. Thus the disaccharide common to these groups gives the bacteria the O:1-antigen specificity. The O:19- and O:37-antigen specificities, which appear concomitantly with O:1 in Groups E4 and G respectively, contain the D-glucose α-1$\rightarrow$6 D-galactose disaccharide in their determinants, but apparently at least two more polysaccharide chain sugars are involved in the specificity of these antigens (Staub and Girard, 1965).

The form variation observed for the O:12_2-antigen is apparently not caused by phage conversion. Although it has not been studied in detail a mechanism such as the flagellar phase variation mechanism is likely (see below). In the O:12_2-positive form the polysaccharide chain is glucosylated. In the O:12_2-negative form the chain lacks the α-1,4-linked D-glucose residue. The frequency with which the O:12_2-antigen is either turned from on to off, or vice versa, seems to be high. We have so far failed to find a fixed O:12_2-positive bacterial culture in our structural studies (Wollin *et al.*, 1981). Consequently, most colonies have either a majority of O:12_2-positive, or O:12_2-negative bacteria, but seldom, if ever, consist exclusively of O:12_2-positive bacteria. Mutants which are O:12_2-negative, e.g. which do not express the O:12_2-antigen, have been selected for (Mäkelä, 1973).

The structural entities in the polysaccharide chain of *S. typhimurium* responsible for the O:12_1- and 12_3-antigens are not known. Kauffmann (1961) reported that the 12 antigens could appear in various combination such as O:12_1, 12_2, 12_3 in *S. typhi* T4, O:12_1, 12_3 in *S. paratyphi* and *S. typhi* T2, and O:12_1, 12_2 in *S. reading*. Whether these specificities are found in the main polysaccharide chain or are constituents of the major chain, e.g. glucosyls, acid or alkali-labile groups (Hellerqvist *et al.*, 1968, 1969a,c) is not known. It is however, likely that an antiserum contains antibodies with more specificities than the O:12_1- and O:12_3-antigen specificities discussed here. It has been found that antibodies directed against the D-mannose α-1$\rightarrow$4 L-rhamnose, L-rhamnose α-1$\rightarrow$3 D-galactose and D-galactose α-1$\rightarrow$2 D-mannose disaccharides are elicited in rabbits (Svenson and Lindberg, 1978). However, most or all of these antibody populations are removed during absorptions for the production of factor sera.

(e) *O:2-antigen.* O:2-antigen is the serogroup determining antigen factor in Group A. Quantitative immunochemical studies have convincingly demonstrated that it is the dideoxyhexosyl paratose which is the immunodominant structure (Staub *et al.*, 1966; Jörbeck *et al.*, 1980). By analogy with studies on the O:4-antigen, the corresponding disaccharide hapten for the

O:2-antigen paratose α-1$\to$3 D-mannose has been synthesized and covalently linked to an immunogenic carrier. Antibodies elicited in rabbits were superior to conventionally prepared antiserum preparations in the enzyme-linked immunosorbent assay (ELISA), immunofluorescence and coagglutination (Svenungsson and Lindberg, 1978a,b).

(f) *O:5-antigen.* The antigen is found in Group B together with O:4-antigen. Some, but not all, of the species in this serogroup express the O:5-antigen. Structurally O:5-antigen specificity is caused by O-acetylation of C2 of the abequose residue (Table II; Hellerqvist *et al.*, 1968). Not all abequose residues are acetylated, figures ranging from 50% to 98% have been estimated Hellerqvist *et al.*, 1968, 1969a). The ester-linked O-acetyl groups are sensitive to alkaline hydrolysis. Therefore O:5-antigen specificity can easily be removed by treating the bacteria under mild alkaline conditions. Synthesis of the 2-OAc-abequose residue followed by covalent linkage to ovalbumin enabled Stellner *et al.* (1972) to elicit antibodies with O:5-antigen specificity in rabbits.

(g) *O:6,7- and O:14-antigens.* The antigens are found in Groups C1 and C4. The structure of the O-antigen polysaccharide chain in this group has still not been completely determined. This is the reason why the antigenic determinants cannot be represented by a specific saccharide structure. Recent studies have corrected the structure proposed earlier (Fuller and Staub, 1968; Fuller *et al.*, 1968; S. B. Svenson *et al.*, unpublished data). Available evidence indicates that the O:antigen-chain consists of D-mannose and *N*-acetyl-D-glucosamine residues in a ratio of 4 to 1, which are linked as shown in Table II. Nuclear magnetic resonance (n.m.r.) studies have shown that there are two α- and three β-linkages in the tentative repeat unit, but only the β-linkage between the amino sugar and the D-mannose residue has been proven.

An oligosaccharide fraction isolated after hydrolysis of the O-antigen polysaccharide chain by phage 14 *endo*-mannosidase was covalently linked to an immunogenic carrier protein (S. B. Svenson *et al.*, unpublished data). Rabbit antibodies elicited by the conjugate showed only O:7-antigen specificity (Ekwall *et al.*, 1982). This strongly suggests that the O:7-antigenic determinant is found in the main polysaccharide chain. The suggestion is further strengthened by the frequently observed reciprocal immunological cross-reaction between O:7-antigen serum and *Candida* species. Yeasts have a cell wall mannan which consists of an α-1,6-linked polymannose backbone with branches of α-1,2- and α-1,3-linked D-mannose residues (Ballou, 1976).

The polysaccharide chain can be substituted by D-glucose linked to one of the D-mannose residues (Fuller and Staub, 1968). Available evidence indicates that this D-glucose residue gives the bacteria O:6-antigen specificity. The O:6-antigen can be subdivided in O:6_1- and O:6_2-antigenic determinants. Some

Group C1 strains express both partial antigens, others only one or the other. Glucosylation of either D-mannose II or III was found to be the likely structural background for the O:6_2-antigenic determinant (Fuller et al., 1968). Lysogenization of an O:6_2 7 strain by phage 14, which leads to O:14-antigen specificity, was accompanied by glucosylation of C3 of the D-mannose IV residue instead of the D-mannose II or III residue (Fuller and Staub, 1968). Strains in Group C2 (O:6,8) have the O:6_1-antigenic determinant. The structural studies on *S. newport* revealed the presence of D-mannose substituted by D-glucose at C3 (Table II; Hellerqvist et al., 1970a). The same structural element is found in Group C1 bacteria (Table II). Furthermore, the D-mannose residue in the Group C2 is in the α-$\rightarrow$2 D-Man α-1$\rightarrow$ linkage, which may provide the clue to the O:6_1-antigenic determinant. There is also an O:6-antigen in Group H (6,14; 6,14,24; 1,6,14,25). Bacteria in this serogroup are not agglutinated by factor O:6_1- or O:6_2-antisera, which indicates a further subdivision of the O:6-antigen. The structure of the O-antigen polysaccharide chain in Group H is, however, not known. The O:6_1-antigen has been found to be subject to form variation (Edwards, 1945) as was described for the O:1- and O:12_2-antigens (see above), where glucosylation/non-glucosylation of the polysaccharide chain is the structural basis for the variation.

(h) *O:8- and O:20-antigens.* The structures of the O-antigen polysaccharide chains in Groups C2 (O:6,8) and C3 (O:8,20) are given in Table II. The specificity of the O:8-antigenic determinant was attributed early on to the dideoxyhexosyl abequose (Lüderitz et al., 1966a; Hellerqvist et al., 1969, 1970). This was confirmed when the disaccharide abequose α-1$\rightarrow$3 L-rhamnose was synthesized, and as a hapten was coupled to bovine serum albumin (BSA), and the antibodies elicited in rabbits were found by ELISA, immunofluorescence and coagglutination to be specific for O:8-antigen (Svenungsson and Lindberg, 1979).

The O:12-antigenic determinant can most likely be attributed to glucosylation of the D-galactose residue at position C4 (Table II). The presence of the O-acetyl at C2 of the D-glucose residue did not influence immunochemical specificity since a deacetylated LPS preparation was no better as inhibitor than the acetylated LPS in a passive haemagglutination assay system (Hellerqvist et al., 1969).

(i) *O:3,10,15,34-antigens.* O:3-antigen is the serogroup-specific antigen in Groups E1 (O:3,10), E2 (O:3,15), E3 (O:3,15,34) and E4 (O:1,3,19). The immunodominant structure of the O:3-antigenic determinant was studied by inhibition with oligosaccharides obtained from acid hydrolysis of the polysaccharide chain (Uchida et al., 1963). The disaccharide D-mannose β-

$1 \rightarrow 4$ L-rhamnose, which is internal in the polysaccharide chain, was the immunodominant structure. The linkage between the D-galactose and D-mannose residues in successive repeat units is α-1,6 in Groups E1 and E4, but β-1,6 in Groups E2 and E3. All four serogroups are O:3-antigen specific, and this linkage apparently does not significantly influence the O:3-antigen specificity. It is likely, however, that a detailed immunochemical study using various saccharides representing the Group E polysaccharides would make it possible to subdivide the O:3-antigen into further antigenic determinants.

O:10 antigen specificity is a consequence of O-acetylation of the D-galactose residue (Table II). This conclusion was arrived at as a result of treatment with alkali, which removes ester-linked acetyl groups, O:10-antigen specificity was destroyed, and that acetylation of the serologically inactive trisaccharide D-galactose α-1$\rightarrow$2 D-mannose β-1$\rightarrow$4 L-rhamnose converted it into a potent inhibitor of a complement fixation system with O:10-antigen specificity (Uchida *et al.*, 1963).

O:15-antigen specificity appears concomitant with lysogenization of Group E1 bacteria (O:3,10) with ε15-like phages (Iseki and Sakai, 1953). Structurally lysogenization is followed by a linkage change between successive repeat units, from α-1,6 to β-1,6, and by disappearance of O-acetyl from the D-galactose residue. Quantitative inhibition experiments with oligosaccharide released by partial acid hydrolysis suggest that the D-Gal β-1$\rightarrow$6 D-Man β-1$\rightarrow$4 L-Rha trisaccharide is the structure responsible for O:15-antigen specificity with the β-1,6-linked D-galactose residue as the immunodominant monosaccharide (Robbins and Uchida, 1962a,b).

O:34-antigen specificity, which appears concomitant with lysogenization of Group E2 bacteria with ε34-like phages, is caused by the simultaneous appearance of D-glucose residues linked α-1,4 to D-galactose of the main chain polysaccharide (Table II) (Robbins and Uchida, 1962a). The disaccharide D-glucose α-1$\rightarrow$4 D-galactose is also found in Groups A, B and D, where it is responsible for O:12_2-antigen specificity. As expected an O:12_2-antigen serum agglutinates O:34-antigen-containing bacteria and vice versa.

In Group E4 the antigenic formula contains O:1- and O:19-antigen specificities in addition to the serogroup-specific O:3-antigen (Table II). The O:1-antigen is as in Groups A, B and D, a consequence of the D-glucose α-1$\rightarrow$6 D-galactose disaccharide. The O:19-antigen specificity was found to contain the same disaccharide, but in addition it also contained the α-$\rightarrow$6 D-Man β-1$\rightarrow$4 L-Rha structural element as an important part of the antigenic determinant (Staub and Girard, 1965).

(j) *Other O-antigen determinants.* Bacteria belonging to Groups A to E are the most frequently isolated. Of the 57 most commonly isolated serotypes only five belonged to a serogroup other than Groups A to E (Table II). Three belong to Group G2 (O:1,13,23), and one each to Groups K (O:18) and O (O:35). Of these only the structure of the O-antigen polysaccharide chain in Group G has been studied and partially elucidated structurally (Table II) (Simmons *et al.*, 1965a). The lack of immunochemical studies makes it impossible to assign a chemical structure to any of the O-antigenic determinants O:13, O:22 or O:23. The occasional presence of O:1-antigen specificity in Group G is certainly caused by D-glucose residues α-1,6-linked to the chain D-galactosyl units.

There are some additional serogroups where the structure of the O-antigen polysaccharide chain has been studied, namely Groups L (O:21; Lüderitz *et al.*, 1981), N (O:30; Simmons *et al.*, 1965b), and U (O:43; Lüderitz *et al.*, 1965a). The proposed partial structures for the polysaccharides of these serogroups are given in Table II. As quantitative immunochemical studies have not been done, the structure responsible for each O-antigenic determinant cannot be given. However, the observed reciprocal cross-reactivity between Groups G (O:13,22; O:13,23) and U (O:43) is understandable since the O-antigen polysaccharide chains contain at least one common region: the D-Gal β-1→3 D-GalNAc disaccharide. Although only one O-antigen specificity is given for each group, the antigenic determinants can be subdivided. Thus two different antigenic determinants have been identified in Group N and five in Group U (Le Minor and Rohde, 1978).

(k) *T:1- and T:2-antigens.* The two serological variants designated as T:1 and T:2 (T for transient) (Kauffmann, 1956, 1957) are unstable and frequently undergo mutations into R-forms. The two variants are characterized by their serological specificity which differentiates them from each other and from smooth and rough *Salmonella* strains.

Chemical analyses have revealed unique structures. The T:1-specific polysaccharide chains are composed of polymers of β-linked D-ribofuranosyl and β-linked D-galactofuranosyl residues (Berst *et al.*, 1969). Both chains are β-1,4 linked to the D-glucose II residue of the core in the same way as the O-polysaccharide chains. It is obvious with the given structures that the T:1-antigen specificity is composed of several determinants.

For the T:2-antigen polysaccharide chain only a partial structure is known (Bruneteau *et al.*, 1974). A N-acyl-D-glucosamine residue is β-1,4 linked to the D-glucose II residue of the core. The N-acyl group and a substituent at positions 3 or 4, which makes the N-acyl-D-glucosamine periodate resistant, have not been identified.

(l) *Vi- and M-antigens.* The Vi-antigen is a homopolysaccharide composed of β-1,4-linked N-acetyl-D-galactosaminuronic acid residues which are acetylated at C3 (Heyns and Kiessling, 1967). Vi-antigens lacking the acetyl group at C3 have been found.

The *Salmonella* M-antigen is a polysaccharide with a carbohydrate backbone which is the same for all *Salmonella* strains, and other Enterobacteriaceae as well (Gorin and Ishikawa, 1966):

$$O\text{AC}$$
$$|2/3$$
$$\rightarrow 4 \text{ L-Fuc}p\ 1 \rightarrow 3 \text{ D-Glc}p\ 1 \rightarrow 3 \text{ L-Fuc}p\ 1 \rightarrow$$
$$\alpha \qquad\quad \beta \qquad_{4}\quad \beta$$
$$\beta\uparrow$$
$$1$$
$$\text{D-Gal}p\ 1 \rightarrow 4 \text{ D-Glc}p\text{A}\ 1 \rightarrow 3 \text{ D-Gal}p$$
$$\beta \qquad\qquad \beta$$

A group linked acetalically either to C3 and C4, or to C4 and C6 of the terminal β-linked galactosyl residue may be pyruvic acid, acetaldehyde or formaldehyde (Garegg *et al.*, 1971a,b, 1980; Bennet and Bishop, 1977; Lvov *et al.*, 1981).

2. *The H-antigens*

The flagellum is a complex structure consisting of a basal body, a hook and a filament (Fig. 4). The basal body anchors the flagellum to the bacterial cell envelope and the hook connects the basal body with the filament which is a passive device rotated by a mechanism in the basal body (Iino, 1969, 1977; Macnab, 1978). When serotyping *Salmonella* only antigen specificity found in the filament is used.

The filament consists of polymerized protein components called flagellin. These have a molecular weight in the range of 51 000 to 57 000 (Macnab, 1978). In *Salmonella* more than 60 different H-antigen specificities have been found so far. The antigenic differences are a consequence of differences in the primary structure of the various flagellin molecules. An early hypothesis was that each H-antigen specificity correlated with an unique flagellin structure. Studies of antigenically different flagellins demonstrated different amino acid compositions, although certain similarities were found (McDonough, 1965). Treatment of flagellin from *S. adelaide*, H:1:*f,g*, with cyanogen bromide resulted in the isolation of four polypeptide fragments (Parish *et al.*, 1969). The largest (18 000 molecular weight) contained all the antigen specificities detectable on intact flagella. Subsequent studies with five antigenically different polymerized flagellin preparations (H:*i*; H:*1,2*; H:*d*; H:*f,g*; H:*g,p*) used as immunogens in mice and rabbits showed both common and variable regions in the flagellin molecule (Fig. 4) (Langman, 1972). This finding contradicted the earlier hypothesis of unique flagellin structures, but has been supported by more recent investigations. The present knowledge indicates

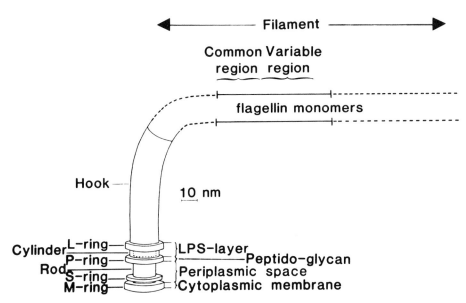

Fig. 4. Model of the basal structure of a flagellum from *Salmonella*.

that the flagellin has a common or essential region and an antigenically variable region (Iino, 1977). Support for this hypothesis besides Langman's results (1972) is provided by genetical (Iino, 1977) and immunochemical studies (Joys, 1976; Smith and Potter, 1976; Smith *et al.*, 1979). The concept of a common region has found strong support in the observation that an IgA myeloma protein M467 binds to flagella from all *Salmonella* species studied, including both the H:1 and H:2 phases of diphasic strains (Smith and Potter, 1976; Smith *et al.*, 1979).

All evidence indicates that the H-antigen specificity of the Kauffmann–White scheme resides in a relatively small variable region of the flagellin molecules. Antibodies against the common essential region of the flagellin molecule are also elicited during immunization (Langman, 1972), but these antibodies are removed during the absorption(s) performed for the preparation of the H-antigen sera.

Attempts to determine the amino acid sequence of flagellin have only been partly successful (Davidson, 1971; Joys and Rankis, 1972). Flagellin *i* from *S. typhimurium* and *f,g* from *S. adelaide* are composed of 470 and 386 amino acids respectively. In flagellin *i* amino acid sequences in 29 tryptic peptides with a total of 217 amino acids were determined. The order of these peptides in the flagellin remains unknown. Flagellin from spontaneous mutants of *S. typhimurium* with altered *i*-antigen specificity have been analysed and found to differ from the wild type in four peptides (Joys and Stocker, 1966; Joys and Rankis, 1972). The serological analyses suggest that the H:*i*-antigen specificity is composed of several antigenic subfactors (Joys and Stocker, 1966). Independent support of the serological results was obtained when H:1-*i*-flagellin from two strains of *S. typhimurium* was studied, and although apparently identical in the antigenically common region the flagellins differed in at least five amino acid sites in one soluble tryptic peptide, presumably in the variable region (Joys *et al.*, 1974). The differences found required transition, transversion and double mutational events. There are several instances where reciprocal cross-reactions between different H-antigens are observed, such as between *i* and *r*, *k* and *z*, *1,5* and z_6. These cross-reactions may rest on shared H-antigenic determinants.

A complete amino acid sequence of a flagellin molecule has so far only been reported for *B. subtilis* strain 168 (De Lange *et al.*, 1976). Comparison of this sequence of 304 amino acids with the known partial amino acid sequences of *Salmonella* flagellin shows partial homology in the common region, but less homology in the variable region (De Lange *et al.*, 1973). Thus the common region seems to be more conservative than the variable region.

Although there has been progress in our understanding of the immuno-chemical background of the H-antigen specificities in the last decade, it is obvious that we still know less about the H-antigens than about the O-antigens. The observation that one H-antigen specificity appears to consist of several antigenic determinants (Joys and Stocker, 1966; Joys *et al.*, 1973) points to the Kauffmann–White scheme not as a method of taking bacterial fingerprints but as a practical means of typing.

3. *Biosynthesis and genetics of O-antigens*

Biosynthesis of the polysaccharide chain containing the O-antigenic determinants is a complex process. It involves reactions which can be grouped into five general classes (Wright and Kanegasaki, 1971; Osborn and Rothfield, 1971; Nikaido, 1973; Osborn, 1979).

1. Synthesis of sugar mucleotides which are high-energy precursors such as UDP-glucose, UDP-galactose, GDP-mannose and TDP-rhamnose.
2. Transfer of the monosaccharide, or other unit, from the precursor to an acceptor, which is a special isoprenol carrier lipid. The transfer occurs either directly to the lipid, such as the formation of the galactosyl-carrier lipid compound in the first step in the biosynthesis of the *S. typhimurium* repeat unit, or to a monosaccharide already linked to the acceptor, such as the addition of L-rhamnose to the galactosyl-carrier lipid compound.

3. Polymerization or transfer of a previously synthesized lipid-borne repeat unit to the distal non-reducing end of a newly synthesized repeat unit.
4. Transfer of the chain of polymerized repeat units from the carrier lipid to the core LPS.
5. Modification of the polysaccharide like glucosylation and addition of O-acetyl groups to the repeat unit.

The genetics of the O-antigen polysaccharide chain biosynthesis has been extensively studied (Mäkelä and Stocker, 1981). Most of the genes for the synthesis of components of the O-antigen chains, and for their assembly, are specified by genes in the main *his*-linked *rfb* cluster at position 44–45 in the *Salmonella* map (Fig. 5). For *Salmonella* Group B at least nine enzymes are known to be required for the synthesis of the TDP-rhamnose, GDP-mannose and CDP-abequose sugar nucleotides. All these enzymes, and at least four transferases, are determined by closely linked genes at *rfb*. At least six of the mapped *rfb* genes are part of a single transcription unit or operon (Levinthal and Nikaido, 1969). The gene for synthesis of galactose, *galE* (for UDP-galactose-epimerase), the fourth essential sugar in the *Salmonella* Group B repeat unit is located separately from the LPS genes (Fig. 5). This is true also for some other monosaccharides which are used for other purposes in the bacteria, such as *galU* for UDP-glucose synthesis. Mutations in a gene resulting either in a blocked synthesis of a precursor, such as TDP-rhamnose, or its transfer to the growing repeat unit, lead to a bacterium unable to synthesize its O-antigen polysaccharide chain. This rough mutant has a LPS which terminates with a complete core oligosaccharide (Fig. 3). Mutations in a gene affecting a precursor such as UDP-galactose, which is used as a component in both the core and the O-antigen polysaccharide chain, results in a bacterium that has an incomplete core (in this case terminating with the D-Glc I unit: Fig. 3).

Polymerization of the repeat units in Group B is determined by a gene function located at the *trp*-linked *rfc* site at position 32 (Fig. 5). Mutants defective in the polymerization step have only one repeat unit linked to the core. Such bacteria are in many characteristics intermediate between smooth (with complete O-antigen-chains) and rough forms (with no O-antigen polysaccharide chains) and were therefore termed semi-rough (SR) (Naide *et al.*, 1965).

Assembly of the O-antigen polysaccharide chain can also occur without preassembly and subsequent polymerization of repeat units. In Groups C1 (O:6,7,14) and L (O:21) (Table II) the synthesis appears to proceed through sequential addition of the monosaccharides to the non-reducing end of the growing chain linked to a carrier lipid (Flemming and Jann, 1978a,b). The O-antigen-chain in Group C1 which appeared to start with glyceraldehyde

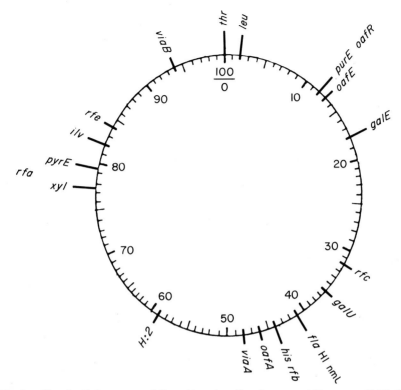

Fig. 5. Circular linkage map of *S. typhimurium* (Sanderson and Hartman, 1978). The map shows the location of some of the genes, or blocks of genes, involved in the formation of O- and H-antigens.

(Gmeiner, 1975), has subsequently been shown to start with D-mannose (Heasley, 1981). For assembly of the C1 chain the function of a gene, or genes, of a *rfe* locus to *ilv* (Fig. 5) is required (Mäkelä and Jahkola, 1970; Schmidt *et al.*, 1976), but its mode of action is not known.

The enzymes responsible for modification of the O-antigen polysaccharide chain are often determined by genes outside *rfb* and *rfc*. Thus O-acetylation of abequose in *S. typhimurium*, the O:5-antigenic determinant (Table II), is determined by the *oafA* gene (Fig. 5; Mäkelä, 1965). Glucosylation of the D-galactose residue in *S. typhimurium* (Group B) and *S. anatum* (Group E) requires the function of the *oafR* and *oafC* genes respectively (Mäkelä, 1973; Plosila and Mäkelä, 1972).

Genes modifying the polysaccharide chain, thereby adding or deleting antigenic determinants, can also be carried by non-chromosomal elements such as plasmids or bacteriophages. The changes brought about are termed antigenic conversion (Uetake *et al.*, 1958). Lysogenization of *S. anatum*

(Group E1; O:3,10) with phage ε15 is accompanied by (i) a block in transacetylase synthesis so that the O-acetyl groups are not transferred to C6 of the D-galactose residue of the repeat unit, which leads to loss of the O:10-antigen specificity (Table II), (ii) production of the phage-directed protein which inhibits the action of the α-1,6 polymerase linking the repeat units and (iii) synthesis of a new polymerase which links the repeat units β-1,6 with the concomitant appearance of a new antigenic determinant, O:15-antigen (Robbins and Uchida, 1965; Table II).

A similar conversion, mediated by phage P27, occurs in Groups A, B and D. The linkage between successive repeat units is changed from α-1,2 to α-1,6 (Lindberg *et al.*, 1978), which leads to the appearance of an O:27 antigenic determinant. The structural change most likely also results in disappearance of antigenic determinants from the native chain, but this has not yet been studied.

Several *Salmonella* phages cause glucosylation of the O-polysaccharide chain. Phage P22 in Groups A, B and D (Mäkelä, 1973), phage ε34 in Group E2 (Wright and Barzilai, 1971) and several others (Le Minor, 1969) add a glucosyl-linked α-1,3, α-1,4 or α-1,6 to the repeat unit (Table II). The glucosylation brought about by P22 is recognized as O:1-antigen specificity, and that by phage ε34 as O:34-antigen specificity. For P22 and ε34 the phages contain the genetic information both for the formation of the glucosyl acceptor lipid and for the transfer of the D-glucose residue to the LPS.

The mechanism for form variation, e.g. the quantitative changes in expression of particular O-antigens such as O:1 and O:12$_2$, are not known. The O:12$_2$-antigenic determinant D-glucose α-1 → 4 D-galactose is determined by *oafR* genes (Fig. 5), which control both the on or off state whether variation occurs or not (Mäkelä and Mäkelä, 1966). It is a commonly held hypothesis that the molecular mechanism is similar to the mechanism of flagellar phase variation (see below).

4. *Biosynthesis and genetics of H-antigens*

Many genes are involved in the formation and function of flagella, e.g. chemotaxis (Iino, 1969, 1977; Macnab, 1978). Here only a brief account of the genetics of the filament, in which the H-antigen specificity resides, will be given.

Most *Salmonella* species have two H-antigen specificities, H phases 1 and 2, specified by genes *H:1* (at position 41, Fig. 5) and *H:2* (at position 59, Fig. 5) respectively. Normally the *H:1* or the *H:2* gene is expressed in an individual bacterium. However, approximately every thousand to ten thousand division causes a bacterium expressing the H:1-antigen to switch to expressing the H:2-antigen instead. The probability for a reverse change is somewhat lower. The molecular basis for this change has recently been determined by Simon and co-workers (Silverman and Simon, 1977; Zieg *et al.*, 1977; Silverman *et al.*, 1979; Simon *et al.*, 1980).

The controlling element is at a site adjacent to the *H:2* gene (Fig. 6). When this site is in the *H:2* on state, the *H:2* gene and another adjacent *rhl* gene are expressed. The product of the *rhl* gene acts as a repressor of the *H:1* gene. Therefore only *H:2* flagellin is produced, and the bacteria agglutinate a phase H:2-antiserum. When the controlling element is in the off state, neither the *H:2* nor the *rhl* genes are expressed. The lack of the *rhl* repressor allows transcription of the *H:1* gene, H:1 flagellin is produced and the bacteria agglutinate in a phase H:1-antiserum.

The controlling element is a 970 base pair long DNA segment adjacent to the *H:2* gene (Fig. 6). This segment contains a promotor region (P) which is necessary for initiation of transcription of the *H:2* gene. When the segment is in the *H:2* on state, the promotor is connected to the *H:2* gene and transcription of the *H:2* and *rhl* genes occurs. In the *H:2* off state, the element

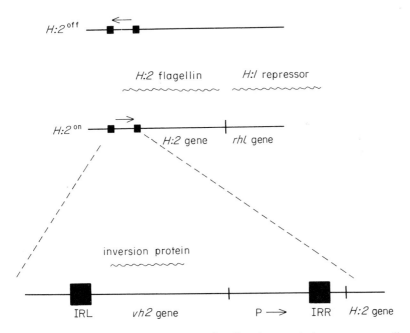

Fig. 6. Components of the *S. typhimurium* flagellar phase variation system according to Simon *et al.* (1980). The top two lines illustrate the mechanism of phase transition. The arrows indicate the orientation of the DNA sequence in the invertible region. P, promotor; IRL, Inverted Repeat Left; IRR, Inverted Repeat Right; *vh2=hin*, inversion protein involved in mediating specific inversion.

is inverted and the promotor disconnected from the *H:2* gene. Therefore transcription of the *H:2* and *rhl* genes does not take place. The frequency of switching determines the relative proportions of phase 1 and phase 2 bacteria in the population under study.

Although the mechanism for the phase switching has been clarified, the molecular events that are involved in the transition from one state to the other are not completely understood (Simon *et al.*, 1980; Mäkelä and Stocker, 1981).

Genetic analyses in *Salmonella* have led to the identification of 16 flagellar (*fla*) genes in the *H:1* cluster, and 10 in the *H:2* cluster (Iino, 1977). The H-antigenic determinants are located in the variable region of the flagellin. Mutations in the essential region leads to non-flagellate bacteria, which suggests that the region may correspond to a section of the polypeptide chain that is important for the conformation and the ability for self-assembly of the filament. Mutations in the variable region lead to changes in antigen specificity (Joys *et al.*, 1974) but seldom to non-flagellate bacteria. A hypothesis is that the variable region may correspond to an exposed part of the flagellin molecule (Iino, 1977). A gene *nml*, closely linked to *H:1* (Fig. 5) (Stocker *et al.*, 1961a), determines N-methylation of the ε-amino residues of lysine in the flagellin. This N-methylation modifies the antigenic character of the flagella.

The process of flagellin assembly from filaments has been clarified by polymerization *in vitro* (Asakura, 1970). In the presence of appropriate concentrations of flagellin monomers and flagellar filaments under controlled temperature, pH and ionic strength conditions the flagellin monomers bind to the end of an existing filament. The flagellin monomers are thought to be arranged in 11 nearly longitudinal rows leaving room for an internal core (O'Brien and Bennett, 1972). The end to which the monomers bind corresponds to the distal end of a flagellar filament on a living bacterium. Next the monomer is firmly incorporated into the filament, with a concomitant conformational change. The incorporated monomer then acts as a part of the nucleus for polymerization of the next monomer. Thus, the conformational change of flagellin which occurs during assembly confers structural polarity to the filaments so that the next monomer can only assemble at the distal end of each filament. The process *in vivo* can occur with the same speed as the *in vitro* process (Iino, 1977). *In vivo* the distal end of each hook serves as the nucleus for polymerization of flagellin. For growth of the flagellar filament *in vivo* the flagellin monomers are most probably transported from the cell body to the tip of the filament through the central core canal inside the filament (Arakura, 1970; Iino, 1977).

III. Technical details

A. Preparation of bacterial suspensions for titration of antibodies in one serum

1. *O-antigen suspension*

Smooth bacteria are cultivated overnight on nutrient agar. A heavy suspension is prepared in saline, and mixed with an equal volume of 96% ethanol whereby the H-antigen is denatured.

Strains used as immunizing antigens are selected for their sensitivity (titre of agglutination as high as possible against homologous antiserum) and specificity (titre as low as possible against heterologous antisera).

2. *Vi-antigen suspension*

The method described by Ando and Shimojo (1953) gives a stable suspension. The strain of *S. typhi* Ty6S does not possess any O-antigen, only Vi- and H-antigens. A heavy suspension of an overnight culture grown on nutrient agar is prepared in saline added with an equal volume of ethanol. After one day at $+4°C$, the bacteria are centrifuged, and the supernatant discarded. The bacteria are suspended in saline added with $CaCl_2$ (0.5 g 100 ml^{-1}) and formalin (0.5 ml 100 ml^{-1}). Such a suspension is stable, and the Vi-antigen remains fixed on the bacteria that are agglutinable only by Vi-agglutinins. Sensitivity is controlled with an anti-Vi serum prepared with a *Citrobacter* Vi+, its specificity must be controlled with H:*d* and O:9,12-antisera.

3. *H-antigen suspension*

The most motile bacteria are selected by culture in U tubes containing soft agar, or by migration on Gard plates. It is generally necessary to prepare H-antigen suspensions of bacteria in the same H-phase. This presents no problem if the strain is monophasic. If the strain is diphasic, however, it is necessary to select the desired phase by passaging the culture in soft agar containing antiserum to the non-wanted phase. Another simpler technique is to use monphasic mutants of usually diphasic serotypes. Highly motile bacteria are subcultivated in nutrient broth, if possible stirred in a water bath. After 8 h, formalin (0.5% v/v final concentration) is added. After overnight incubation at $37°C$, the sensitivity and specificity are controlled.

The sterility of each suspension is controlled.

For tube agglutinations are used saline suspensions of about 5×10^8 bacteria per millilitre.

A dilution series of the sera are prepared, distributed in tubes, and the bacterial suspension is added (e.g. 0.1 ml serum dilution + 0.9 ml suspension). Tubes are incubated either at 37°C or at 50°C (the results are similar). After 2 h H-antigen agglutinations can be read and the morphology of agglutinates are typical.

For the determination of the O-antigen agglutination titre, it is necessary to wait 18 h, keeping the tubes either in the incubator or at room temperature.

4. *The Widal reaction*

Various techniques use the same principle. We describe one tube agglutination technique. The serum to be studied is diluted 1/10, 1/20 and 1/160 in saline. Of each dilution 0.1 ml is distributed in Kahn tubes, in numbers corresponding to the number of the bacterial suspension used, e.g. eight tubes if one uses the suspension O and H of *S. typhi*, *S. paratyphi* Groups A, B and C. Then 0.9 ml of the bacterial suspensions is added. The tubes are incubated for 2 h at 37°C for the H-agglutination and overnight at room temperature for the O-agglutination. A concave mirror may make it easier to read the results.

The choice of cells to be used routinely must be adapted according to local requirements. For example, if *S. paratyphi* A does not exist in a particular country or geographical area inclusion of this serotype would not be required except in imported cases. A suspension of *S. paratyphi* C, O and H may be useful for work on the agglutinins of Group C, or if agglutinins for H:c are of interest.

Sera of patients with typhoid syndromes generally exhibit agglutinins at about day 10. O-agglutinin titres increase to 1/400 and disappear with an interval of one to three months after recovery. H-agglutinin titres rise to 1/800 or 1/1600 and remain increased after the recovery (low titres of e.g. 1/200 can remain for years). Cross-reactions are often observed between the O-antigens of the Group B and the O-antigens of the Group D. This is based on the O:12-antigen common to both serogroups (Table II).

Titres such as *S. typhi* O 1/400, *S. paratyphi* BO 1/100 and *S. typhi* H 1/800 are observed after the second week in patients with typhoid.

An elevated *S. typhi* H titre of 1/200 may be observed in individuals which formerly had typhoid fever, or which have been vaccinated with a typhoid vaccine (during the weeks after vaccination, the H titre is still higher and O-agglutinins are present, as during illness). If the vaccine used also contains *S. paratyphi* B, its H-agglutinins will also be found.

If only O-agglutinins for *S. typhi* are found, possibly also agglutininating the O-antigens of *S. paratyphi* B, the illness may be caused by any *Salmonella* within the Group D, also with antigen specificities other than in the cell antigen used in the set up, e.g. *S. enteritidis*. This possibility is strengthened if the H-agglutinins of *S. enteritidis* are found.

The Widal reaction can only serve as a diagnostic indication and never be interpreted as proof of a certain species diagnosis. It is of lower value than the isolation and identification of the pathogenic agent. Cross-reactions may be observed also with bacteria belonging to a genus other than *Salmonella*. For example, *Y. pseudotuberculosis* type II cross-reacts with *Salmonella* of Group D and type IV with Group B. Hence, agglutination of an O-antigen suspension of *Salmonella* (H-antigens are more specific) is no proof of a *Salmonella* infection.

Otherwise it is necessary to remember that more than 2000 serotypes of *Salmonella* are known. Among these 12 are the most frequent and 14 somewhat less frequent (Kelterborn, 1979). In one particular country about 300 serotypes may be found (Le Minor and Le Minor, 1981). It is necessary to know the principle of the Kauffmann–White scheme in order to interpret the results of the Widal reaction. For example, it is normal to find high O-agglutinin titre for *S. paratyphi* B and high H-agglutinin titre for *S. typhi* when the illness is due to *S. stanley*, which is a serotype of Group B that possesses H:*d* as does *S. typhi*.

B. Preparation of antisera

Here is an abstract of the more detailed document BAC 78/1 published by WHO (Le Minor and Rohde, 1978).

1. *General*

Sera are prepared using rabbits. These animals produce high titre agglutinins.

Some rabbits spontaneously have anti-O *Salmonella* agglutinins, most often anti-group D antibodies. It is therefore advisable, before immunization, to take a blood sample of about 2 ml from the marginal vein of the ear and to determine whether the serum contains agglutinins against a *S. typhi* O-antigen suspension commonly employed in the Widal test. Rabbits with spontaneous agglutinin titres of 1/100 or more should not be used because these agglutinins decrease the specificity of the immune serum and must subsequently be absorbed.

To obtain uniform results, the sera are prepared, as far as possible, by inoculating standard strains distributed by the Centre for Reference and Research on *Salmonella* (Pasteur Institute, Paris).

Designation of antigenic factors is conventional. An antigen factor is identified by means of an antiserum from a rabbit which has been immunized with a specific culture, and absorbed by other hetero-specific cultures. This preparation, too, is conventional and an agreement on modes of preparation, the strains which should be and those which should not be agglutinated by this antiserum, is essential to achieve reproducibility between laboratories. That is why WHO has published the methods used in two laboratories that have been cooperating for twelve years in the field of the *Salmonella*. These methods are not fixed and permanent; they will be continually improved as our knowledge of *Salmonella* serology advances.

Control tests are most important; they should not be overlooked, whereas immunization and absorption methods can be modified.

(a) *O-antisera.* A perfectly smooth colony is plated on nutrient agar. After 18 h incubation at 37°C, the culture is harvested in saline (0.85% NaCl) and kept for 2.5 h in boiling water in order to destroy the H-antigen. The suspension is centrifuged and the bacterial pellet resuspended in saline, titrated photometrically and adjusted to 2×10^9 bacteria per millilitre. At intervals of four days the rabbits receive intravenously 1, 2, 4 and 6×10^9 bacteria (0.5, 1, 2 and 3 ml). They are bled eight to ten days after the last injection. The titres obtained are usually in the range 1/1600–1/6400.

Special case of O:5- and O:2-antisera. A culture of a non-flagellated strain should be used. It is harvested in saline with the addition of 0.5% (v/v) formalin (formalin is an aqueous solution of 30% formaldehyde). A promising method is the use of synthetic *Salmonella* O-antigens to have specific antisera (Svenungsson and Lindberg, 1978, 1979; Jörbeck *et al.*, 1979a).

(b) *Vi-antisera.* To prepare Vi-antisera a *Citrobacter* strain with the Vi-antigen (Ballerup, 7851) is normally employed. This culture has no O- or H-antigens common to the *Salmonella*. Consequently, any agglutination of a *Salmonella* in this serum will most likely be due to the presence of a Vi-antigen. Conversely, to detect the presence of Vi-antigen in *Citrobacter* strains, this serum should not be employed, but rather a serum prepared with a *S. typhi* rich in Vi-antigen should be used (e.g. strain Ty 6S which has no O-antigen, little H-antigen and a considerable amount of Vi-antigen).

The Ballerup culture is streaked on agar. The opaque colonies (richest in Vi) are collected separately and suspended in saline. Formalin (final concentration 0.5% v/v) is added to the suspension which is titrated photometrically and then adjusted to 2×10^9 bacteria per millilitre. Rabbits are immunized as described for the preparation of O-antisera. Titration is carried out using a standard Vi-antigen suspension (see below). The agglutination titres are usually in the range 1/800–1/1600.

For slide agglutination this anti-Vi-serum is diluted to 1/20–1/50, and the controls tests are performed with Vi-positive and Vi-negative *Salmonella*.

(c) *H-antisera.* To avoid having O-agglutinins in H-antisera it is best to use rough (R) variants of *Salmonella*, i.e. variants that have lost their O-antigen.

The problem is simple if monophasic *Salmonella* are used; it is easy to prepare a *d*-antiserum with an R-form of *S. typhi*. Motility needs just to be enhanced by passage through a U tube containing semi-solid agar followed by a broth culture. After 8 h incubation at 37°C in a water bath, formalin should be added to a final concentration of 0.5% (v/v). It is more complicated when diphasic *Salmonells* must be used. The strain must be passaged several times on semi-solid agar supplemented with serum containing the agglutinin corresponding to the phase which is to be suppressed. Even when working with the utmost care and patience, it is sometimes difficult to get rid of one phase completely.

The sufficiently motile culture is inoculated by flooding on swarm agar [a broth supplemented with 0.4–0.8 g agar (depending upon the quality of batches) 100 ml^{-1}] to which the serum corresponding to the unwanted phase has been added. After 8 h incubation at 37°C, the culture is harvested in saline, treated with formalin, adjusted to 2×10^9 bacteria per millilitre and injected into rabbits as described above. Agglutination titres are commonly about 1/25 000.

Titration of H-antisera is carried out against specific H-antigen suspensions corresponding to each of the phases and against the homologous O-antigen suspension.

If the antiserum is purely anti-H and specific for one phase only, it will be ready for slide agglutination after suitable dilution. Otherwise (and this is the most frequent case), the O-agglutinins (if the strain was not completely rough) and the H-agglutinins corresponding to the undesired phase must be absorbed.

Other methods for extraction of the H-antigen can be employed, but they are more tedious, e.g. see Aleksic and Rohde (1972) and Fey (1979).

(d) *Absorption and control methods.* The amount of bacteria needed for adsorption can only be estimated by experience. Bacteria are conveniently grown on nutrient agar in a Roux bottle when a fairly large volume of serum needs be absorbed.

After 18 h the culture is harvested in saline supplemented with an antiseptic, e.g.:

A. Concentrated solution for storage

Mercuric iodide	5 g
Potassium iodide	20 g
Water	500 ml

This should be kept in a stoppered bottle at room temperature, and in the dark.

B. Diluted solution to be used for harvesting the cultures
    Solution A   100 ml
    Formalin (30% formaldehyde)   1 ml
    Normal saline   1000 ml

The bacteria are centrifuged in screw-capped tubes. The supernatant is discarded and the pellets are stored at $+4°C$. They can be used over a period of at least six months.

The antiserum to be absorbed is diluted 1/5, and carefully mixed with the bacteria. The mixture is kept for 2 h at $37°C$ (water bath) and then centrifuged. A slide agglutination test is made to determine whether it still agglutinates the absorbing strain. If the result with the absorbed serum is still positive, absorption is repeated. If it is negative then other tests are made or, if necessary, absorptions continued with the other strain.

When the absorptions are finished and the control tests are satisfactory, the antiserum should be diluted to the highest dilution (in a geometric series) which still agglutinates a culture harbouring the corresponding antigen factor(s) well (within a minute). Since these antisera are intended for slide agglutination the control tests should be made by this method.

Strains used for immunization and absorption of the most commonly used antisera are listed in Tables IV–V (see Section III.B.2.a for an explanation of Tables IV–VII and their symbols). These are examples of methods which are at present satisfactory. Other methods can be used provided that the antisera be properly controlled for sensitivity and specificity.

Specificity controls are limited to groups with which cross-agglutinations are known to occur. The extent of such cross-agglutinations varies. For example, the cross-reactions of O:30 antiserum with Group I and J strains can be of a low titre. In such cases a mere dilution of the serum generally eliminates their interference with other antigens in slide agglutination. If cross-agglutinins cannot be eliminated by dilution, absorption(s) is necessary.

Absorptions and control testing of antisera corresponding to accessory O-antigen factors (i.e. antigen determinants other than those defining O-antigen groups) can be restricted to the O-antigen group considered. For example, it is not essential to absorb the O:27-antiserum (an accessory Group B determinant) by a Group E4 culture (O:1,3,19) with which there is a cross-reaction, as O:27-antiserum is only used with Group B cultures (i.e. cultures having O:4-antigen factor). It should be kept in mind that all antigenic relationships are not detailed in the Kauffmann–White scheme. The latter contains only factors of diagnostic interest.

It is essential to control antisera very carefully at the time of preparation. Subsequently, especially for highly absorbed antisera, it is wise to check their sensitivity periodically with a positive strain in order to avoid false-negative agglutinations caused by a drop in titre.

Each H-antiserum should be subjected to control tests to ensure that it is free from O-agglutinins. If present, O-agglutinins must be removed by absorption with bacteria heated for 2 h at 100°C, using as far as possible, a culture of the same serotype as the one employed to immunize the animals.

The control tests should be carried out with strains having the antigens shown in Tables IV–V. The antigenic relationships may either be apparent in the Kauffmann–White scheme (e.g. *g,m* and *g,p*) or they may not be indicated by the usual symbol notations (e.g. *i* and z_6).

It is not practical to check each antiserum against all strains; the choice of absorptions and control tests is based on experience. There may be individual variations in the response of rabbits immunized at the same time with the same antigen suspension. We have indicated in brackets (Tables IV–V) certain absorptions which are sometimes necessary and on other occasions unnecessary. For example, to prepare a z_{36}-antiserum, absorption by *S. lille* (z_{38}) is generally sufficient, but if *S. loenga* (its z_{38}-factor is not identical with that of *S. lille*) is still agglutinated, absorption must be carried out with that strain. The same symbol used to designate an antigen possessed by different serotypes signifies that these antigens are closely related. However, it cannot be assumed that they are identical. Conversely, extensive cross-reactions between antigens designated by different symbols in the Kauffmann–White scheme may be found.

Very motile cultures on swarm agar may be used if stored at +4°C. Before stored cultures are used for control tests, it should be assured that they are still strongly agglutinable by the homologous antiserum. As with the O-antigen factors, a certain logical order in looking for the H-antigen factors should be followed. The major H-antigen factors should be identified before looking for the accessory ones. For example, it must be confirmed that a strain has factor z_4 before testing for z_{23}, z_{24} or z_{32}. This avoids unnecessary absorptions in preparing these factor-specific antisera.

(i) *Comments on absorption.* O-antigens are heat-stable whereas H-antigens are heat-labile. A suspension of O- + H-antigen bacteria, which has been used to absorb H-antibodies, can be reused for absorbing O-antibodies; it is merely necessary to resuspend the bacteria in saline, heat them for 2 h at 100°C, and again centrifuge them. If required, the same bacteria, after heating for 2 h at 100°C, can be used for two successive absorptions of O-antibodies.

It is also possible to render lipopolysaccharides insoluble by adding glutaraldehyde and to use them for repeated absorptions. All that is needed is to remove the fixed antibodies each time by trypsin digestion (Eskenazy *et al.*, 1975).

TABLE IV
Group-specific O-antisera

O-antiserum	Immunization	Absorption	Controls (O-antigen group) −	Controls (O-antigen group) +
A/ 1, 2[a]	S. paratyphi A 205 (1, 2, 12: −:−)	S. typhi O 901 (9, 12) + S. typhi-murium SL 501 (4, 12)	B, D_1 (1^-)	A, E_4
B/ 4	S. essen (4, 12)	S. typhi O 901 (9, 12) + S. kentucky (8, 20)	A, C_2, C_3, D_1, D_2, E_4	B
4, 5[b]	S. paratyphi B 1 (4, 5, 12)	S. typhi O 901 (9, 12) + S. kentucky (8, 20)	A, C_2, C_3, E_4, M	B
C/($C_1 + C_2 + C_3$)	S. thompson (6_1, 6_2, 7) + S. newport (6_1, 8) + S. kentucky (8, 20)	S. westerstede (1, 3, 19) + S. waycross (41)	B, D_1, E_1, E_4, F, $G(1^+)$, S	C_1, C_2, C_3, C_4, H
D/ 9	S. typhi T_2 (9, 12_1, 12_3)	S. paratyphi A (1, 2, 12_1, 12_3)	A, B, E_4	D_1, D_2, D_3
E/ 3, 10	S. london (3, 10)	0	B (27^+), F, S	D_2, E_1, E_2, E_3, E_4
3, 15	S. newington (3, 15)	0	D, C, F	D_2, E_1, E_2, E_3, E_4 [S] [54][c]
1, 3, 19[d]	S. senftenberg (1, 3, 19)	0	D_1 (1^-), C_1, F	E_1, E_2, E_3, E_4 [$G(1^+)$]
F/ 11	S. aberdeen (11)	S. egusi (41) + S. vietnam (41) + S. anatum (3, 10)	C_1, E_1, E_2, E_4, S	11
G/ 13, 22	S. poona (13, 22, 36)	S. enteritidis (9, 12) + S. senftenberg (1, 3, 19)	D, E_1, E_4, 51	G_1, G_2, [U][e]

13, 23	*S. grumpensis* (13, 23, 36)	*S. senftenberg* (1, 3, 19) + S_{IV} *houten* (43_1, 43_3, 43_4) + *S. milwaukee* (43_1, 43_2, 43_3)	E_1, E_4, U. 51	G_1, G_2
H/ 6, 14, 24	*S. carrau* (6, 14, 24)	*S. bredeney* (4, 12, 27) + *S. beloha* (18)	B (27^+), K	C_1, C_2, C_4, H
I/ 16	*S. hvittingfoss* (16)	*S. godesberg* (30_1, 30_2)	N, Q, R	I
J/ 17	*S. kirkee* (17)	*S. midhurst* (53)	M, 53	J (SG I, II, III)
K/ 18	*S. cerro* (18)	*S. boecker* (6, 14) + *S. onderstepoort* (1, 6, 14, 25)	H	K
L/ 21	*S. minnesota* (21)	*S. anatum* (3, 10) [+*S. uccle* (54)]	E_1, E_2, E_4 [54]	L
M/ 28	*S. tel-aviv* (28_1, 28_2) + *S. dakar* (28_1, 28_3)	*S. champaign* (39) + S_{II} *mondeor* (39)	D (12_2^+) N, O, Q	M
N/ 30	*S. urbana* (30_1, 30_2)	0	H, I, J, R	N
O/ 35	*S. adelaide* (35)	0	M	O
P/ 38	*S. inverness* (38)	0	57, S_{III} 57	P
Q/ 39	*S. champaign* (39)	*S. dakar* (28_1, 28_3) + *S. pomona* (28_1, 28_2) + *S. frankfurt* (16)	I, M, J, 51	Q
R/ 40	*S. riogrande* (40_1, 40_2) + *S. bulawayo* (l, 40_1, 40_3)	*S. urbana* (30_1, 30_2) + *S. angola* (30) + *S. landau* (30) + *S. senftenberg* (1, 3, 19)	A (1^+), N, E_4	R, S_{III} 40
S/ 41	*S. waycross* (41)	*S. etterbeck* (11) + *S. rossleben* (54)	C_1, E_1, F	S. 54

TABLE IV — *continued*

O-antiserum	Immunization	Absorption	Controls (O-antigen group) −	Controls (O-antigen group) +
T/42	*S. weslaco* (42)	*S. uccle* (54)	54	T
U/43	*S. milwaukee* (43_1, 43_2, 43_3)	*S. worthington* (*1*, 13, 23, 37) + *S. grumpensis* (13, 23, 36) + *S. poona* (13, 22, 36) + *S. tione* (51)	G, 51	U
V/44	*S. niarembe* (44_1, 44_2) + *S. clovelly* (44_1, 44_3)	*S. treforest* (*1*, 51) + *S. senftenberg* (1, 3, 19)	E_4, 51	V
W/45	*S. deversoir* (45_1, 45_2) + *S. dugbe* (45_1, 45_3)	0	Z, S_{III} 50	W
X/47	*S. kaolak* (47_1, 47_3) + *S. bergen* (47_1, 47_2)	0	Y	X
Y/48^f	*S. dahlem* (48_1, 48_2) + S_{III} 48_1, 48_3, 48_4:i:z_{35}	S_{III} 61:c:z_{35} + S_{III} 61:k:1,5,7	X, 61	Y (48_1, 48_2—48_1, 48_3 and S_{III} 48_1, 48_3, 48_4)
Z/50	*S. greenside* (50)	*S. adelaide* (35)	G, V, W	Z
51	*S. treforest* (51)	*S. vleuten* (44)	E_4, V, R, U	51
52	*S. utrecht* (52)	0		52
53	*S. humber* (53)	*S. bleadon* (17)	J, 59	53
54	*S. uccle* [(7), 3, 42_3, 54]	*S. weslaco* (42) + *S. anatum* (3, 10)	E_1, E_2, S, T	54
55	*S. tranaroa* (55)	0		55

56	S. artis (56)	0		56
57	S. locarno (57)	0	P	57
58	S. basel (58)	0		58
59	S. betiocky (58)	0	53	59
60	S. luton (60)	0		60
61	S_{III} $61_1, 61_2$:k:1, 5, 7 $+S_{III}$ $61_1, 61_3$:l:v:1, 5, 7	S. dahlem ($48_1, 48_2$) + S. djakarta ($48_1, 48_2, 48_3$) + S_{III} $48_1, 48_3, 48_4$:i:z_{35}	X ($47_1, 47_2-47_1, 47_3$) Y ($48_1, 48_2-48_1, 48_2, 48_3 -48_1, 48_3, 48_4$)	61 ($61_1, 61_2-61_1-63_3$)
62	S_{III} 62:g:z_{51}:—	0		62
63	S_{III} 63:z_4, z_{32}:—	[S. luton 0 60]	60	63
64 ($=48_4$)	combined with 48f			
65	S_{III} 65:i:e, n, x, z_{15}	[S_{III} $61_1, 61_3$:i:z_{53}]	61	65
66	S. brookfield	[S. niarembe ($44_1, 44_2$)]	V (44)	66
67	S. crossness	0		67

---

[a] This antiserum, which is easier to prepare than O:2-antiserum, is suitable for routine diagnosis of Group A serotypes. Agglutination with other serotypes having the O:1-antigen occurs in particular with those of Group E4 (O:1,3,19). However, the latter are agglutinated by O:3,10- and O:3,15-antisera, unlike those of Group A.

[b] With this antiserum, Group B strains can be easily diagnosed. Only strains with factor O:4 can have factor O:5. When the latter is strongly agglutinable, the factor O:4 characteristic of the group is less strongly agglutinated than in the O:5-strains. It is therefore often more interesting to use an O:4,5-antiserum directly rather than an O:4:antiserum. Detection of the O:5-antigen can proceed subsequently.

[c] Agglutination with O:54 causes no trouble in practice: this group is rare and agglutinates only with O:3,15-antiserum and not with the other Group E antisera.

[d] The same antiserum, but more concentrated (dilution 1/10), is recommended for detection of O:1-antigen in groups other than Group E.

[e] Agglutination of Group U strains in this antiserum is not a problem in practice: strains O:13,22 are agglutinated by O:13,23-antiserum and so are strains O:13,23 by O:13,22-antiserum. Only one of these antisera needs to be absorbed by Group U cultures.

[f] Winkle (1976).

TABLE V
Accessory O-antisera

O-antiserum	Immunization	Absorption	Controls (O-antigen group)	
			−	+
IA	S. paratyphi A (1, 2, 12)	S. paratyphi A subsp. durazzo (2, 12)	A, B, D (1⁻)	A, B, D (1⁺), E₄
I for group other than E	S. senftenberg (1, 3, 19)	0	A, B, D₁, R, T, 51(1⁻)	A, B, D₁, R, T 51 (1⁺)
2	S. paratyphi A subsp. durazzo (2, 12) (formolized)	S. typhi (9, 12) + S. enteritidis (9, 12) + S. essen (4, 12) + S. abortus equi (4, 12)	B, D, E₄	A
5	S. paratyphi B (4, 5, 12) (formolized)	S. typhi-murium (4, 12) + S. paratyphi B (4, 12) [+ S. schwarzengrund (4, 12, 27)]	A, B (5⁻), E₄	B (5⁺)
6, 7ᵃ	S. thompson (6₁, 6₂, 7)	S. typhi O 901 (9, 12)	C₃, D₁, E₁, F, S, 54	C₁, C₂, H
6, 8ᵃ	S. newport (6, 8)	0	D₁	C₁, C₂, C₃, H
7	S. lomita (6₂, 7)	S. onderstepoort (1, 6, 14, 25) [+ S. anatum (3, 10) + S. aberdeen (11)]	C₂, C₃, H [E₁, F]	C₁
8	S. virginia or S. amherstiana (8)	0	C₁, C₄, H	C₂, C₃

10	*S. london* (3, 10)	*S. newington* (3, *15*) +*S. kinshasa* (3, *15*) +*S. niloese* (1, 3, 19) [+*S. minnesota* (21)]	E_2, E_3, E_4, L	E_1, D_2[b]
<u>14</u>	*S. boecker* (6, 14)	*S. muenchen* (6, 8) +*S. thompson* (6, 7) +*S. mbandaka* (6, 7) +*S. cholerae suis* 5210 N^+ (6_2, 7, N) +*S. langenhorn* (18)	C_1, C_2, K S_{III} 6, 7 (*Ar.* 27) +S_{III} 18 (*Ar.* 7a, 7b)	C_1 (14^+) ($=C_4$) H, K (6, 14^+) S_{III} 6, 14 (*Ar.* 7a, 7c)
15	*S. newington*	*S. london* (3, 10) +*S. orion* (3, 10)	E_1, E_4	E_2, E_3, D_2 (15^+)
19	*S. senftenberg* (1, 3, 19)	*S. london* (3, 10) +*S. schwarzengrund* (*1*, 4, 12, 27)	A (1^+), B (27^+) E_1, E_2, E_3	E_4
20	*S. kentucky* (8, 20)	*S. newport* (6, 8) +*S. glostrup* (6, 8) +*S. cholerae suis* 5210 (6, 7)	C_1, C_2, C_3 (20^-), C_4	C_3 (20^+)
22	*S. poona* (13, 22, 36)	*S. grumpensis* (13, 23, 36) [+*S. fanti* (13, 23, 36)]	G_2 (1^+ and 1^-)	G_1 (1^+ and 1^-)
23	*S. grumpensis* (13, 23, 36)	*S. poona* (13, 22, 36) +*S. clifton* (13, 22, 36)	G_1 (1^+ and 1^-)	G_2 (1^+ and 1^-)
24[d]	*S. carrau* (6, 14, 24)	*S. ondersteptoort* (1, 6, 14, 25) +*S. harburg* (1, 6, 14, 25) +*S. bornum* (6, 7, 14)]	C_4 ($=C_1$ 14^+), H (25^+) [K (14^+)]	H (24^+)
25[e]	*S. ondersteptoort* (1, 6, 14, 25)	*S. senftenberg* (1, 3, 19) [+*S. carrau* (6, 14, 24)] +*S. bahrenfeld* (6, 14, 24) +*S. cholerae suis* 5210 N^{e+} (6, 7 N) +*S. surat* 25^- (6, 14)	E_4, H (25^-) [C_4 and C_1 N^+]	H (25^+)

TABLE V —continued

O-antiserum	Immunization	Absorption	Controls (O-antigen group)	
			−	+
27	S.II 1, 9, 12, (46), 27:y:z$_{39}$	*S. typhi* (9, 12) + *S. dublin* (*1*, 9, 12) + *S. paratyphi A* (*1, 2, 12*) + *S. strasbourg* (9, 46) + *S. baildon* (9, 46)	A, B, D (27⁻), E$_4$	B (27⁺), D$_3$
34	*S. illinois* (3, *15*, 34)	*S. newington* (3, *15*) + *S. senftenberg* (1, 3, 19) [+ *S. typhi* (9, 12)]	E$_1$, E$_2$, E$_4$ [D]	D$_2$ (*34⁺*), E$_3$
46	*S. strasbourg* (3, 10, 9, 46)	*S. anatum* (3, 10) + *S. enteritidis* (9, 12)	D, E$_1$	D$_2$ [D$_3$]
T$_1^f$	*S. paratyphi B* T$_1$ form (T$_1$:b:1, 2)		B	T$_1$ forms

[a] The presence of factor O:6 in *Salmonella* Group C is shown by agglutination in O:6 14, 24-antisera. This factor is always 6$_1$ in *Salmonella* Group C2 (6, 8). A Group C1 (6, 7) culture agglutinated by O:6.8-antiserum therefore has the O:6$_1$-factor, either alone or together with factor O:6$_2$. Factor O:6$_1$ may be subject to form variation differences in intensity of agglutination according to the particular colony. Factor O:6 of Group H is different from both factors O:6$_1$ and O:6$_2$.

[b] Unconverted Group D2 strains (e.g. *S. strasbourg*) have O:10-antigen. Consequently, they can be agglutinated by O:10-antiserum. This antigen can turn into O:*15*-antigen after conversion by phage ε15, and then into O:*15, 34, 12$_2$*-antigen after subsequent conversion by phage ε34.

[c] The antigen provisionally designated as *N* may exist in Group C1 strains (Le Minor, 1968). To avoid complicating the Kauffmann–White scheme it has not been assigned a number, but its existence should be known to specialists, since it cross-reacts with factors O:1, 14 and O:25. N antiserum is prepared with an antiserum to *S. livingstone N*+ which is absorbed by *S. livingstone 14*+, *S. onderstepoort* and *S. mikawasima*.

[d] O:24-antigen is subject to form variation. The colonies which are most strongly agglutinable by antiserum to O:24 should be used for immunizing rabbits.

[e] O:1- and O:25-antigens are subject to simultaneous form variation. The colonies most strongly agglutinable by O:1-antiserum should be selected for immunization.

[f] Factors connected with phage conversion are set in italic and those to which this does not apply (6$_1$ of Group C$_1$) are underlined with a broken line.

TABLE VI
Major H-antisera

O-antiserum	Immunization	Absorption	Controls (O-antigen group) −	Controls (O-antigen group) +
a	S. paratyphi A	S. chittagong (z_{35}) + S_{III} 48:k:z_{35}	z_{35} (SG I, II, III), z_6, z_{10}	a (SG I, II, III)
b	S. paratyphi B ph. 1	—	1, 2	b
c	S. cholerae suis ph. 1	—	1, 5	c
d	S. typhi	—	—	d
e, h	S. newport ph. 1	—	—	e, h e, n, x e, n, z_{15}
E mixture	S. abortus equi (e, n, x)	—	—	e, h e, n, x e, n, z_{15}
G mixture	S. oranienburg (m, t) + S. derby (f, g) + S. dublin (g, p)	—	—	g, m g, p g, t g, q g, m, s g, z_{51} m, t f, g g, p, u
g, m	S. enteritidis	—	—	
g, p	S. dublin	—	m, t	
r^a	S. typhi-murium ph. 1	—	r z_6 1, 2	i

TABLE VI – *continued*

O-antiserum	Immunization	Absorption	Controls (O-antigen group)	
			−	+
k standard[b]	S. thompson ph. 1	S. mura (l, w)	1, 5 L complex (l, v / l, w / l, z_{13} / l, z_{28}) z	k
L mixture	S. worthington ph. 2 (l, w)	S. neumuenster (k)	k e, n, z_{15} 1, 2 1, 5	L complex
r	S. virchow ph. 1	S. typhi-murium (i)	1, 2 i (SG I, II, III)	r (SG I, II, III)
y	S. bareilly ph. 1	—	1, 5	y
z standard[c]	S. poona ph. 1	S. orientalis (k) +S_{III} 41:k:− +S. chittagong (z_{35}) +S. weslaco (z_{36}) +S. cairina (z_{35}) +S. quinhon (z_{44})	a k (SG I, II) e, n, x e, n, z_{15} l, v l, w z_{35} z_{36} z_{44}	z (SG I, II, III)
Z mixture	S. cerro (z_4, z_{23}) or S. tallahassee (z_4, z_{32})	—	—	z_4, z_{23} z_4, z_{24} z_4, z_{32}

z_6	S. kentucky ph. 2	S. bovis morbificans (1, 5)	1, 5 a i	z_6
z_{10}	S. glostrup ph. 1 or S. guinea monoph. (z_{10})	[S. paratyphi A (a) + S. miami (a)]	a (SG I, II, III) z_{53} e, n, z_{15} z z_6 z_{35}	z_{10}
z_{29}	S. tennessee	—		z_{29}
z_{35}	S. chittagong ph. 2	S. paratyphi A (a) + S. arechavaleta (a) + S. poona (z)	a (SG I, III) z (SG I, II) b z_6 z_{36} z_{38} z_{44}	z_{35} (SG I, II, III)
z_{36}^d	S. weslaco	S. lille (z_{38})	z_{38} z_{35}	z_{36}(SG I, II, III, IV) z_{36}, z_{38}
z_{38}^d	S. lille	S. weslaco (z_{36}) S_{IV} argentina (z_{36})	z_{36} (SG I, II, III, IV)	z_{38} z_{36}, z_{38}
z_{39}	S_{II} springs ph. 2	S. paratyphi A (a) + S_{II} bloemfontein (z_{42}) + S_{II} elsiesrivier (z_{42})	a z_{42}	z_{39}
z_{41}	S. karamoja ph. 1	S_{II} springs (z_{39}) + S. quinhon (z_{44})	z_{39} z_{44} 1, 2	z_{41}
z_{42}	S_{II} bunnik ph. 2	S_{II} springs (z_{39}) [+ S_{II} soutpan (z_{39})]	z_{39}	z_{42}

TABLE VI — *continued*

O-antiserum	Immunization	Absorption	Controls (O-antigen group)	
			−	+
z_{44}	S. quinhon	S_{II} greenside (z) + S. karamoja (z_{41}) [+ S. westerstede (l, z_{13})]	z z_{35} z_{36} z_{37} z_{41} e, n, x e, n, z_{15} l, z_{13} l, z_{40}	z_{44} (coagglutinates with L complex, in particular l, z_{40} and l, z_{13})
z_{60}	S. aesch ph. 1 (z_{60})	[S. newport ph. 2 (1, 2)]	$1, 2$	z_{60}
z_{64}	S. aarhus ph. 2 (z_{64})	S. sternschange (z_{59})	z_{59} z_4, z_{23}	z_{64}
z_{65}	S. malawi (z_{65})			z_{65}
1 mixture	1, 2 (S. paratyphi B ph.2) +1, 5 (S. cholerae suis ph. 2) +z_6 (S. kentucky ph. 2)	S. neukoelln (l, z_{28}) +S. vredelust (l, z_{28}) +S. vrom (l, z_{13}, z_{28})	l, z_{13} l, z_{28} l, z_{13}, z_{28}	$1, 2$ $1, 5$ $1, 6$ $1, 7$ z_6

[a] Antigenic relationships exist between *i* and *r*. In general with a high titre *i*-antiserum dilution may eliminate anti-agglutinins. Such a diluted antiserum is adequate for routine work. An absorbed antiserum should be used in special cases, e.g. for checking the *r*, *i*-complex which does not contain the complete *i*-factor.

[b] This antiserum is suitable for all *Salmonella* subgenus I cultures.

[c] H:z-antigen is complex and shows considerable differences in the three subgenera. For that reason the use of a z-antiserum with a wider spectrum (polyvalent z) for study of the strains of subgenera II and III, is recommended (WHO, 1978).

[d] Because of the antigenic relationship between H:z_{36} and H:z_{38} it is necessary to use absorbed antisera.

TABLE VII
Minor H-antisera

O-antiserum	Immunization	Absorption	Controls (O-antigen group) −	Controls (O-antigen group) +
f	S. derby (4, 12:f, g: −)	S. essen (g, m) + S. budapest (g, t) + S. dublin (g, p) + S. congo (g, t) [+ S. rostock (g, p, u)]	g, m g, p g, t g, q g, m, s g, z_{51} m, t	f, g f, g, t (SG I, II)
h	S. newport ph. 1 (6,8:e.h:1,2)	S_{11} dar-es-salaam (e, n, x) + S. potsdam (e, n, z_{15}) + S. abortus equi (e, n, x)	e, n, x e, n, z_{15}	e, h
m	S. oranienburg (6, 7:m.t:—)	S. senftenberg (g, s, t) + S. berta (f, g, t) + S_{11} foulpointe (g, t) + S. moscow (g, q) + S. rostock (g, p, u) [+ S. naestved (g, p, s)] + S. budapest (g, t) + S. wayne (g, z_{51})	g, s, t f, g, t g, t g, q f, g g, z_{51}	g, m g, m, s m, t
n, x, z_{16}	S. abortus equi (4, 12: − :e, n, x, z_{16})	S. reading (e, h) [+ S. eastbourne (e, h)] [+ S. onderstepoort (e, h)] [+ S. mara (e, h)] [+ S_{11} greenside (z)]	e, h	e, n, x e, n, z_{15}
n, z_{15}	S. san-diego (4, 5, 12:e, h:e, nz_{15}, z_{17})	S. reading (e, h) [+ S. eastbourne (e, h)] [+ S. mara (e, h)]	e, h	e, n, x e, n, z_{15}

TABLE VII – *continued*

O-antiserum	Immunization	Absorption	Controls (O-antigen group)	
			−	+
p	*S. dublin* ($1, 9, 12$:g, p:−)	*S. enteritidis* (g, m) +*S. montevideo* (g, m, s) +*S. moscow* (g, q) +S_{II} *fremantle* (f, g, t) +*S. senftenberg* (g, s, t) +*S. wayne*	$f, g, t - g, t$ $g, m - g, m, s$ $g, m, s, t - g, s, t$ $g, q - g, m, t$	$g, p - g, p, s$ $g, p, u - m, p, t, u$ ($S.$ *haelsingborg*) g, z_{51} (*SG I, III, IV*) g, m, p, s (*S. montevideo* subsp. $p+$)
q	*S. moscow* ($9, 12$:g, q:−)	*S. enteritidis* (g, m) [+*S. montevideo* (g, m, s)] *S. wayne* (g, z_{51}) [+*S. rostock* (g, p, u)]	$f, g - f, g, t$ $g, t - g, s, t,$ $g, m - g, m, s$ $g, m, s, t - g, p$ m, p, t, u $g, p, s, - m, t$ g, z_{51}	$g, q - g, m, q$
s	*S. montevideo* ($6, 7$:g, m, s:−)	*S. enteritidis* (g, m) +*S. essen* (g, m) +*S. oranienburg* (m, t) +*S. budapest* (g, t) [+*S. dublin* (g, p)] +*S. moscow* (g, q) +*S. wayne* (g, z_{51}) +*S. derby* (f, g)	$g, m - g, m, t$ $g, t, - g, p, - g, p, s$ $g, p, u - g, q$ $m, p, t, u -$ ($S.$ *haelsingborg*) $m, t - f, g - f, g, t$ ε, z_{51} (*SG I, III, IV*)	$g, m, s - g, m, s, t$ $g, s, t - g, p, s$ f, g, s
t	*S. oranienburg* ($6, 7$:m, t:−)	*S. montevideo* (g, m, s) +*S. rostock* (g, p, u) +*S. wayne* (g, z_{51}) [+S_{IV} *wassenaar* (g, z_{51})] [+S_{IV} *marina* (g, z_{51})] +*S. blegdam* (g, m, q)	$g, m - g, m, s - g, p$ $g, p, u - g, p, s$ $f, g - f, g, s$ $g, z_{51} - g, m, q$	$m, t - g, m, s, t - g, m, t$ $g, t - g, s, t, - f, g, t$ g, t

u	S. rostock (1, 9, 12:g, p, u:—)	S. dublin (g, p) [+S. moscow (g, q)] [+S. adelaide (f, g)]	g, p − g, q − g, m g, m, s, t − g, s, t f, g, t − g, z_{51} g, p, s, − m, p, t	g, p, u − m, p, t, u
v	S. london monoph. (3, 10:l, v:—)	S. westerstede (l, z_{13}) + S. makiso (l, z_{13}, z_{28}) + S. basel (l, z_{13}, z_{28}) + S. eimsbuettel (l, w) + S_{II} dar-es-salaam (l, w) + S. rutgers (l, z_{40})	l, w l, z_{13} − l, z_{13}, z_{28} l, z_{28} − l, z_{40}	l, v − l, (v), w [S_{II} 17:l, (v), w l, v SG III (Ar. H_{23})]
w	S. livingstone ph. 2 (6, 7:d:l, w)	S. london (l, v) + S. westerstede (l, z_{13}) + S. javiana monoph. (l, z_{28}) + S. niloese (d)	l, v (SG I, II, III) (Ar. H_{25}) l, z_{13}−l, z_{28} (SG I, II) l, z_{13}, z_{28} (SG I, II) l, z_{40} − d	l, w
x, z_{16} [a]	S. abortus equi (4, 12:—:e, n, x, z_{16})	S. san-diego (e, n, z_{15}) + S. tel-aviv (e, n, z_{15}) + S. poona (z) + S_{II} bulawayo (z)	e, n, z_{15} e, h z (SG I, II, III) (Ar. H_{31})	e, n, x, z_{16} e, n, x, z_{17} e, n, x, z_{15} (= e, n, z_{15}, z_{16})
x [b]	S. chester (4, 5, 12:e, h, e, n, x, z_{17})	S_{II} dar-es-salaam (e, n, z_{16}, z_{18}) + S. glostrup (e, n, z_{15}, z_{17}) + S. san-diego (e, n, z_{15}, z_{17}) + S_{111} 48_1, 48_2:k:e, n, x, z_{15} (e, n, x, z_{15} = e, n, z_{15}, z_{16}) + S. mara (e, h)	e, h e, n, x (SG III) (= e, n, z_{16}) e, n, x, z_{16}) = e, n, z_{15}, z_{16}	e, n, x, z_{16} e, n, x, z_{17}
z_{15} [b]	S. 4, 12:—:e, n, z_{15}	S. chester (e, n, x, z_{17}) + S_{II} helsinki (e, n, x = e, n, z_{16})	e, n, x, z_{17} (e.g. S. bonariensis) e, n, x, z_{16} (e.g. S. abortus-equi) e, n, x (SG 11) (= e, n, z_{16}) (e.g. S_{II} dar-es-salaam) e, h	e, n, z_{15} e, n, z_{15}, z_{16} (SG II, III) (Ar. H_{28})

TABLE VII — *continued*

O-antiserum	Immunization	Absorption	Controls (O-antigen group) −	Controls (O-antigen group) +
z_{13}	S. uganda (3, 10:l, z_{13}:1, 5)	S. dar-es-salaam (l, w) + S. panama (l, v)	l, v (SG I, II, III)	l, z_{13} l, z_{13}, z_{28}
		+ S. javiana (l, z_{28}) + S. astraal (l, z_{28}) + S. rutgers (l. z_{40}) [+ S. shangani (1, 5)] [+ S. thompson (l, 5)]	l, w − l, z_{28} (SG I, II) l, z_{40} 1, 5	
z_{23}	S. cerro (z_4, z_{23})	S. tallahassee (z_4, z_{32})	z_4, z_{24} z_4, z_{32}	z_4, z_{23}
z_{24}	S. duesseldorf (z_4, z_{24})	S. tallahassee (z_4, z_{32})	z_4, z_{23} z_4, z_{32}	z_4, z_{24}
z_{28}	S. javiana monoph. (1, 9, 12:1, z_{28}:−)	S. london (l, v) + S. panama (l, w) + S_{II} dar-es-salaam (l, w) + S. livingstone (l, w) + S. westerstede (l, z_{13})	l, v l, w l, z_{13} l, z_{40}	l, z_{28} l, z_{13}, z_{28}
z_{32}	S. tallahassee (6, 8:z_4, z_{32}:−)	S. cerro (z_4, z_{23}) + S. duesseldorf (z_4, z_{24}) + S. wangata (z_4, z_{23}) + S. gera (z_4, z_{23})	z_4, z_{23} (SG I, II, III IV) z_4, z_{24} (SG I, II, III IV)	z_4, z_{32} (SG I, II, III, IV)
z_{51} (global)	S. wayne (30:g, z_{51}:−)	S. dublin (g, p) + S. budapest (g, t) + S_{II} mobeni (g, m, s, t) + S. moscow (g, q) + S. berta (f, g, t)	G complex except g, z_{51}	g, z_{51} (SG I, III, IV)

	or S_{II} 4, 12:g, z_{62}:-	+ S. kulsrvver (g, m, s, t)	g, t	g, z_{62}
		+ S. congo (g, m, s, t)	g, m	
		+ S. bechuana (g, t)	g, m, s, t	
			g, p	
			g, q	
			g, m, q	
			g, z_{51}	
z_{63}	S. antarctica (9, 12:g, z_{63}:—)	S. enteritidis (g, m)	G complex except g, z_{62}	g, z_{62}
2	S. paratyphi B ph. 2 (1, 2)	S. arechavaleta (1, 7)	1, 5	1, 2
		+ S. bovis morbificans (1, 5)	1, 5, 7	1, (2), 7
		+ S. panama (1, 5)	1, 6	1, 2, 3
		+ S. anatum (1, 6)	1, 7	(e.g. S. typhi-murium
			z_6	S. muenchen
			l, v	S. newport)
5	S. cholerae-suis subsp. kunz. (1, 5)	S. anatum (1, 6)	1, 2	1, 5
		+ S. newport (1, 2)	1, 2, 3	1, 5, 7 (SG II, III)
		+ S. westerstede (1, z_{13})	1, 6	
			1, 7	
			z_6	
			l, v	
6	S. london ph. 2 (1, 6)	S. thompson (1, 5)	1, 2	1, 6
		+ S. muenster (1, 5)	1, 2, 3	1, 6, 7 (S. oakland)
		+ S. cholerae-suis var. kunz. (1, 5)	1, 5	
			1, 5, 7	
			1, 7	
			z_6	
7	S. bredeney ph. 2 (1, 7)	S. newport (1, 2)	1, 2	1, 7
		+ S. typhi-murium (1, 2)	1, 5	1, 5, 7 (SG II, III)
		+ S. panama (1, 5)	1, 6	
		+ S. muenster (1, 5)	z_6	
		+ S. illinois (1, 5)	l, v	
		+ S. potsdam (1, v)		

[a] The x-antiserum advocated by Kauffmann for diagnosis is an H:x, z_{16} antiserum.

[b] Most of the e, n, x phases of subgenus I contain both H:x- and :z_{16}-antigens and their full formula is consequently H:e, n, x, z_{16}. Exceptionally, the e, n, x, z_{17} phase may be found. Pure x-factor never exists in the e, n, x phases of SG_{II} and e, n, z_{15} phases of SG III. They have z_{16} (H:e, n, z_{16} or H:e, n, z_{15}, z_{16}), whereas the factor shown as x in their formula is in fact z_{16}.

(e) *Storage of antisera*

(i) *Crude antisera.* The simplest method consists of bleeding rabbits under aseptic conditions. The antiserum is separated from the clot (e.g. by placing a metal cylinder on the clot more antiserum can be collected), centrifuged and the supernatant dispensed under sterile conditions in ampoules in a suitable volume, and stored at $+4°C$. Loss of titre is negligible even after several years.

(ii) *Absorbed or diluted antisera for use in slide agglutination.* Filter if necessary through membrane filters and add merthiolate (final concentration of 1/10 000) or sodium azide (final concentration of 1/1000: e.g. 0.5 ml of a solution containing 20 g 100 ml^{-1} added to 100 ml of antiserum). The sera are stored at $+4°C$ for routine use.

(f) *Media for storage of strains.* *Salmonella* cultures remain viable for several years if they are stored in well-stoppered (screw-capped) tubes and in the dark. Simple media are suitable.

Meat extract	5 g
Peptone	10 g
NaCl	3 g
$Na_2HPO_4 \times 12H_2O$	2 g
Agar	10 g
Distilled water	1000 ml
pH 7.4	

This medium is dispensed in small tubes, sterilized for 20 min at 110°C and allowed to set. Small tubes measuring 9 × 9 mm and hermetically sealed by caps, or paraffined cork, or rubber stoppers (in the latter two cases the tubes are plugged with cotton wool for sterilization and the cotton wool replaced by sterile stoppers after inoculation). Inoculation is carried out by a central stab. After overnight incubation, the tubes are labelled and stored in the dark at room temperature. There is no need, at least in temperate climates, to store them in the refrigerator.

When the stored strains need be used again a portion of the culture is transferred to a sterile broth with an inoculating loop. After overnight incubation, it is streaked on nutrient agar. When isolated colonies develop on the agar, a smooth or a rough colony is selected as needed.

2. *The main O- and H-antisera*

For information on production and controls of specific typing sera of *Salmonella*, see the Tables IV–VII.

(a) *Polyvalent O- and H-antisera*

(i) *General comments.* Antigenic analysis cannot be carried out with these antisera, but in the absence of antisera prepared for that purpose they may be useful in some cases. They are easy to prepare, but the indications for their use are limited to the few serotypes isolated most frequently from blood cultures in adults, especially *S. typhi.*

(ii) *Presentation of tables and symbols.* Tables IV–VII comprise five columns. The first shows the agglutinins required, the second the strain employed for immunization with, for H-antisera, the phase selected when there is no monophasic variant. The third column shows, following the same principle, the strains to be used for absorptions. There may be variation of intensity in the agglutinin response corresponding to certain antigenic factors in rabbits immunized with the same bacterial suspension or suspensions prepared in an identical way after an interval of a few months. Certain additional absorptions are sometimes necessary. These are indicated in brackets.

When it is necessary to use strains of subgenus III (*S. arizonae*), for control tests these are shown in the Tables IV–VII.

The last two columns indicate the O-antigen groups or H phases of strains which should be utilized for control tests of specificity and activity. It is wise to use at least two strains of the group indicated for these controls, belonging to different serotypes.

For simplification, we adopt the following symbols for designating the subgenera.

1. The serotypes of subgenus II are designated by S_{II} followed by the specific name (e.g. S_{II} *rowbarton*). Unnamed serotypes (described after the IXth International Congress of Microbiology, Moscow, 1966) are designated by S_{II} and the formula (e.g. S_{II} 58:b:1,5).
2. Serotypes of subgenus III (*S. arizonae*) are designated by S_{III} followed by the *Salmonella* formula in accordance with the Kauffmann–White scheme as edited by the International Reference and Research Centre on *Salmonella* and published by WHO (1974) and Appendices I and II.
3. Serotypes of subgenus IV (*S. houtenae*) are designated according to the same principle for subgenus II (e.g. S_{IV} *bonaire*, S_{IV} $50:z_4:z_{24}:-$).

Preparation. A broth inoculated with a typical *S. typhi* strain that has been passaged on semi-solid agar (U tube or swarm agar in a Petri dish) to enhance motility. After 8 h incubation at 37°C, formalin is added to the broth (0.5% v/v final concentration). A sterility test is carried out the next day. The formalin-treated broth can be used, even without titration of bacterial density, for immunizing rabbits intravenously by doses of 0.5, 1, 2 and 3 ml injected at intervals of four days. Rabbits are bled ten days after the last injection.

This antiserum, intended principally for slide agglutination, is diluted to 1/10, 1/20, 1/40 and 1/640. Each dilution is submitted to a slide test against a *S. typhi* strain cultured on agar medium. The maximal dilution which gives very distinct agglutination in a few seconds should be employed. Sterility is maintained by addition of merthiolate (0.01% w/v) or sodium azide (0.1% w/v).

This antiserum may give granular agglutinations with bacteria sharing the O-antigens of *S. typhi* (e.g. *S. enteritidis*) and flocculent agglutination by bacteria having the H:*d*-antigen in common with *S. typhi*. The probability of finding blood cultures with *Salmonella* serotypes other than *S. typhi* possessing H:*d* is low. Moreover, their biochemical characteristics are different from that of *S. typhi* (gas$^-$, little or no H_2S, LDC$^+$, ODC$^-$, Simmons citrate test$^-$).

In regions where typhoid fever is endemic it is advisable to have a Vi-antiserum available (see above), also because a high incidence of Vi-strains which are O-non-agglutinable but not very motile, and, consequently, possess less H-antigen. Such strains will be poorly agglutinated by polyvalent O- + H-antiserum, but strongly agglutinated by Vi-antiserum.

N.B. The titres usually obtained are about O: 3200; H: 25 000; Vi: 800.

In the same way, antisera for *S. paratyphi* A, *S. paratyphi* B or *S. paratyphi* C should be prepared, if these bacteria are frequent in the region concerned.

Although these antisera have limited indications, it is better to have them available than to have no antiserum at all.

(b) *O-antisera*

(i) *Mixed O-antisera.* These antisera are prepared by mixing the un-absorbed constituents antisera in a final dilution that gives a strong agglutination with the corresponding strains.

These mixed O-antisera are intended to guide diagnosis, but definite conclusions can be reached only by agglutination in group-specific O-antiserum. Any O-agglutinable *Salmonella* culture will immediately agglutinate in one of these mixed O-antisera. A later delayed and less intense agglutination may occur in one or more of the other mixed antisera which, it will be remembered, are not absorbed.

The very great majority (of the order of 98%) of *Salmonella* strains isolated from man and warm-blooded animals have an O-antigen corresponding to the agglutinins present in the OMA- (O Mixture A) and OMB-antisera.

OMA = O-agglutinins of Groups A, B, D, E, L
   (O:1,2,12 + 4,5,12 + 9,12 + (9),46 + 3,10 + 1,3,19 + 21)
OMB = O-agglutinins of Groups C, F, G, H
   (O:6,7 + 6,8 + 11 + 13,22 + 13,23 + 6,14,24 + 8,20)
OMC = O-agglutinins of Groups I, J, K, M, N, O, P
   (O:16 + 17 + 18 + 28 + 30 + 35 + 38)
OMD = O-agglutinins of Groups Q, R, S, T, U, V, W
   (O:39 + 40 + 41 + 43 + 44 + 45)
OME = O-agglutinins of Groups X, Y, Z, 51 to 53
   (O:47 + 48 + 50 + 51 + 52 + 53)
OMF = O-agglutinins of Groups 54 + 55 + 56 + 57 + 58 + 59
OMG = O-agglutinins of Groups 60 + 61 + 63 + 65 + 66 + 67

We do not feel that it is advisable to recommend a single O-antiserum corresponding to all the *Salmonella* O-antigen factors, because such sera coagglutinate with many cultures not belonging to the genus *Salmonella*. Diagnosis of the genus *Salmonella* is based on biochemical characteristics and a polyvalent *Salmonella* O-antiserum is of little value and causes considerable confusion. It seems preferable rather to use the Felix O1 phage for screening purposes. An antigen study should only be undertaken when it is certain that the culture being studied is a *Salmonella*. This study should begin by identifying the O-antigen group. The mixed O-antisera can serve as a guide, but they are not essential.

(ii) *Group-specific O-antisera*. Group specific O-antisera are the most important because they should be used first for identifying the *Salmonella* culture O-antigen group.

Only when the O-antigen group has been determined, should other antigenic determinants be determined because they have a lesser discriminating value, i.e. for differentiation within the group. For example, in Group B the important factor, characteristic of the group, is O:4. Factor O:5 exists only if the O:4-antigen is present. If the abequose is not O-acetylated at C2 the strain will not have the O:5-antigen, but the serotype is the same.

O:1-antigen exists in the Group B only if the bacteria are lysogenized by special converting phages. When the phage genome is lost, the bacterium loses the O:1-antigen, but the diagnosis of the serotype will not change. The same applies to O:27-antigen, determined by another phage. Consequently, the diagnostic value of such O-antigens is only supplementary.

Finally, identification of factors with very limited discriminating value is omitted in everyday work. This is the case of O:12-antigen for example.

(c) *H-antisera.* *Salmonella* H-antigens are more specific for the bacteria of this genus (whose diagnosis should be based on biochemical and not on antigenic characteristics). Some food bacteriologists use a polyvalent H-antiserum, sometimes conjugates with fluorescein, for detecting *Salmonella*. Such an antiserum can not be used to establish antigenic formulae.

(i) *Mixed H-antisera for purposes of orientation.* These antisera labelled HMA (H Mixture A) are not absorbed. They are made up by mixing the antisera indicated below, each of these being in the final dilution as in the monospecific antisera. They are useful for orientation purposes in laboratories identifying *Salmonella* of very diverse serotypes (e.g. in food inspection). A diagnosis can be based only on agglutination in the specific antisera.

Composition
$HMA = a + b + c + d + i + z_{10} + z_{29}$
$HMB = e,h + e,n,x + G$
$HMC = k + y + z + L + z_4 + r$
$HMD = z_{35} + z_{36} + z_{38} + z_{39} + z_{41} + z_{42} + z_{44} + z_{60}$

Another HM-antiserum may be prepared mixing H-antisera corresponding to H-antigens only found in strains belonging to the subgenus III.

(ii) *Mixed H-antisera for "phase reversal".* By means of these mixed antisera labelled SG1 (Sven Gard 1), the number of antisera necessary for reverting the phases of the H-antigen can be reduced.

Composition
$SG1 = $anti-$a + b + c + z_{10}$
$SG2 = $anti-$d + i + e,h$
$SG3 = $anti-$k + y + l,w$ (convenient to immobilize also
$$l,v - l,z_{13} - l,z_{28})$$
$SG4 = $anti-$r + z$
$SG5 = $anti-$e,n,x$ (convenient to immobilize also e,n,z_{15})
$SG6 = $anti-$1,2 + 1,5,7 + z_6$ (convenient to immobilize also
$$1,5 - 1,6 - 1,7)$$

Use
It is first necessary to specify the agar content of the nutrient agar intended for H-phase reversal, for which a Petri dish is commonly used. It may vary according to the batch of agar from 0.4 to 0.8%. The agar must be sufficiently

soft for a spot-inoculated motile *Salmonella* to spread on the surface of the medium after overnight incubation. Preliminary trials, without addition of antiserum, are therefore necessary, employing either a nutrient agar specially prepared for the purpose or ordinary laboratory nutrient agar with the addition of the necessary amount of broth to give it the requisite qualities. The volume required for a 10-cm diameter Petri dish is 30 ml.

Phase reversal technique

1. Add one drop of SG-antiserum corresponding to the unwanted phase (i.e. the phase already determined) to melted agar brought to about 45°C.
2. Mix serum and medium by a gentle rotatory movement and then pour it into a Petri dish.
3. After setting (make sure that no drops of water remain on the surface, otherwise dry with the lid partly open) spot inoculate in the centre (one loopful of culture taken from the agar medium).
4. Incubate at 37°C for 18 h.
5. Scratch with an inoculating loop at the periphery of the swarming zone and look for the other phase H.

Example: O:4,5- and H:*1,2*-antigens have been determined in a *Salmonella* under study. SG6-serum should be used for inverting the phase. The next day agglutination will be found in H:*i*-antiserum in the case of *S. typhimurium* and in H:*b*-antiserum in the case of *S. paratyphi* B.

The same antisera can be employed for phase inversions in a U tube containing semi-solid nutrient agar 0.2–0.3% agar.

Acknowledgements

This work was supported by the Swedish Medical Research Council (Grant No. 16X–656) and the Swedish Board for Technical Development (to A. A. Lindberg).

References

Aleksic, S. and Rohde, R. (1972). *Ann. Inst. Pasteur* **123**, 363–377.
Anderson, E. S. (1962). *Ann. Inst. Pasteur* **102**, 379–388.
Ando, K. and Shimojo, J. (1953). *Bull. W.H.O.* **9**, 575–577.
Andrewes, F. W. (1922). *J. Pathol. Bacteriol.* **25**, 505–521.
Andrewes, F. W. (1925). *J. Pathol. Bacteriol.* **28**, 345–349.
Arakatsu, Y., Ashwell, G. and Kabat, E. (1966). *J. Immunol.* **97**, 858–866.
Asakura, S. (1970). *Adv. Biophys.* **1**, 99–155.

Ballou, C. E. (1976). *Adv. Microb. Physiol.* **14**, 93–158.

Bennet, L. G. and Bishop, C. T. (1977). *Immunochemistry* **14**, 693–696.

Berst, M., Hellerqvist, C. G., Lindberg, B., Lüderitz, O., Svensson, S. and Westphal, O. (1969). *Eur. J. Biochem.* **11**, 353–359.

Bruneteau, M., Volk, W. A., Singh, P. P. and Lüderitz, O. (1974). *Eur. J. Biochem.* **43**, 501–508.

Cherry, W. B., Davis, B. R., Edwards, P. R. and Hogan, R. B. (1954). *J. Lab. Clin. Med.* **44**, 51–55.

Craigie, J. and Yen, C. H. (1938). *Can. J. Public Health* **29**, 448–463.

Crosa, J. F., Brenner, D. J., Ewing, W. H. and Falkow, S. (1973). *J. Bacteriol.* **115**, 307–315.

Davidson, B. E. (1971). *Eur. J. Biochem.* **18**, 524–529.

DeLange, R. J., Chang, J. Y., Shaper, J. H., Martinez, R. J., Komatsu, S. K. and Glaser, A. N. (1973). *Proc. Nat. Acad. Sci. USA* **70**, 3428–3431.

DeLange, R. J., Chang, J. Y., Shaper, J. H. and Glaser, A. N. (1976). *J. Biol. Chem.* **251**, 705–711.

Edwards, P. R. (1945). *Proc. Soc. Exp. Biol. Med.* **59**, 49–52.

Ekborg, G., Eklind, K., Garegg, P. J., Gotthammar, B., Carlsson, H. E., Lindberg, A. A. and Svenungsson, B. (1977). *Immunochemistry* **14**, 153–157.

Ekwall, E., Svenson, S. B. and Lindberg, A. A. (1982). *J. Med. Microbiol.* **15**, 173–180.

Eskenazy, M., Strahilov, D., Ivanova, R. and Kalenova, R. (1975). *J. Clin. Microbiol.* **2**, 368–372.

Espersen, F., Høiby, N. and Hertz, J. B. (1980). *Acta Pathol. Microbiol. Scand. Sect. B* **88**, 243–248.

Felix, A. (1955). *Bull. W.H.O.* **13**, 109–170.

Fey, H. (1979). *Zentralbl. Bakteriol. Parasitenkd. Infektionskr. Hyg. Abt. Orig.* **245**, 55–66.

Flemming, H.-C. and Jann, K. (1978a). *Eur. J. Biochem.* **83**, 47–52.

Flemming, H.-C. and Jann, K. (1978b). *FEMS Microbiol. Lett.* **4**, 203–205.

Fuller, N. A. and Staub, A.-M. (1968). *Eur. J. Biochem.* **4**, 286–300.

Fuller, N. A., Etievant, M. and Staub, A.-M. (1968). *Eur. J. Biochem.* **6**, 525–533.

Fumahara, Y. and Nikaido, H. (1980). *J. Bacteriol.* **141**, 1463–1465.

Garegg, P. J., Lindberg, B., Onn, T. and Holme, T. (1971a). *Acta Chem. Scand.* **25**, 1185–1194.

Garegg, P. J., Lindberg, B., Onn, T. and Sutherland, I. W. (1971b). *Acta Chem. Scand.* **25**, 2103–2108.

Garegg, P. J., Jansson, P. E., Lindberg, B., Lindh, F., Lönngren, J., Kvarnström, I. and Nimmich, W. (1980). *Carbohydr. Res.* **78**, 127–132.

Gmeiner, J. (1975). *Eur. J. Biochem.* **51**, 449–457.

Gmeiner, J., Simon, M. and Lüderitz, O. (1971). *Eur. J. Biochem.* **21**, 355–356.

Gorin, P. A. J. and Ishikawa, T. (1967). *Can. J. Chem.* **45**, 521–532.

Heasley, F. A. (1981). *J. Bacteriol.* **145**, 624–627.

Hellerqvist, C.-G. and Lindberg, A. A. (1971). *Carbohydr. Res.* **16**, 39–48.

Hellerqvist, C.-G., Lindberg, B., Svensson, S., Holme, T. and Lindberg, A. A. (1968). *Carbohydr. Res.* **8**, 43–55.

Hellerqvist, C.-G., Lindberg, B., Svensson, S., Holme, T. and Lindberg, A. A. (1969a). *Carbohydr. Res.* **9**, 237–241.

Hellerqvist, C.-G., Lindberg, B., Svensson, S., Holme, T. and Lindberg, A. A. (1969b). *Acta Chem. Scand.* **23**, 1588–1596.

Hellerqvist, C.-G., Larm, O., Lindberg, B., Holme, T. and Lindberg, A. A. (1969c). *Acta Chem. Scand.* **23**, 2217–2222.
Hellerqvist, C.-G., Lindberg, B., Svensson, S., Holme, T. and Lindberg, A. A. (1970a). *Carbohydr. Res.* **14**, 17–26.
Hellerqvist, C.-G., Lindberg, B., Pilotti, Å. and Lindberg, A. A. (1970b). *Acta Chem. Scand.* **24**, 1168–1174.
Hellerqvist, C.-G., Lindberg, B., Samuelsson, K. and Lindberg, A. A. (1971a). *Acta Chem. Scand.* **25**, 955–961.
Hellerqvist, C.-G., Lindberg, B., Lönngren, J. and Lindberg, A. A. (1971b). *Acta Chem. Scand.* **25**, 601–606.
Hellerqvist, C.-G., Lindberg, B., Lönngren, J. and Lindberg, A. A. (1971c). *Carbohydr. Res.* **16**, 289–296.
Hellerqvist, C.-G., Lindberg, B., Pilotti, Å. and Lindberg, A. A. (1971d). *Carbohydr. Res.* **16**, 297–302.
Hellerqvist, C.-G., Lindberg, B., Lönngren, J. and Lindberg, A. A. (1971e). *Acta Chem. Scand.* **25**, 939–944.
Hellerqvist, C.-G., Hoffmann, J., Lindberg, B., Pilotti, Å. and Lindberg, A. A. (1971f). *Acta Chem. Scand.* **25**, 1512–1513.
Hellerqvist, C.-G., Hoffmann, J., Lindberg, A. A., Lindberg, B. and Svensson, S. (1972). *Acta Chem. Scand.* **26**, 3282–3286.
Heyns, K. and Kiessling, G. (1967). *Carbohydr. Res.* **3**, 340–353.
Iino, T. (1964). *In* "Salmonella: Genetics Today", pp. 731–740. Pergamon, London.
Iino, T. (1969). *Bacteriol. Rev.* **33**, 454–475.
Iino, T. (1977). *Annu. Rev. Genet.* **11**, 161–182.
Iino, T. and Lederberg, J. (1964). *In* "The World Problem of Salmonellosis" (E. Van Oye, Ed.), pp. 111–142. Junk Publishers, The Hague.
Iseki, S. and Kasiwagi, K. (1957). *Proc. Jpn. Acad.* **33**, 481–485.
Iseki, S. and Sakai, T. (1953). *Proc. Jpn. Acad.* **29**, 121–126.
Jörbeck, H., Carlsson, H. E., Svenson, S. B., Lindberg, A. A., Alfredsson, G., Garegg, P. J., Svensson, S. and Wallin, N. H. (1979a). *Int. Arch. Allergy Appl. Immunol.* **58**, 11–19.
Jörbeck, H. J. A., Svenson, S. B. and Lindberg, A. A. (1979b). *J. Immunol.* **123**, 1376–1381.
Jörbeck, H., Svenson, S. B. and Lindberg, A. A. (1980). *Int. Arch. Allergy Appl. Immunol.* **61**, 55–64.
Joys, T. M. (1976). *Microbios* **15**, 221–228.
Joys, T. M. and Rankis, V. (1972). *J. Biol. Chem.* **247**, 5180–5193.
Joys, T. M. and Stocker, B. A. D. (1966). *J. Gen. Microbiol.* **44**, 121–138.
Joys, T. M., Martin, J. F., Wilson, H. L. and Rankis, V. (1974). *Biochim. Biophys. Acta* **351**, 301–305.
Kabat, E. A. (1961). *In* "Experimental Immunochemistry" (E. A. Kabat and Mayer, Eds). Thomas, Springfield, Illinois.
Kabat, E. A. (1966). *J. Immunol.* **97**, 1–11.
Kauffmann, F. (1956). *Acta Pathol. Microbiol. Scand.* **39**, 299–304.
Kauffmann, F. (1957). *Acta Pathol. Microbiol. Scand.* **40**, 343–344.
Kauffmann, F. (1961). "Die Bakteriologie der *Salmonella* species", pp. 1–257. Munksgaard, Copenhagen.
Kauffmann, F. (1966). "The Bacteriology of Enterobacteriaceae", pp. 1–400. Munskgaard, Copenhagen.
Kauffmann, F. (1978). "Das Fundament", pp. 1–57. Munksgaard, Copenhagen.

Kelterborn, E. (1979). *Zentralbl. Bakteriol. Parasitenkd. Infektionskr. I. Orig.* **243**, 289–307.
Köhler, G. and Milstein, C. (1975). *Nature (London)* **256**, 495–497.
Kutsukate, K. and Iino, T. (1980). *Nature (London)* **284**, 479–481.
Langman, R. E. (1972). *Eur. J. Immunol.* **2**, 582–586.
Lederberg, J. and Iino, T. (1956). *Genetics* **41**, 743–757.
Le Minor, L. (1963). *Ann. Inst. Pasteur* **105**, 879–896.
Le Minor, L. (1966). *Ann Inst. Pasteur* **110**, 562–567.
Le Minor, L. (1968). *Ann. Inst. Pasteur* **114**, 49–62.
Le Minor, L. (1969). "Colloques Internationaux du Centre National de la Recherche Scientifique", No. 174, pp. 155–160. Ed. Du Centre National de la Recherche Scientifique, Paris.
Le Minor, L. and Le Minor, S. (1981). *Rev. Epid. Sant. Publ.* **29**, 45–55.
Le Minor, L. and Rohde, R. (1978). "WHO Guidelines for the Preparation of *Salmonella* Antisera", BAC 78/1.
Le Minor, L., Ackerman, H. W. and Nicolle, P. (1963). *Ann. Inst. Pasteur* **104**, 469–476.
Levinthal, M. and Nikaido, H. (1969). *J. Mol. Biol.* **42**, 511–520.
Lindberg, A. A. (1973). *Annu. Rev. Microbiol.* **27**, 205–241.
Lindberg, A. A., Hellerqvist, C.-G., Bagdian-Motta, G. and Mäkelä, P. H. (1978). *J. Gen. Microbiol.* **107**, 279–287.
Lüderitz, O., Galanos, C., Risse, H. J., Ruschmann, E., Schlecht, S., Schmidt, G., Schulte-Holthausen, H., Wheat, R., Westphal, O. and Schlosshardt, J. (1966). *Ann. N.Y. Acad. Sci.* **133**, 349–374.
Lüderitz, O., Freudenberg, M. A., Galanos, C., Lehmann, V., Rietschel, E. T. and Shaw, D. W. (1982). *In* "Microbial Membrane Lipids" (S. Razin and S. Rottem, Eds), pp. 79–151. Current Topics in Membranes and Transport, Vol. 2. Academic Press, New York and London.
Lvov, V. L., Tochtamysheva, N. V., Dmitriev, B. A., Kochetkov, N. K. and Hofman, I. L. (1981). *Bioorg. Chem.* **6**, 1842–1850.
McDonough, M. W. (1965). *J. Mol. Biol.* **12**, 342–355.
Macnab, R. M. (1978). *CRC Crit. Rev. Biochem.* **5**, 291–340.
Mäkelä, P. H. (1965). *J. Gen. Microbiol.* **41**, 57–65.
Mäkelä, P. H. (1973). *J. Bacteriol.* **116**, 847–856.
Mäkelä, P. H. and Jahkola, M. (1970). *J. Gen. Microbiol.* **60**, 91–106.
Mäkelä, P. H. and Mäkelä, O. (1966). *Ann. Med. Exp. Biol. Fenn.* **44**, 310–317.
Mäkelä, P. H. and Stocker, B. A. D. (1969). *Annu. Rev. Genet.* **3**, 291–322.
Mäkelä, P. H. and Stocker, B. A. D. (1981). *In* "Genetics as a Tool in Microbiology" (S. W. Glover and D. A. Hopwood, Eds), pp. 219–264.
Naide, Y., Nikaido, H., Mäkelä, P. H., Wilkinson, R. G. and Stocker, B. A. D. (1965). *Proc. Natl. Acad. Sci. U.S.A.* **53**, 147–153.
Nikaido, H. (1973). *In* "Bacterial Membranes and Walls" (L. Leive, Ed.), pp. 131–180. Dekker, New York.
Nikaido, H. and Nakae, T. (1979). *Adv. Microbiol. Physiol.* **20**, 163–250.
O'Brien, E. J. and Bennett, P. M. (1972). *J. Mol. Biol.* **70**, 133–152.
Ørskov, F. and Ørskov, I. (1978). *In* "Methods in Microbiology" (T. Bergan and J. R. Norris, Eds), Vol. 11, pp. 1–77. Academic Press, London and New York.
Osborn, M. J. (1979). *In* "Bacterial Outer Membranes" (M. Inouye, Ed.), pp. 15–34. Wiley, New York.
Osborn, M. J. and Rothfield, L. (1971). *In* "Microbial Toxins" (G. Weinbaum, S.

Kadis and S. J. Ajl, Eds), Vol. 4, pp. 331–350. Academic Press, New York and London.

Parish, C. R., Wistar, Jr, R. and Ada, G. L. (1969). *Biochem. J.* **113**, 501–506.

Plosila, M. and Mäkelä, P. H. (1972). *Scand. J. Clin. Lab. Invest.* **29**, 55.

Robbins, P. W. and Uchida, T. (1962a). *Fed. Am. Soc. Exp. Biol.* **21**, 702–710.

Robbins, P. W. and Uchida, T. (1962b). *Biochemistry* **1**, 323–335.

Robbins, P. W. and Uchida, T. (1965). *J. Biol. Chem.* **240**, 375–383.

Rohde, R. (1979). *Zentralbl. Bakteriol. Parisitenkd. Infektionskr. Abt. Orig. A.* **243**, 148–176.

Rohde, R., Aleksic, S., Muller, G., Plasovic, S. and Aleksic, V. (1975). *Zentralbl. Bakteriol. Parasitenkd. Infektionskr. Hyg. Abt. Orig.* **230**, 38–50.

Sanderson, K. E. and Hartman, P. E. (1978). *Microbiol. Rev.* **42**, 471–519.

Schmidt, G., Mayer, H. and Mäkelä, P. H. (1976). *J. Bacteriol.* **127**, 755–762.

Sertic, V. and Boulgakov, N. A. (1936). *C.R. Soc. Biol.* **123**, 887–888.

Silverman, M. and Simon, M. (1977). *Annu. Rev. Microbiol.* **31**, 397–419.

Silverman, M., Zieg, J., Hilman, M. and Simon, M. (1979a). *Proc. Natl. Acad. Sci. U.S.A.* **76**, 391–395.

Silverman, M., Zieg, J. and Simon, M. (1979b). *J. Bacteriol.* **137**, 517–523.

Simmons, D. A. R., Lüderitz, O. and Westphal, O. (1965a). *Biochem. J.* **97**, 807–814.

Simmons, D. A. R., Lüderitz, O. and Westphal, O. (1965b). *Biochem. J.* **97**, 815–819.

Simon, M., Zieg, J., Silverman, M., Mandel, G. and Doolittle, R. (1980). *Science* **209**, 1370–1374.

Smith, A. M. and Potter, M. (1976). *J. Immunol.* **114**, 1847–1850.

Smith, A. M., Miller, J. S. and Whitehead, D. S. (1979). *J. Immunol.* **123**, 1715–1720.

Staub, A.-M. (1961). *Pathol. Microbiol.* **24**, 890–909.

Staub, A.-M. and Girard, R. (1965). *Bull. Soc. Chim. Biol.* **47**, 1245–1268.

Staub, A.-M., Tinelli, R., Lüderitz, O. and Westphal, O. (1959). *Ann. Inst. Pasteur* **96**, 303–332.

Staub, A.-M., Stirm, S., Le Minor, L., Lüderitz, O. and Westphal, O. (1966). *Ann. Inst. Pasteur* **111**, 47–58.

Stellner, K., Lüderitz, O., Westphal, O., Staub, A.-M., Leluc, B., Coynault, C. and Le Minor, L. (1972). *Ann. Inst. Pasteur* **123**, 43–54.

Stocker, B. A. D. (1949). *J. Hyg.* **47**, 308–413.

Stocker, B. A. D. and Mäkelä, P. H. (1971). *In* "Microbial Toxins" (G. Weinbaum, S. Kadis and S. J. Ajl, Eds), Vol. 4, pp. 369–438. Academic Press, New York and London.

Stocker, B. A. D., McDonough, M. W. and Ambler, R. P. (1961a). *Nature (London)* **189**, 556–558.

Stocker, B. A. D., Staub, A.-M., Tinelli, R. and Kopacka, B. (1961b). *Ann. Inst. Pasteur* **98**, 505–523.

Svenson, S. B. and Lindberg, A. A. (1978). *J. Immunol.* **120**, 1750–1757.

Svenungsson, B. and Lindberg, A. A. (1977). *Med. Microbiol. Immunol.* **163**, 1–11.

Svenungsson, B. and Lindberg, A. A. (1978a). *Acta Pathol. Microbiol. Scand. Sect. B.* **86**, 35–40.

Svenungsson, B. and Lindberg, A. A. (1978b). *Acta Pathol. Microbiol. Scand. Sect. B* **86**, 283–290.

Svenungsson, B. and Lindberg, A. A. (1979). *Acta Pathol. Microbiol. Scand. Sect. B* **87**, 29–36.

Svenungsson, B., Jörbeck, H. and Lindberg, A. A. (1979). *J. Infect. Dis.* **140**, 927–936.

Uchida, T., Robbins, P. W. and Luria, S. E. (1963). *Biochemistry* **2**, 663–668.

Uetake, H., Luria, S. E. and Burrows, J. W. (1958). *Virology* **5**, 68–91.
WHO (1980). "WHO Collaborating Centre for Reference and Research on *Salmonella*", BD/72–1 Rev.
Winkle, I. (1976). *Ann. Microbiol.* **127B**, 463–472.
Wollin, R., Eriksson, U. and Lindberg, A. A. (1981). *J. Virol.* (in press).
Wright, A. and Barzilai, N. (1971). *J. Bacteriol.* **105**, 937–939.
Wright, A. and Kanegasaki, S. (1971). *Physiol. Rev.* **51**, 749–783.
Zieg, J., Silverman, M., Hilmen, M. and Simon, M. (1977). *Science* **196**, 170–172.
Zinder, N. (1957). *Science* **126**, 1237.

Appendix I

Kauffmann–White scheme supplemented by the formulae approved up to 31 December 1977 and including those for *S. arizonae* (*Arizona*)

[ ] = may be absent
( ) = not very developed (weakly agglutinable)

The symbols for somatic factors whose presence is connected with phage conversion are in italic (e.g. *6*, *14*, 18). They are present only if the culture is lysogenized by the corresponding converting phage. These factors are mentioned in the table for serotypes in which they were found. It is probable that most, if not all, types in a group could be converted by these bacteriophages.

Groups C_4, E_2 and E_3 are retained in this edition, although it has been shown that the serotypes belonging to them are, respectively, those of Groups C_1, lysogenized by phage 14 (6, 7), and E_1, lysogenized by ε_{15} or $\varepsilon_{15} + \varepsilon_{34}$. No further serotypes have been added to these groups, which are retained provisionally.

All the subgenous I serotypes bear a name.

S. II = subgenus II *Salmonella*

S. (II) = atypical subgenus II

The serotypes belonging to this subgenus and described before the Moscow International Congress (1966) bear a name (e.g. *S.* II *sofia*). Those described subsequently are designated solely by their formula (e.g. *S.* II *1*, 4, 12, *27*; z:1, 5).

S. III = subgenus III *Salmonella* (= *Arizona*)

The serotypes of this subgenus appear in the table with the name *Salmonella arizonae*, followed by the formula according to the symbols used in the Kauffmann–White scheme and in parentheses, the formula according to Edwards, Fife and Ewing. The extent to which these two formulae correspond has been established by R. Rohde.

S. IV = subgenus IV *Salmonella*

Type	Somatic O-antigen	Flagellar H-antigen	
		Phase 1	Phase 2
Group O:2 (A)			
S. paratyphi A	1, 2, *12*	a	[1, 5]
S. nitra	2, 12	g, m	—
S. kiel	*1*, 2, 12	g, p	—

| | | Flagellar H-antigen | |
Type	Somatic O-antigen	Phase 1	Phase 2
	Group O:4 (B)		
S. kisangani	*1*, 4, [5], 12	a	1, 2
S. hessarek	4, 12, *27*	a	1, 5
S. fulica	*4*, [5], 12	a	1, 5
S. arechavaleta	4, [5], 12	a	[1, 7]
S. bispebjerg	*1*, 4, [5], 12	a	e, n, x
S. tinda	*1*, 4, 12, *27*	a	e, n, z_{15}
S. II *makoma*	4, [5], 12	a	—
S. nakuru	*1*, 4, 12, *27*	a	z_6
S. paratyphi B[a]	*1*, 4, [5], 12	b	1, 2
S. limete	*1*, 4, 12, *27*	b	1, 5
S. canada	4, 12	b	1, 6
S. uppsala	4, 12, *27*	b	1, 7
S. abony	*1*, 4, [5], 12, *27*	b	e, n, x
S. abortusbovis	*1*, 4, 12, *27*	b	e, n, x
S. II *sofia*	*1*, 4, 12, *27*	b	[e, n, x]
S. wagenia	*1*, 4, 12, *27*	b	e, n, z_{15}
S. wien	*1*, 4, 12, *27*	b	l, w
S. schleissheim	4, 12, *27*	b	—
S. legon	*1*, 4, 12, *27*	c	1, 5
S. abortusovis	4, 12	c	1, 6
S. altendorf	4, 12, *27*	c	1, 7
S. jericho	*1*, 4, 12, *27*	c	e, n, z_{15}
S. hallfold	*1*, 4, 12, *27*	c	l, w
S. bury	4, 12, *27*	c	z_6
S. stanley	*1*, 4, [5], 12, *27*	d	1, 2
S. eppendorf	*1*, 4, 12, *27*	d	1, 5
S. brezany	*1*, 4, 12, *27*	d	1, 6
S. schwarzengrund	*1*, 4, 12, *27*	d	1, 7
S. II *kluetjenfelde*	4, 12	d	e, n, x
S. sarajane	4, [5], 12, *27*	d	e, n, x
S. duisburg	*1*, 4, 12, *27*	d	e, n, z_{15}
S. salinatis	4, 12	d, e, h	d, e, n, z_{15}
S. mons	*1*, 4, 12, *27*	d	l, w
S. ayinde	*1*, 4, 12, *27*	d	z_6
S. saintpaul	*1*, 4, [5], 12	e, h	1, 2
S. reading	*1*, 4, [5], 12	e, h	1, 5
S. eko	4, 12	e, h	1, 6
S. kaapstad	4, 12	e, h	1, 7
S. chester	*1*, 4, [5], 12	e, h	e, n, x
S. san-diego	4, [5], 12	e, h	e, n, z_{15}
S. II *makumira*	*1*, 4, 12, *27*	e, n, x	q, [5], 7

[a] Variety D-tartrate-positive is often called variety java.

Type	Somatic O-antigen	Flagellar H-antigen	
		Phase 1	Phase 2
S. derby	*1*, 4, [5], 12	f, g	[1, 2]
S. agona	*1*, 4, 12	f, g, s	—
S. II	*1*, 4, [5], 12	f, g, t	$z_6:z_{42}$
S. essen	4, 12	g, m	—
S. hato	4, [5], 12	g, m, s	—
S. II *caledon*	*1*, 4, 12, 27	g, m, [s], t	e, n, x
S. II *bechuana*	*1*, 4, 12, 27	g, [m], t	[1, 5]
S. II	4, 12	g, m, t	z_{39}
S. california	4, 12	g, m, t	—
S. kingston[b]	*1*, 4, [5], 12, 27	g, s, t	[1, 2]
S. budapest	*1*, 4, 12, 27	g, t	—
S. travis	4, [5], 12	g, z_{51}	1, 7
S. tennyson	4, 5, 12	g, z_{51}	e, n, z_{15}
S. II	4, 12	g, z_{62}	—
S. banana	4, [5], 12	m, t	1, 5
S. typhimurium	*1*, 4, [5], 12	i	1, 2
S. lagos	*1*, 4, [5], 12	i	1, 5
S. agama	4, 12	i	1, 6
S. tsevie	4, 12	i	e, n, z_{15}
S. gloucester	*1*, 4, 12, 27	i	l, w
S. massenya	*1*, 4, 12, 27	k	1, 5
S. neumuenster	*1*, 4, 12, 27	k	1, 6
S. II	*1*, 4, 12, 27	k	1, 6
S. ljubljana	4, 12, 27	k	e, n, x
S. texas	4, [5], 12	k	e, n, z_{15}
S. fyris	4, [5], 12	l, v	1, 2
S. azteca	4, [5], 12, 27	l, v	1, 5
S. clackamas	4, 12	l, v	1, 6
S. bredeney[c]	*1*, 4, 12, 27	l, v	1, 7
S. kimuenza	*1*, 4, 12, 27	l, v	e, n, x
S. II	*1*, 4, 12, 27	l, v	e, n, x
S. brandenburg	*1*, 4, 12	l, v	e, n, z_{15}
S. II	*1*, 4, 12, 27	l, v	z_{39}
S. mono	4, 12	l, w	1, 5
S. togo	4, 12	l, w	1, 6
S. II *kilwa*	4, 12	l, w	e, n, x
S. ayton	*1*, 4, 12, 27	l, w	z_6
S. kunduchi	*1*, 4, [5], 12, 27	1, z_{13}, z_{28}	1, 2
S. tyresoe	4, 12	1, [z_{13}], z_{28}	1, 5
S. kubacha	*1*, 4, 12, 27	1, z_{13}, z_{28}	1, 7
S. kano	*1*, 4, 12, 27	1, z_{13}, z_{28}	e, n, x
S. vom	*1*, 4, 12, 27	1, z_{13}, z_{28}	e, n, z_{15}

[b] May possess a R-phase H-antigen: z_{43}.

[c] May possess a R-phase H-antigen: z_{40}.

		Flagellar H-antigen	
Type	Somatic O-antigen	Phase 1	Phase 2
S. reinickendorf	4, 12	$1, z_{28}$	e, n, x
S. II	4, 12	$1, z_{28}$	—
S. heidelberg	*1*, 4, [5], 12	r	1, 2
S. bradford	4, 12, 27	r	1, 5
S. remo	*1*, 4, 12, 27	r	1, 7
S. bochum	4, [5], 12	r	l, w
S. southampton	*1*, 4, 12, 27	r	z_6
S. drogana	*1*, 4, 12, 27	r, i	e, n, z_{15}
S. africana	4, 12	r, i	l, w
S. coeln	4, [5], 12	y	1, 2
S. trachau	4, 12, 27	y	1, 5
S. teddington	*1*, 4, 12, 27	y	1, 7
S. ball	*1*, 4, [5], 12, 27	y	e, n, x
S. jos	*1*, 4, 12, 27	y	e, n, z_{15}
S. kamoru	4, 12, 27	y	z_6
S. shubra	4, [5], 12	z	1, 2
S. kiambu	4, 12	z	1, 5
S. II	*1*, 4, 12, 27	z	1, 5
S. indiana	*1*, 4, 12	z	1, 7
S. neftenbach	4, 12	z	e, n, x
S. II *nordenham*	*1*, 4, 12, 27	z	e, n, x
S. koenigstuhl	*1*, 4, 12	z	e, n, z_{15}
S. preston	*1*, 4, 12	z	l, w
S. entebbe	*1*, 4, 12, 27	z	z_6
S. stanleyville	*1*, 4, [5], 12, 27	z_4, z_{23}	[1, 2]
S. kalamu	4, [5], 12	z_4, z_{24}	[1, 5]
S. haifa	*1*, 4, [5], 12	z_{10}	1, 2
S. ituri	*1*, 4, 12	z_{10}	1, 5
S. tudu	4, 12	z_{10}	1, 6
S. albert	4, 12	z_{10}	e, n, x
S. tokoin	4, 12	z_{10}	e, n, z_{15}
S. mura	*1*, 4, 12	z_{10}	l, w
S. fortune	*1*, 4, 12, 27	z_{10}	z_6
S. vellore	*1*, 4, 12, 27	z_{10}	z_{35}
S. brancaster	*1*, 4, 12, 27	z_{29}	—
S. II *helsinki*	*1*, 4, 12	z_{29}	[e, n, x]
S. pasing	4, 12	z_{35}	1, 5
S. tafo	*1*, 4, 12, 27	z_{35}	1, 7
S. sloterdijk	*1*, 4, 12, 27	z_{35}	z_6
S. yaounde	*1*, 4, 12, 27	z_{35}	e, n, z_{15}
S. tejas	4, 12	z_{36}	—
S. wilhelmsburg	*1*, 4, [5], 12, 27	z_{38}	—
S. II *durbanville*	*1*, 4, 12, 27	z_{39}	1, [5], 7
S. thayngen	*1*, 4, 12, 27	z_{41}	1, (2), 5
S. abortusequi	4, 12	—	e, n, x

Type	Somatic O-antigen	Flagellar H-antigen	
		Phase 1	Phase 2

Group O:6, 7 (C_1)
(the strains of this group may be lysogenized by phage $14 \rightarrow 6, 7, 14$)

Type	Somatic O-antigen	Phase 1	Phase 2
S. san-juan	6, 7	a	1, 5
S. umhlali	6, 7	a	1, 6
S. austin	6, 7	a	1, 7
S. oslo	6, 7	a	e, n, x
S. denver	6, 7	a	e, n, z_{15}
S. coleypark	6, 7	a	l, w
S. II	6, 7	a	z_6
S. II *calvinia*	6, 7	a	z_{42}
S. brazzaville	6, 7	b	1, 2
S. edinburg	6, 7	b	1, 5
S. adime	6, 7	b	1, 6
S. koumra	6, 7	b	1, 7
S. georgia	6, 7	b	e, n, z_{15}
S. II *bloemfontein*	6, 7	b	[e, n, x]:z_{42}
S. ohio	6, 7	b	l, w
S. leopoldville	6, 7	b	z_6
S. kotte	6, 7	b	z_{35}
S. II	6, 7	b	z_{39}
S. paratyphi C	6, 7, [Vi]	c	1, 5
S. choleraesuis	6, 7	[c]	1, 5
S. typhisuis	6, 7	c	1, 5
S. birkenhead	6, 7	c	1, 6
S. kisii	6, 7	d	1, 2
S. isangi	6, 7	d	1, 5
S. kivu	6, 7	d	1, 6
S. kambole	6, 7	d	1, 7
S. II	6, 7	d	1, 7
S. amersfoort	6, 7	d	e, n, x
S. gombe	6, 7	d	e, n, z_{15}
S. livingstone	6, 7	d	l, w
S. wil	6, 7	d	l, z_{13}, z_{28}
S. larochelle	6, 7	e, h	1, 2
S. lomita	6, 7	e, h	1, 5
S. norwich	6, 7	e, h	1, 6
S. braenderup	6, 7	e, h	e, n, z_{15}
S. rissen	6, 7	f, g	—
S. eingedi	6, 7	f, g, t	1, 2, 7
S. afula	6, 7	f, g, t	e, n, x
S. montevideo	6, 7	g, m, [p], s	[1, 2, 7]
S. II	6, 7	g, m, [s], t	e, n, x
S. II	6, 7	(g), m, [s], t	1, 5

Type	Somatic O-antigen	Flagellar H-antigen	
		Phase 1	Phase 2
S. II	6, 7	g, m, s, t	z_{42}
S. othmarschen	6, 7	g, m, [t]	—
S. menston	6, 7	g, s, t	[1, 6]
S. II	6, 7	g, t	e, n, x :z_{42}
S. riggil	6, 7	g, t	—
S. alamo	6, 7	g, z_{51}	1, 5
S. haelsingborg	6, 7	m, p, t, [u]	—
S. oranienburg	6, 7	m, t	—
S. augustenborg	g, 7	i	1, 2
S. oritamerin	6, 7	i	1, 5
S. garoli	6, 7	i	1, 6
S. lika	6, 7	i	1, 7
S. athinai	6, 7	i	e, n, z_{15}
S. norton	6, 7	i	l, w
S. galiema	6, 7	k	1, 2
S. thompson	6, 7	k	1, 5
S. daytona	6, 7	k	1, 6
S. baiboukoum	6, 7	k	1, 7
S. singapore	6, 7	k	e, n, x
S. escanaba	6, 7	k	e, n, z_{15}
S. III *arizonae* (Ar. 27:22:31:37)	6, 7	(k)	z:[z_{55}]
S. II	6, 7	k	[z_6]
S. concord	6, 7	l, v	1, 2
S. irumu	6, 7	l, v	1, 5
S. mkamba	6, 7	l, v	1, 6
S. kortrijk	6, 7	l, v	1, 7
S. bonn	6, 7	l, v	e, n, x
S. potsdam	6, 7	l, v	e, n, z_{15}
S. gdansk	6, 7	l, v	z_6
S. III *arizonae* (Ar. 27:23:25)	6, 7	l, v	z_{53}
S. gabon	6, 7	l, w	1, 2
S. colorado	6, 7	l, w	1, 5
S. II	6, 7	l, w	1, 5, 7
S. nessziona	6, 7	1, z_{13}	1, 5
S. kenya	6, 7	l, z_{13}	e, n, x
S. neukoelln	6, 7	l, z_{13}[z_{28}]	e, n, z_{15}
S. makiso	6, 7	l, z_{13}, z_{28}	z_6
S. II *heilbron*	6, 7	l, z_{28}	1, 5:[z_{42}]
S. virchow	6, 7	r	1, 2
S. infantis[a]	6, 7	r	1, 5
S. nigeria	6, 7	r	1, 6

[a] May possess a R-phase H-antigen: 1, 11; z_{37}; z_{49}.

Type	Somatic O-antigen	Flagellar H-antigen	
		Phase 1	Phase 2
S. colindale	6, 7	r	1, 7
S. papuana	6, 7	r	e, n, z_{15}
S. grampian	6, 7	r	l, w
S. richmond	6, 7	y	1, 2
S. bareilly	6, 7	y	1, 5
S. oyonnax	6, 7	y	1, 6
S. gatow	6, 7	y	1, 7
S. hartford[b]	6, 7	y	e, n, x
S. mikawasima[c]	6, 7	y	e, n, z_{15}
S. II *tosamanga*	6, 7	z	1, 5
S. oakland	6, 7	z	1, 6, [7]
S. cayar	6, 7	z	e, n, x
S. businga	6, 7	z	e, n, z_{15}
S. bruck	6, 7	z	l, w
S. II	6, 7	z	z_6
S. II	6, 7	z	z_{39}
S. II *oysterbeds*	6, 7	z	z_{42}
S. obogu	6, 7	z_4, z_{23}	1, 5
S. aequatoria	6, 7	z_4, z_{23}	e, n, z_{15}
S. goma	6, 7	z_4, z_{23}	z_6
S. IV *roterberg*	6, 7	z_4, z_{23}	—
S. somone	6, 7	z_4, z_{24}	—
S. IV *kralendyk*	6, 7	z_4, z_{24}	—
S. II *cape*	6, 7	z_6	1, 7
S. menden	6, 7	z_{10}	1, 2
S. inganda	6, 7	z_{10}	1, 5
S. eschweiler	6, 7	z_{10}	1, 6
S. ngili	6, 7	z_{10}	1, 7
S. djugu	6, 7	z_{10}	e, n, x
S. mbandaka	6, 7	z_{10}	e, n, z_{15}
S. redba	6, 7	z_{10}	z_6
S. II	6, 7	z_{10}	z_{35}
S. tennessee	6, 7	z_{29}	[1, 2, 7]
S. II	6, 7	z_{29}	—
S. palime	6, 7	z_{35}	e, n, z_{15}
S. II *bacongo*	6, 7	z_{36}	z_{42}
S. IV *argentina*	6, 7	z_{36}	—
S. rumford	6, 7	z_{38}	1, 2
S. lille	6, 7	z_{38}	—
S. II *gilbert*	6, 7	z_{39}	1, 5, 7
S. II	6, 7	z_{41}	1, 7
S. hillsborough	6, 7	z_{41}	l, w

[b] May possess a R-phase H-antigen: z_{50}.
[c] May possess a R-phase H-antigen: z_{47}: z_{50}.

Type	Somatic O-antigen	Flagellar H-antigen	
		Phase 1	Phase 2
S. tamilnadu	6, 7	z_{41}	z_{35}
S. II *sullivan*	6, 7	z_{42}	1, 7
S. II	6, 7	z_{42}	e, n, x : 1, 6
S. III *arizonae* (Ar. 27:–30)	6, 7	—	1, 6

Group O:6, 8 (C₂)

Type	Somatic O-antigen	Phase 1	Phase 2
S. doncaster	6, 8	a	1, 5
S. curacao	6, 8	a	1, 6
S. nordufer	6, 8	a	1, 7
S. narashino	6, 8	a	e, n, x
S. II	6, 8	a	e, n, x
S. leith	6, 8	a	e, n, z_{15}
S. II *tulear*	6, 8	a	z_{52}
S. skansen	6, 8	b	1, 2
S. nagoya	6, 8	b	1, 5
S. stourbridge	6, 8	b	1, 6
S. eboko	6, 8	b	1, 7
S. gatuni	6, 8	b	e, n, x
S. presov	6, 8	b	e, n, z_{15}
S. bukuru	6, 8	b	l, w
S. banalia	6, 8	b	z_6
S. wingrove	6, 8	c	1, 2
S. utah	6, 8	c	1, 5
S. bronx	6, 8	c	1, 6
S. belfast	6, 8	c	1, 7
S. belem	6, 8	c	e, n, x
S. quiniela	6, 8	c	e, n, z_{15}
S. muenchen	6, 8	d	1, 2
S. manhattan	6, 8	d	1, 5
S. sterrenbos	6, 8	d	e, n, x
S. herston	6, 8	d	e, n, z_{15}
S. II	6, 8	d	$z_6 : z_{42}$
S. newport[a]	6, 8	e, h	1, 2
S. kottbus	6, 8	e, h	1, 5
S. cremieu	6, 8	e, h	1, 6
S. tshiongwe	6, 8	e, h	e, n, z_{15}
S. sandow	6, 8	f, g	e, n, z_{15}
S. chincol	6, 8	g, m, [s]	[e, n, x]
S. II	6, 8	g, m, t	[e, n, x]
S. nanergou	6, 8	g, s, t	—
S. II *baragwanath*	6, 8	m, t	1, 5

[a]May possess a R-phase H-antigen: z_{50}

Type	Somatic O-antigen	Flagellar H-antigen	
		Phase 1	Phase 2
S. II *germiston*	6, 8	m, t	e, n, x
S. bassa	6, 8	m, t	—
S. lindenburg	6, 8	i	1, 2
S. takoradi	6, 8	i	1, 5
S. warnow	6, 8	i	1, 6
S. malmoe	6, 8	i	1, 7
S. bonariensis	6, 8	i	e, n, x
S. aba	6, 8	i	e, n, z_{15}
S. cyprus	6, 8	i	l, w
S. blockley	6, 8	k	1, 5
S. schwerin	6, 8	k	e, n, x
S. charlottenburg	6, 8	k	e, n, z_{15}
S. litchfield	6, 8	l, v	1, 2
S. loanda	6, 8	l, v	1, 5
S. manchester	6, 8	l, v	1, 7
S. holcomb	6, 8	l, v	e, n, x
S. II	6, 8	l, v	e, n, x
S. edmonton	6, 8	l, v	e, n, z_{15}
S. fayed	6, 8	l, w	1, 2
S. hiduddify	6, 8	l, z_{13}, z_{28}	1, 5
S. breukelen	6, 8	l, z_{13}, [z_{28}]	e, n, z_{15}
S. bovismorbificans	6, 8	r	1, 5
S. akanji	6, 8	r	1, 7
S. hidalgo	6, 8	r	e, n, z_{15}
S. goldcoast	6, 8	r	l, w
S. tananarive	6, 8	y	1, 5
S. bulgaria	6, 8	y	1, 6
S. II	6, 8	y	1, 6:z_{42}
S. inchpark	6, 8	y	1, 7
S. praha	6, 8	y	e, n, z_{15}
S. mowanjum	6, 8	z	1, 5
S. II	6, 8	z	1, 5
S. kalumburu	6, 8	z	e, n, z_{15}
S. kuru	6, 8	z	l, w
S. lezennes	6, 8	z_4, z_{23}	1, 7
S. chailey	6, 8	z_4, z_{23}	e, n, z_{15}
S. duesseldorf	6, 8	z_4, z_{24}	—
S. tallahassee	6, 8	z_4, z_{32}	—
S. zerifin	6, 8	z_{10}	1, 2
S. mapo	6, 8	z_{10}	1, 5
S. cleveland	6, 8	z_{10}	1, 7
S. hadar	6, 8	z_{10}	e, n, x
S. glostrup	6, 8	z_{10}	e, n, z_{15}
S. wippra	6, 8	z_{10}	z_6
S. II	6, 8	z_{29}	1, 5

Type	Somatic O-antigen	Flagellar H-antigen	
		Phase 1	Phase 2
S. uno	6, 8	z_{29}	[e, n, z_{15}]
S. yarm	6, 8	z_{35}	1, 2
S. aesch	6, 8	z_{60}	1, 2

Group O:8 (C$_3$)

S. be	8, 20	a	—
S. djelfa	8	b	1, 2
S. korbol	8, 20	b	1, 5
S. sanga	8	b	1, 7
S. konstanz	8	b	e, n, x
S. shipley	8, 20	b	e, n, z_{15}
S. tounouma	8, 20	b	z_6
S. alexanderpolder	8	c	l, w
S. santiago	8, 20	c	e, n, x
S. tado	8, 20	c	z_6
S. virginia	8	d	1, 2
S. yovokome	8	d	1, 5
S. labadi	8, 20	d	z_6
S. bardo	8	e, h	1, 2
S. ferruch	8	e, h	1, 5
S. atakpame	8, 20	e, h	1, 7
S. rechovot	8, 20	e, h	z_6
S. emek	8, 20	g, m, s	—
S. reubeuss	8, 20	g, m, t	—
S. alminko	8, 20	g, s, t	—
S. yokoe	8	m, t	—
S. bargny	8, 20	i	1, 5
S. kentucky	8, 20	i	z_6
S. haardt	8	k	1, 5
S. pakistan	8	l, v	1, 2
S. amherstiana	8	l, v	1, 6
S. hindmarsh	8, 20	r	1, 5
S. cocody	8, 20	r, i	e, n, z_{15}
S. brikama	8, 20	r, i	l, w
S. altona	8, 20	r, [i]	z_6
S. giza	8, 20	y	1, 2
S. brunei	8, 20	y	1, 5
S. alagbon	8	y	1, 7
S. sunnycove	8	y	e, n, x
S. kralingen	8, 20	y	z_6
S. bellevue	8	z_4, z_{23}	1, 7

Type	Somatic O-antigen	Flagellar H-antigen	
		Phase 1	Phase 2
S. dabou	8, 20	z_4, z_{23}	l, w
S. corvallis	8, 20	z_4, z_{23}	$[z_6]$
S. albany[a]	8, 20	z_4, z_{24}	—
S. bazenheid	8, 20	z_{10}	1, 2
S. paris	8, 20	z_{10}	1, 5
S. istanbul	8	z_{10}	e, n, x
S. chomedey	8	z_{10}	e, n, z^{60}
S. molade	8, 20	z_{10}	z_6
S. II	8	z_{29}	e, n, x :z_{42}
S. tamale	8, 20	z_{29}	$[e, n, z_{15}]$
S. angers	8, 20	z_{35}	z_6
S. apeyeme	8, 20	z_{38}	—
S. diogoye	8, 20	z_{41}	z_6

[a] May possess a R-phase H-antigen: z_{45}.

Group O:6, 7, 14(C_4)
($=Salmonella$ of Group C_1 lysogenized by phage 14)

S. lockleaze	6, 7, 14	b	e, n, x
S. nienstedten	6, 7, 14	b	[l, w]
S. hissar	6, 7, 14	c	1, 2
S. kaduna	6, 7, 14	c	e, n, z_{15}
S. omderman	6, 7, 14	d	e, n, x
S. eimsbuettel	6, 7, 14	d	l, w
S. nieukerk	6, 7, 14	d	z_6
S. ardwick	6, 7, 14	f, g	—
S. thielallee	6, 7, 14	m, t	—
S. gelsenkirchen	6, 7, 14	l, v	z_6
S. jerusalem	6, 7, 14	z_{10}	l, w
S. bornum	6, 7, 14	z_{38}	—
S. III arizonae (Ar. 27:45:30)	6, 7, 14	z_{39}	1, 2

Group O:9, 12 (D_1)

S. sendai	1, 9, 12	a	1, 5
S. miami	1, 9, 12	a	1, 5
S. II	9, 12	a	1, 5
S. os	9, 12	a	1, 6
S. saarbruecken	1, 9, 12	a	1, 7
S. lomalinda	1, 9, 12	a	e, n, x
S. II	1, 9, 12	a	e, n, x

Type	Somatic O-antigen	Flagellar H-antigen	
		Phase 1	Phase 2
S. durban	9, 12	a	e, n, z_{15}
S. II	9, 12	a	z_{39}
S. onarimon	1, 9, 12	b	1, 2
S. frintrop	1, 9, 12	b	1, 5
S. II mjimwema	1, 9, 12	b	e, n, x
S. II blankenese	1, 9, 12	b	z_6
S. II suederelbe	1, 9, 12	b	z_{39}
S. goeteborg	9, 12	c	1, 5
S. ipeko	9, 12	c	1, 6
S. elokate	9, 12	c	1, 7
S. alabama	9, 12	c	e, n, z_{15}
S. ridge	9, 12	c	z_6
S. ndolo	1, 9, 12	d	1, 5
S. tarshyne	9, 12	d	1, 6
S. II rhodesiense	9, 12	d	e, n, x
S. zega	9, 12	d	z_6
S. jaffna	1, 9, 12	d	z_{35}
S. typhi[a]	9, 12 [Vi]	d	—
S. bournemouth	9, 12	e, h	1, 2
S. eastbourne	1, 9, 12	e, h	1, 5
S. israel	9, 12	e, h	e, n, z_{15}
S. II lindrick	9, 12	e, n, x	1, [5], 7
S. II	9, 12	e, n, x	1, 6
S. berta	1, 9, 12	f, g, t	—
S. enteritidis	1, 9, 12	g, m	[1, 7]
S. blegdam	9, 12	g, m, q	—
S. II	1, 9, 12	g, m, [s], t	[1, 5]:[z_{42}]
S. II kuilsrivier	1, 9, 12	g, m, s, t	e, n, x
S. dublin	1, 9, 12 [Vi]	g, p	—
S. naestved	1, 9, 12	g, p, s	—
S. rostock	1, 9, 12	g, p, u	—
S. moscow	9, 12	g, q	—
S. II neasden	9, 12	g, s, t	e, n, x
S. newmexico	9, 12	g, z_{51}	1, 5
S. II	1, 9, 12	g, z_{62}	—
S. antarctica	9, 12	g, z_{63}	—
S. II	9, 12	m, t	e, n, x
S. pensacola	1, 9, 12	m, t	—
S. seremban	9, 12	i	1, 5
S. claibornei	1, 9, 12	k	1, 5
S. goverdhan	9, 12	k	1, 6
S. mendoza	9, 12	l, v	1, 2
S. panama	1, 9, 12	l, v	1, 5

[a] May possess a R-phase H-antigen: j; z_{66}.

Type	Somatic O-antigen	Flagellar H-antigen Phase 1	Phase 2
S. kapemba[b]	9, 12	1, v	1, 7
S. II	9, 12	1, v	e, n, x
S. goettingen	9, 12	1, v	e, n, z_{15}
S. II	9, 12	1, v	z_{39}
S. victoria	1, 9, 12	1, w	1, 5
S. II daressalaam	1, 9, 12	1, w	e, n, x
S. itami	9, 12	1, z_{13}	1, 5
S. miyazaki	9, 12	1, z_{13}	1, 7
S. napoli	1, 9, 12	1, z_{13}	e, n, x
S. javiana[c]	1, 9, 12	1, z_{28}	1, 5
S. II	9, 12	1, z_{28}	e, n, x
S. jamaica	9, 12	r	1, 5
S. camberwell	9, 12	r	1, 7
S. campinense	9, 12	r	e, n, z_{15}
S. lome	9, 12	r	z_6
S. lawndale	1, 9, 12	z	1, 5
S. kimpese	9, 12	z	1, 6
S. II stellenbosch	1, 9, 12	z	1, 7
S. II angola	1, 9, 12	z	z_6
S. II hueningen	9, 12	z	z_{39}
S. wangata	1, 9, 12	z_4, z_{23}	[1, 7]
S. portland	9, 12	z_{10}	1, 5
S. II canastel	9, 12	z_{29}	1, 5
S. II	1, 9, 12	z_{29}	e, n, x
S. penarth	9, 12	z_{35}	z_6
S. elomrane	1, 9, 12	z_{38}	—
S. II wynberg	1, 9, 12	z_{39}	1, 7
S. ottawa	1, 9, 12	z_{41}	1, 5
S. gallinarum-pullorum	1, 9, 12	—	—

[b] May possess a R-phase H-antigen: z_{40}.
[c] May possess a R-phase H-antigen: 1, 13.

Group O:9, 46 (D$_2$)[a]			
S. baildon	9, 46	a	e, n, x
S. doba	9, 46	a	e, n, z_{15}
S. zadar	9, 46	b	1, 6
S. worb	9, 46	b	e, n, x
S. II lundby	9, 46	b	e, n, x
S. bamboye	9, 46	b	1, w

[a] The serotypes of this group also contain factors O:3 and (10); the latter is not very well developed. They can be lysogenized by phages ε_{15} and ε_{34}, and in the case of double lysogenization become strongly agglutinable, as strains of Group E$_3$, by antisera against O:34 and O:12$_2$.

Type	Somatic O-antigen	Flagellar H-antigen	
		Phase 1	Phase 2
S. linguere	9, 46	b	z_6
S. itutaba	9, 46	c	z_6
S. ontario	9, 46	d	1, 5
S. quentin	9, 46	d	1, 6
S. strasbourg	9, 46	d	1, 7
S. olten	9, 46	d	e, n, z_{15}
S. plymouth	9, 46	d	z_6
S. bergedorf	9, 46	e, h	1, 2
S. guerin	9, 46	e, h	z_6
S. II	9, 46	e, n, x	1, 5, 7
S. wernigerode	9, 46	f, g	—
S. hillingdon	9, 46	g, m	—
S. II *duivenhoks*	9, 46	g, m, s, t	e, n, x
S. gateshead	9, 46	g, s, t	—
S. II	9, 46	m, t	e, n, x
S. sangalkam	9, 46	m, t	—
S. mathura	9, 46	i	e, n, z_{15}
S. potto	9, 46	i	z_6
S. marylebone	9, 46	k	1, 2
S. cochin	9, 46	k	1, 5
S. ceyco	9, 46	k	z_{35}
S. india	9, 46	l, v	1, 5
S. geraldton	9, 46	l, v	1, 6
S. toronto	9, 46	l, v	e, n, x
S. shoreditch	9, 46	r	e, n, z_{15}
S. sokode	9, 46	r	z_6
S. benin	9, 46	y	1, 7
S. mayday	9, 46	y	z_6
S. II *haarlem*	9, 46	z	e, n, x
S. bambylor	9, 46	z	e, n, z_{15}
S. ekotedo	9, 46	z_4, z_{23}	—
S. II *maarssen*	9, 46	z_4, z_{24}	$z_{39}:z_{42}$
S. lishabi	9, 46	z_{10}	1, 7
S. inglis	9, 46	z_{10}	e, n, x
S. louisiana	9, 46	z_{10}	z_6
S. II	9, 46	z_{10}	z_6
S. II	9, 46	z_{10}	z_{39}
S. ouakam[b]	9, 46	z_{29}	—
S. hillegersberg	9, 46	z_{35}	1, 5
S. basingstoke	9, 46	z_{35}	e, n, z_{15}
S. trimdon	9, 46	z_{35}	z_6
S. fresno	9, 46	z_{38}	—
S. II	9, 46	z_{39}	1, 7
S. wuppertal	9, 46	z_{41}	—

[b] May possess a R-phase H-antigen: z_{45}.

| Type | Somatic O-antigen | Flagellar H-antigen | |
		Phase 1	Phase 2
Group O:1, 9, 12, (46), 27, (D$_3$)			
S. II *zuerich*	1, 9, 12, (46), 27	c	z_{39}
S. II	9, 12, (46), 27	g, t	e, n, x
S. II	1, 9, 12, (46), 27	l, z_{13}, z_{28}	z_{39}
S. II	1, 9, 12, (46), 27	y	z_{39}
S. II	1, 9, 12, (46), 27	z_4, z_{24}	1, 5
S. II	1, 9, 12, (46), 27	z_{10}	e, n, x
S. II	1, 9, 12, (46), 27	z_{10}	z_{39}
Group O:3, 10 (E$_1$)			
S. aminatu	3, 10	a	1, 2
S. goelzau	3, 10	a	1, 5
S. oxford	3, 10	a	1, 7
S. masembe	3, 10	a	e, n, x
S. II *matroosfontein*	3, 10	a	e, n, x
S. galil	3, 10	a	e, n, z_{15}
S. II	3, 10	a	z_{39}
S. kalina	3, 10	b	1, 2
S. butantan	3, 10	b	1, 5
S. allerton	3, 10	b	1, 6
S. huvudsta	3, 10	b	1, 7
S. benfica	3, 10	b	e, n, x
S. II	3, 10	b	e, n, x
S. yaba	3, 10	b	e, n, z_{15}
S. epicrates	3, 10	b	l, w
S. II	3, 10	b	z_{39}
S. gbadago	3, 10	c	1, 5
S. ikayi	3, 10	c	1, 6
S. pramiso	3, 10	c	1, 7
S. agege	3, 10	c	e, n, z_{15}
S. anderlecht	3, 10	c	l, w
S. okefoko	3, 10	c	z_6
S. stormont	3, 10	d	1, 2
S. shangani	3, 10	d	1, 5
S. lekke	3, 10	d	1, 6
S. onireke	3, 10	d	1, 7
S. souza	3, 10	d	e, n, x
S. II	3, 10	d	e, n, x
S. madjorio	3, 10	d	e, n, z_{15}
S. birmingham	3, 10	d	l, w
S. weybridge	3, 10	d	z_6

Type	Somatic O-antigen	Flagellar H-antigen	
		Phase 1	Phase 2
S. *maron*	3, 10	d	z_{35}
S. *vejle*	3, 10	e, h	1, 2
S. *muenster*[a]	3, 10	e, h	1, 5
S. *anatum*	3, 10	e, h	1, 6
S. *nyborg*	3, 10	e, h	1, 7
S. *newlands*	3, 10	e, h	e, n, x
S. *meleagridis*	3, 10	e, h	l, w
S. *sekondi*	3, 10	e, h	z_6
S. II *chudleigh*	3, 10	e, n, x	1, 7
S. *regent*	3, 10	f, g, [s]	[1, 6]
S. *alfort*	3, 10	f, g	e, n, x
S. *suberu*	3, 10	g, m	—
S. *amsterdam*	3, 10	g, m, s	—
S. II	3, 10	g, m, s, t	—
S. *westhampton*[b]	3, 10	g, s, t	—
S. II *islington*	3, 10	g, t	—
S. *southbank*	3, 10	m, t	[1, 6]
S. II *stikland*	3, 10	m, t	e, n, x
S. *cukmere*	3, 10	i	1, 2
S. *amounderness*	3, 10	i	1, 5
S. *truro*	3, 10	i	1, 7
S. *bessi*	3, 10	i	e, n, x
S. *falkensee*	3, 10	i	e, n, z_{15}
S. *yeerongpilly*	3, 10	i	z_6
S. *wimborne*	3, 10	k	1, 2
S. *zanzibar*	3, 10	k	1, 5
S. *yundum*	3, 10	k	e, n, x
S. *marienthal*	3, 10	k	e, n, z_{15}
S. *newrochelle*	3, 10	k	l, w
S. *nchanga*	3, 10	l, v	1, 2
S. *sinstorf*	3, 10	l, v	1, 5
S. *london*	3, 10	l, v	1, 6
S. *give*	3, 10	[d], l, v	1, 7
S. II	3, 10	l, v	e, n, x
S. *ruzizi*	3, 10	l, v	e, n, z_{15}
S. II *fuhlsbuettel*	3, 10	l, v	z_6
S. *sinchew*	3, 10	l, v	z_{35}
S. *assinie*[c]	3, 10	l, w	z_6
S. *freiburg*	3, 10	l, z_{13}	1, 2
S. *uganda*	3, 10	l, z_{13}	1, 5
S. *fallowfield*	3, 10	l, z_{13}, z_{28}	e, n, z_{15}

[a] May possess a R-phase H-antigen: z_{48}.
[b] May possess a R-phage H-antigen: z_{37}.
[c] May possess a R-phage H-antigen: z_{45}.

Type	Somatic O-antigen	Flagellar H-antigen Phase 1	Phase 2
S. hoghton	3, 10	$1, z_{13}, z_{28}$	z_6
S. II	3, 10	$1, z_{28}$	1, 5
S. joal	3, 10	$1, z_{28}$	1, 7
S. lamin	3, 10	$1, z_{28}$	e, n, x
S. II *westpark*	3, 10	$1, z_{28}$	e, n, x
S. II	3, 10	$1, z_{28}$	z_{39}
S. ughelli	3, 10	r	1, 5
S. elisabethville	3, 10	r	1, 7
S. simi	3, 10	r	e, n, z_{15}
S. weltevreden	3, 10	r	z_6
S. seegefeld	3, 10	r, i	1, 2
S. dumfries	3, 10	r, i	1, 6
S. amager^d	3, 10	y	1, 2
S. orion	3, 10	y	1, 5
S. mokola	3, 10	y	1, 7
S. ohlstedt	3, 10	y	e, n, x
S. bolton	3, 10	y	e, n, z_{15}
S. langensalza	3, 10	y	l, w
S. stockholm	3, 10	y	z_6
S. fufu	3, 10	z	1, 5
S. II *alexander*	3, 10	z	1, 5
S. huddinge	3, 10	z	1, 7
S. II *finchley*	3, 10	z	e, n, x
S. clerkenwell	3, 10	z	l, w
S. landwasser	3, 10	z	z_6
S. II *tafelbaai*	3, 10	z	z_{39}
S. adabraka	3, 10	z_4, z_{23}	[1, 7]
S. florian	3, 10	z_4, z_{24}	—
S. II	3, 10	z_4, z_{24}	—
S. okerara	3, 10	z_{10}	1, 2
S. lexington^e	3, 10	z_{10}	1, 5
S. coquilhatville	3, 10	z_{10}	1, 7
S. kristianstad	3, 10	z_{10}	e, n, z_{15}
S. biafra	3, 10	z_{10}	z_6
S. II	3, 10	z_{29}	e, n, x
S. jedburgh	3, 10	z_{29}	—
S. zongo	3, 10	z_{35}	1, 7
S. shannon	3, 10	z_{35}	l, w
S. cairina	3, 10	z_{35}	z_6
S. macallen	3, 10	z_{36}	—
S. bolombo	3, 10	z_{38}	[z_6]
S. II *mpila*	3, 10	z_{38}	z_{42}
S. II *winchester*	3, 10	z_{39}	1, 7

^{d} May possess a R-phase H-antigen: z_{45}.
^{e} May possess a R-phase H-antigen: z_{49}.

		Flagellar H-antigen	
Type	Somatic O-antigen	Phase 1	Phase 2

Group O:3, 15 (E_2)
(= *Salmonella* of Group E_1 lysogenized by phage ε_{15})

Type	Somatic O-antigen	Phase 1	Phase 2
S. clichy	3, 15	a	1, 5
S. rosenthal	3, 15	b	1, 5
S. westminster	3, 15	b	z_{35}
S. pankow	3, 15	d	1, 5
S. eschersheim	3, 15	d	e, n, x
S. goerlitz	3, 15	e, h	1, 2
S. newhaw	3, 15	e, h	1, 5
S. newington	3, 15	e, h	1, 6
S. selandia	3, 15	e, h	1, 7
S. cambridge	3, 15	e, h	l, w
S. drypool	3, 15	g, m, s	—
S. II parow	3, 15	g, m, s, t	—
S. halmstad	3, 15	g, s, t	—
S. nancy	3, 15	l, v	1, 2
S. portsmouth	3, 15	l, v	1, 6
S. newbrunswick	3, 15	l, v	1, 7
S. kinshasa	3, 15	l, z_{13}	1, 5
S. lanka	3, 15	r	z_6
S. tuebingen	3, 15	y	1, 2
S. binza	3, 15	y	1, 5
S. tournai	3, 15	y	z_6
S. manila	3, 15	z_{10}	1, 5

Group O:3, 15, 34 (E_3)
(= *Salmonella* of Group E_1 lysogenized by phages ε_{15} and ε_{34})

Type	Somatic O-antigen	Phase 1	Phase 2
S. khartoum	3, 15, 34	a	1, 7
S. arkansas	3, 15, 34	e, h	1, 5
S. minneapolis	3, 15, 34	e, h	1, 6
S. wildwood	3, 15, 34	e, h	l, w
S. canoga	3, 15, 34	g, s, t	—
S. menhaden	3, 15, 34	l, v	1, 7
S. thomasville	3, 15, 34	y	1, 5
S. illinois	3, 15, 34	z_{10}	1, 5
S. harrisonburg	3, 15, 34	z_{10}	1, 6

Group O:1, 3, 19 (E_4)

Type	Somatic O-antigen	Phase 1	Phase 2
S. juba	1, 3, 19	a	1, 7
S. gwoza	1, 3, 19	a	e, n, z_{15}

| Type | Somatic O-antigen | Flagellar H-antigen | |
		Phase 1	Phase 2
S. gnesta[a]	1, 3, 19	b	1, 5
S. visby	1, 3, 19	b	1, 6
S. tambacounda	1, 3, 19	b	e, n, x
S. kande	1, 3, 19	b	e, n, z_{15}
S. broughton	1, 3, 19	b	l, w
S. accra	1, 3, 19	b	z_6
S. madiago	1, 3, 19	c	1, 7
S. ahmadi	1, 3, 19	d	1, 5
S. liverpool	1, 3, 19	d	e, n, z_{15}
S. tilburg	1, 3, 19	d	l, w
S. niloese	1, 3, 19	d	z_6
S. vilvoorde	1, 3, 19	e, h	1, 5
S. sanktmarx	1, 3, 19	e, h	1, 7
S. sao	1, 3, 19	e, h	e, n, z_{15}
S. calabar	1, 3, 19	e, h	l, w
S. rideau	1, 3, 19	f, g	—
S. maiduguri	1, 3, 19	f, g, t	e, n, z_{15}
S. kouka	1, 3, 19	g, m, [t]	—
S. senftenberg[b]	1, 3, 19	g, [s], t	—
S. cannstatt	1, 3, 19	m, t	—
S. stratford	1, 3, 19	i	1, 2
S. machaga	1, 3, 19	i	e, n, x
S. avonmouth	1, 3, 19	i	e, n, z_{15}
S. zuilen	1, 3, 19	i	l, w
S. taksony	1, 3, 19	i	z_6
S. ngor	1, 3, 19	l, v	1, 5
S. parkroyal	1, 3, 19	l, v	1, 7
S. westerstede	1, 3, 19	l, z_{13}	[1, 2]
S. winterthur	1, 3, 19	l, z_{13}	1, 6
S. lokstedt	1, 3, 19	l, z_{13}, z_{28}	1, 2
S. stuivenberg	1, 3, 19	1, [z_{13}] z_{28}	1, 5
S. bedford	1, 3, 19	l, z_{13}, z_{28}	e, n, z_{15}
S. tomelilla	1, 3, 19	l, z_{28}	1, 7
S. yalding	1, 3, 19	r	e, n, z_{15}
S. fareham	1, 3, 19	r, i	l, w
S. gatineau	1, 3, 19	y	1, 5
S. krefeld	1, 3, 19	y	l, w
S. korlebu	1, 3, 19	z	1, 5
S. lerum	1, 3, 19	z	1, 7
S. schoeneberg	1, 3, 19	z	e, n, z_{15}
S. carno	1, 3, 19	z	l, w
S. sambre	1, 3, 19	z_4, z_{24}	—
S. dallgow	1, 3, 19	z_{10}	e, n, z_{15}

[a] May possess a R-phase H-antigen: z_{37}.
[b] May possess a R-phase H-antigen: z_{27}; z_{34}; z_{43}; z_{45}; z_{46}.

Type	Somatic O-antigen	Flagellar H-antigen	
		Phase 1	Phase 2
S. llandoff	1, 3, 19	z_{29}	$[z_6]$
S. chittagong	1, 3, 10, 19	b	z_{35}
S. bilu	1, 3, 10, 19	f, g, t	1, (2), 7
S. ilugun	1, 3, 10, 19	z_4, z_{23}	z_6
S. dessau	1, 3, *15*, 19	g, s, t	—
S. cannonhill	1, 3, *15*, 19	y	e, n, x

Group O:11 (F)

Type	Somatic O-antigen	Flagellar H-antigen	
S. gallen	11	a	1, 2
S. marseille	11	a	1, 5
S. toowong	11	a	1, 7
S. luciana	11	a	e, n, z_{15}
S. epinay	11	a	l, z_{13}, z_{28}
S. II glencairn	11	a	$z_6:z_{42}$
S. atento	11	b	1, 2
S. leeuwarden	11	b	1, 5
S. wohlen	11	b	1, 6
S. II	11	b	1, 7
S. II srinagar	11	b	e, n, x
S. pharr	11	b	e, n, z_{15}
S. chiredzi	11	c	1, 5
S. gustavia	11	d	1, 5
S. chandans	11	d	e, n, x
S. II montgomery	11	d, (a)	d, e, n, z_{15}
S. findorff	11	d	z_6
S. chingola	11	e, h	1, 2
S. adamstua	11	e, h	1, 6
S. redhill	11	e, h	l, z_{13}, z_{28}
S. II grabouw	11	g, m, s, t	z_{39}
S. IV mundsburg	11	g, z_{51}	—
S. II lincoln	11	m, t	e, n, x
S. aberdeen	11	i	1, 2
S. brijbhumi	11	i	1, 5
S. heerlen	11	i	1, 6
S. veneziana	11	i	e, n, x
S. pretoria	11	k	1, 2
S. abaetetuba	11	k	1, 5
S. sharon	11	k	1, 6
S. colobane	11	k	1, 7
S. kisarawe	11	k	e, n, x, $[z_{15}]$
S. amba	11	k	l, z_{13}, z_{28}
S. III arizonae (Ar. 17:29:25)	11	k	z_{53}
S. stendal	11	l, v	1, 2

Type	Somatic O-antigen	Flagellar H-antigen	
		Phase 1	Phase 2
S. maracaibo	11	l, v	1, 5
S. fann	11	l, v	e, n, x
S. bullbay	11	l, v	e, n, z_{15}
S. III arizonae (Ar. 17:23:31)	11	l, v	z
S. III arizonae (Ar. 17:23:25)	11	l, v	z_{53}
S. glidji	11	l, w	1, 5
S. osnabrueck	11	l, z_{13}, z_{28}	e, n, x
S. II huila	11	l, z_{28}	e, n, x
S. senegal	11	r	1, 5
S. rubislaw	11	r	e, n, x
S. volta	11	r	l, z_{13}, z_{28}
S. solt	11	y	1, 5
S. jalisco	11	y	1, 7
S. herzliya	11	y	e, n, x
S. nyanza	11	z	z_6
S. II soutpan	11	z	z_{39}
S. remete	11	z_4, z_{23}	1, 6
S. etterbeek	11	z_4, z_{23}	e, n, z_{15}
S. III arizonae (Ar. 17:1, 2, 5:—)	11	z_4, z_{23}	—
S. IV parera	11	z_4, z_{23}	—
S. yehuda	11	z_4, z_{24}	—
S. IV	11	z_4, z_{32}	—
S. wentworth	11	z_{10}	1, 2
S. straengnaes	11	z_{10}	1, 5
S. telhashomer	11	z_{10}	e, n, x
S. lene	11	z_{38}	—
S. maastricht	11	z_{41}	1, 2
S. II	11	—	1, 5

Group O:13, 22 (G₁)

S. mim	13, 22	a	1, 6
S. marshall	13, 22	a	l, z_{13}, z_{28}
S. ibadan	13, 22	b	1, 5
S. oudwijk	13, 22	b	1, 6
S. rottnest	1, 13, 22	b	1, 7
S. vaertan	13, 22	b	e, n, x
S. bahati	13, 22	b	e, n, z_{15}
S. II	1, 13, 22	b	z_{42}
S. haouaria	13, 22	c	e, n, x, z_{15}
S. friedenau	13, 22	d	1, 6

Type	Somatic O-antigen	Flagellar H-antigen	
		Phase 1	Phase 2
S. diguel	1, 13, 22	d	e, n, z_{15}
S. willemstad	1, 13, 22	e, h	1, 6
S. raus	13, 22	f, g	e, n, x
S. II	13, 22	(f), g, t	—
S. bron	13, 22	g, m	$[e, n, z_{15}]$
S. II *limbe*	1, 13, 22	g, m, t	[1, 5]
S. newyork	13, 22	g, s, t	—
S. II *rotterdam*	1, 13, 22	g, t	1, 5
S. washington	13, 22	m, t	—
S. II	13, 22	k	$1, 5:z_{42}$
S. lovelace	13, 22	l, v	1, 5
S. borbeck	13, 22	l, v	1, 6
S. II	13, 22	l, z_{28}	1, 5
S. tanger	1, 13, 22	y	1, 6
S. poona[a]	1, 13, 22	z	1, 6
S. bristol	13, 22	z	1, 7
S. tanzania	1, 13, 22	z	e, n, z_{15}
S. ried	1, 13, 22	z_4, z_{23}	$[e, n, z_{15}]$
S. III *arizonae* (Ar. 18:1, 2, 5)	13, 22	z_4, z_{23}	—
S. roodepoort	1, 13, 22	z_{10}	1, 5
S. II *clifton*	13, 22	z_{29}	1, 5
S. II *goodwood*	13, 22	z_{29}	e, n, x
S. agoueve	13, 22	z_{29}	—
S. mampong	13, 22	z_{35}	1, 6
S. nimes	13, 22	z_{35}	e, n, z_{15}
S. leiden	13, 22	z_{38}	—
S. II	13, 22	z_{39}	1, 5, (7)
S. III *arizonae* (Ar. 18:— :—)	13, 22	—	—

[a] May possess a R-phase H-antigen: z_{59}.

Group O:13, 23 (G$_2$)

Type	Somatic O-antigen	Phase 1	Phase 2
S. chagoua	1, 13, 23	a	1, 5
S. wyldegreen	13, 23	a	1, w
S. II *tygerberg*	1, 13, 23	a	z_{42}
S. mississippi	1, 13, 23	b	1, 5
S. II *acres*	1, 13, 23	b	$[1, 5]:z_{42}$
S. bracknell	13, 23	b	1, 6
S. ullevi	1, 13, 23	b	e, n, x
S. durham	13, 23	b	e, n, z_{15}
S. handen	1, 13, 23	d	1, 2

Type	Somatic O-antigen	Flagellar H-antigen Phase 1	Phase 2
S. mishmarhaemek	*1*, 13, 23	d	1, 5
S. wichita[a]	*1*, 13, 23	d	[1, 6]
S. grumpensis	13, 23	d	1, 7
S. II	13, 23	d	e, n, x
S. telelkebir	13, 23	d	e, n, z_{15}
S. putten	13, 23	d	l, w
S. isuge	13, 23	d	z_6
S. tschangu	*1*, 13, 23	e, h	1, 5
S. II *epping*	*1*, 13, 23	e, n, x	1, 7
S. havana	*1*, 13, 23	f, g, [s]	—
S. agbeni	13, 23	g, m	—
S. II	13, 23	g, m, s, t	1, 5
S. II *luanshya*	*1*, 13, 23	g, m, [s], t	[e, n, x]
S. congo	13, 23	g, m, s, t	—
S. okatie	13, 23	g, s, t	—
S. II *gojenberg*	*1*, 13, 23	g, t	1, 5
S. II	*1*, 13, 23	g, t	z_{42}
S. III *arizonae* (Ar. 18:13, 14:−)	*1*, 13, 23	g, z_{51}	—
S. II *katesgrove*	*1*, 13, 23	m, t	1, 5
S. II *worcester*	*1*, 13, 23	m, t	e, n, x
S. II *boulders*	*1*, 13, 23	m, t	z_{42}
S. kintambo	13, 23	m, t	—
S. idikan	*1*, 13, 23	i	1, 5
S. jukestown	13, 23	i	e, n, z_{15}
S. kedougou	*1*, 13, 23	i	l, w
S. II	13, 23	k	z_{41}
S. nanga	*1*, 13, 23	l, v	e, n, z_{15}
S. II	13, 23	l, z_{28}	1, 5
S. II	13, 23	l, z_{28}	z_6
S. II *vredelust*	*1*, 13, 23	l, z_{28}	z_{42}
S. adjame	13, 23	r	1, 6
S. linton	13, 23	r	e, n, z_{15}
S. yarrabah	13, 23	y	1, 7
S. ordonez	*1*, 13, 23	y	l, w
S. tunis	*1*, 13, 23	y	z_6
S. II *nachshonim*	*1*, 13, 23	z	1, 5
S. farmsen	13, 23	z	1, 6
S. worthington	*1*, 13, 23	z	l, w
S. ajiobo	13, 23	z_4, z_{23}	—
S. III *arizonae* (18:1, 6, 7:−)	13, 23	z_4, z_{23}, z_{32}	—
S. romanby	13, 23	z_4, z_{24}	—

[a] May possess a R-phase H-antigen: z_{37}.

Type	Somatic O-antigen	Flagellar H-antigen	
		Phase 1	Phase 2
S. III *arizonae* (Ar. 18:1, 3, 11:−)	*1*, 13, 23	z_4, z_{24}	—
S. demerara	13, 23	z_{10}	l, w
S. II	*1*, 13, 23	z_{29}	e, n, x
S. cubana[b]	*1*, 13, 23	z_{29}	—
S. anna	13, 23	z_{35}	e, n, z_{15}
S. fanti	13, 23	z_{38}	—
S. II *stevenage*	*1*, 13, 23	$[z_{42}]$	1, [5], 7
S. II	13, 23	—	1, 6

[b] May possess a R-phase H-antigen: z_{37}; z_{43}.

Group O:6, 14 (H)			
S. garba	1, 6, 14, 25	a	1, 5
S. ferlac	1, 6, 14, 25	a	e, n, x
S. banjul	1, 6, 14, 25	a	e, n, z_{15}
S. ndjamena	1, 6, 14, 25	b	1, 2
S. tucson	[1], 6, 14, [25]	b	[1, 7]
S. II	6, 14	b	e, n, x, z_{15}
S. blijdorp	1, 6, 14, 25	c	1, 5
S. kassberg	1, 6, 14, 25	c	1, 6
S. runby	1, 6, 14, 25	c	e, n, x
S. minna	1, 6, 14, 25	c	l, w
S. heves	6, 14, 24	d	1, 5
S. finkenwerder	[1], 6, 14, [25]	d	1, 5
S. midway	6, 14, 24	d	1, 7
S. florida	[1], 6, 14, [25]	d	1, 7
S. lindern	6, 14, 24	d	e, n, x
S. charity	1, 6, 14, 25	d	e, n, x
S. teko	1, 6, 14, 25	d	e, n, z_{15}
S. encino	1, 6, 14, 25	d	1, z_{13}, z_{28}
S. albuquerque	1, 6, 14, 24	d	z_6
S. bahrenfeld	6, 14, 24	e, h	1, 5
S. onderstepoort	1, 6, 14, [25]	e, h	1, 5
S. magumeri	1, 6, 14, 25	e, h	1, 6
S. beaudesert	[1], 6, 14, [25]	e, h	1, 7
S. warragul	1, 6, 14, 25	g, m	—
S. caracas	[1], 6, 14, [25]	g, m, s	—
S. catanzaro	6, 14	g, s, t	—
S. II *rooikrantz*	1, 6, 14	m, t	1, 5
S. II *emmerich*	6, 14	[m, t]	e, n, x
S. kaitaan	1, 6, 14, 25	m, t	—
S. mampeza	1, 6, 14, 25	i	1, 5

Type	Somatic O-antigen	Flagellar H-antigen	
		Phase 1	Phase 2
S. buzu	1, 6, 14, 25	i	1, 7
S. schalkwijk	6, 14, 24	i	e, n, . . .
S. moussoro	1, 6, 14, 25	i	e, n, z_{15}
S. harburg	1, 6, 14, 25	k	1, 5
S. II	6, 14	k	[e, n, x]
S. III arizonae (Ar. 7a, 7c:29:31)	6, 14	k	z
S. II	1, 6, 14	k	$z_6 : z_{42}$
S. III arizonae (Ar. 7a, 7c:29:25)	6, 14	k	z_{53}
S. boecker	[1], 6, 14, [25]	l, v	1, 7
S. horsham	1, 6, 14, [25]	l, v	e, n, x
S. III arizonae (Ar. 7a, 7c:23:31)	(6), 14	l, v	z
S. III arizonae (Ar. 7a, 7c:23:21)	(6), 14	l, v	z_{35}
S. aflao	1, 6, 14, 25	1, z_{28}	e, n, x
S. III arizonae (Ar. 7a, 7c:24:31)	(6), 14	r	z
S. surat	[1], 6, 14, [25]	r, [i]	e, n, z_{15}
S. carrau	6, 14, [24]	y	1, 7
S. madelia	1, 6, 14, 25	y	1, 7
S. fischerkietz	1, 6, 14, 25	y	e, n, x
S. mornington	1, 6, 14, 25	y	e, n, z_{15}
S. homosassa	1, 6, 14, 25	z	1, 5
S. soahanina	6, 14, 24	z	e, n, x
S. sundsvall	1, 6, 14, 25	z	e, n, x
S. poano	1, 6, 14, 25	z	1, z_{13}, z_{28}
S. bousso	1, 6, 14, 25	z_4, z_{23}	[e, n, z_{15}]
S. IV	6, 14	z_4, z_{23}	—
S. chichiri	6, 14, 24	z_4, z_{24}	—
S. uzaramo	1, 6, 14, 25	z_4, z_{24}	—
S. nessa	1, 6, 14, 25	z_{10}	1, 2
S. II bornheim	1, 6, 14, 25	z_{10}	1, (2), 7
S. II simonstown	1, 6, 14	z_{10}	1, 5
S. III arizonae (Ar. 7a, 7c:27:28)	(6), 14	z_{10}	e, n, x, z_{15}
S. III arizonae (Ar. 7a, 7c:27:[31]:[38])	(6), 14	z_{10}	[z] [z_{56}]
S. II slangkop	1, 6, 14	z_{10}	$z_6 : z_{42}$
S. potosi	6, 14	z_{36}	1, 5
S. sara	1, 6, 14, 25	z_{38}	[e, n, x]
S. II	1, 6, 14	z_{42}	1, 6
S. III arizonae (Ar. 7a, 7c . . .:26:21)	1, 6, 14, 25	z_{52}	z_{35}

Type	Somatic O-antigen	Flagellar H-antigen Phase 1	Flagellar H-antigen Phase 2
		Group O:16 (I)	
S. *hannover*	16	a	1, 2
S. *brazil*	16	a	1, 5
S. *amunigun*	16	a	1, 6
S. *nyeko*	16	a	1, 7
S. *togba*	16	a	e, n, x
S. *fischerhuette*	16	a	e, n, z_{15}
S. *heron*	16	a	z_6
S. *hull*	16	b	1, 2
S. *wa*	16	b	1, 5
S. *glasgow*	16	b	1, 6
S. *hvittingfoss*	16	b	e, n, x
S. II	16	b	e, n, x
S. *sangera*	16	b	e, n, z_{15}
S. *malstatt*	16	b	z_6
S. II	16	b	z_{39}
S. II	16	b	z_{42}
S. *vancouver*	16	c	1, 5
S. *gafsa*	16	c	1, 6
S. *shamba*	16	c	e, n, x
S. *hithergreen*	16	c	e, n, z_{15}
S. *oldenburg*	16	d	1, 2
S. II	16	d	1, 5
S. *sherbrooke*	16	d	1, 6
S. *gaminara*	16	d	1, 7
S. *barranquilla*	16	d	e, n, x
S. *nottingham*	16	d	e, n, z_{15}
S. *caen*	16	d	l, w
S. *barmbek*	16	d	z_6
S. *malakal*	16	e, h	1, 2
S. *saboya*	16	e, h	1, 5
S. *rhydyfelin*	16	e, h	e, n, x
S. *weston*	16	e, h	z_6
S. II *bellville*	16	e, n, x	1, (5), 7
S. *tees*	16	f, g	—
S. *adeoyo*	16	g, m	—
S. *nikolaifleet*	16	g, m, s	—
S. II *mobeni*	16	g, [m], [s], t	[e, n, x]
S. II *merseyside*	16	g, t	[1, 5]
S. II	16	m, t	e, n, x
S. II *rowbarton*	16	m, t	$[z_{42}]$
S. *mpouto*	16	m, t	—
S. *amina*	16	i	1, 5
S. *wisbech*	16	i	1, 7

Type	Somatic O-antigen	Flagellar H-antigen	
		Phase 1	Phase 2
S. frankfurt	16	i	e, n, z_{15}
S. pisa	16	i	l, w
S. abobo	16	i	z_6
S. III *arizonae* (Ar. 25:33:21)	16	i	z_{35}
S. szentes	16	k	1, 2
S. nuatja	16	k	e, n, x
S. orientalis	16	k	e, n, z_{15}
S. III *arizonae* (Ar. 25:29:31)	16	k	z
S. III *arizonae* (Ar. 25:22:21)	16	(k)	z_{35}
S. III *arizonae* (Ar. 25:29:25)	16	k	z_{53}
S. III *arizonae* (Ar. 25:23:30)	16	l, v	1, 5, 7
S. shanghai[a]	16	l, v	1, 6
S. welikade	16	l, v	1, 7
S. salford	16	l, v	e, n, x
S. burgas	16	l, v	e, n, z_{15}
S. III *arizonae* (Ar. 25:23:31:[41])	16	l, v	$z:[z_{61}]$
S. losangeles	16	l, v	z_6
S. III *arizonae* (Ar. 25:23:21)	16	l, v	z_{35}
S. III *arizonae* (Ar. 25:23:25)	16	l, v	z_{53}
S. westeinde	16	l, w	1, 6
S. lomnava	16	l, w	e, n, z_{15}
S. II *noordhoek*	16	l, w	z_6
S. mandera	16	l, z_{13}	e, n, z_{15}
S. enugu	16	$1, [z_{13}], z_{28}$	[1, 5]
S. battle	16	l, z_{13}, z_{28}	1, 6
S. ablogame	16	l, z_{13}, z_{28}	z_6
S. II *sarepta*	16	l, z_{28}	z_{42}
S. rovaniemi	16	r, i	1, 5
S. annedal	16	r, i	e, n, x
S. zwickau	16	r, i	e, n, z_{15}
S. saphra	16	y	1, 5
S. akuafo	16	y	1, 6
S. kikoma	16	y	e, n, x
S. avignon	16	y	e, n, z_{15}
S. fortlamy	16	z	1, 6

[a] May possess a R-phase H-antigen: z_{45}.

| Type | Somatic O-antigen | Flagellar H-antigen | |
		Phase 1	Phase 2
S. lingwala	16	z	1, 7
S. II *louwbester*	16	z	e, n, x
S. brevik	16	z	e, n, z_{15}
S. II	16	z	z_{42}
S. kibi	16	z_4, z_{23}	—
S. II *haddon*	16	z_4, z_{23}	—
S. IV *ochsenzoll*	16	z_4, z_{23}	—
S. IV *chameleon*	16	z_4, z_{32}	—
S. II	16	z_6	1, 6
S III *arizonae*	16	z_{10}	1, 5, 7
(Ar. 25:27:30)			
S. lisboa	16	z_{10}	1, 6
S. III *arizonae*	16	z_{10}	e, n, x, z_{15}
(Ar. 25:27:28)			
S. redlands	16	z_{10}	e, n, z_{15}
S. angouleme	16	z_{10}	z_6
S. saloniki	16	z_{29}	—
S. II *jacksonville*	16	z_{29}	—
S. dakota	16	z_{35}	e, n, z_{15}
S. naware	16	z_{38}	—
S. II *wookstock*	16	z_{42}	1, (5), 7
S. II *elsiesrivier*	16	z_{42}	1, 6
S. III *arizonae*	16	z_{52}	z_{35}
(Ar. 25:26:21)			

Group O:17 (J)

Type	Somatic O-antigen	Phase 1	Phase 2
S. bonames	17	a	1, 2
S. jangwani	17	a	1, 5
S. kinondoni	17	a	e, n, x
S. kirkee	17	b	1, 2
S. II *hillbrow*	17	b	e, n, x, z_{15}
S. bignona	17	b	e, n, z_{15}
S. II	17	b	z_6
S. victoriaborg	17	c	1, 6
S. II *woerden*	17	c	z_{39}
S. berlin	17	d	1, 5
S. niamey	17	d	l, w
S. jubilee	17	e, h	1, 2
S. II *verity*	17	e, n, x, z_{15}	1, 6
S. II	17	e, n, x, z_{15}	1, 7
S. II *bleadon*	17	(f), g, t	[e, n, x, z_{15}]
S. II	17	g, t	z_{39}
S. bama	17	m, t	—

Type	Somatic O-antigen	Flagellar H-antigen	
		Phase 1	Phase 2
S. II	17	m, t	—
S. ahanou	17	i	1, 7
S. III *arizonae* (Ar. 12:33:21)	17	i	z_{35}
S. irenea	17	k	1, 5
S. matadi	17	k	e, n, x
S. II	17	k	—
S. morotai	17	l, v	1, 2
S. michigan	17	l, v	1, 5
S. carmel	17	l, v	e, n, x
S. III *arizonae* (Ar. 12:23:28)	17	l, v	e, n, x, z_{15}
S. III *arizonae* (Ar. 12:23:21)	17	l, v	z_{35}
S. granlo	17	l, z_{28}	e, n, x
S. lode	17	r	1, 2
S. III *arizonae* (Ar. 12:24:31)	17	r	z
S. II	17	y	—
S. gori	17	z	1, 2
S. warengo	17	z	1, 5
S. tchamba	17	z	e, n, z_{15}
S. II *constantia*	17	z	l, w : z_{42}
S. III *arizonae* (Ar. 12:1, 2, 5:—) (Ar. 12:1, 2, 6:—)	17	z_4, z_{23}	—
S. III *arizonae* (Ar. 12:1, 6, 7, 9:—)	17	z_4, z_{23}, z_{32}	—
S. III *arizonae* (Ar. 12:1, 3, 11:—)	17	z_4, z_{24}	—
S. III *arizonae* (Ar. 12:1, 6, 6:—) (Ar. 12:1, 7, 8:—)	17	z_4, z_{32}	—
S. djibouti	17	z_{10}	e, n, x
S. III *arizonae* (Ar. 12:27:28:[38])	17	z_{10}	e, n, x, z_{15} : [z_{56}]
S. III *arizonae* (Ar. 12:27:31)	17	z_{10}	z
S. kandla	17	z_{29}	—
S. III *arizonae* (Ar. 12:16, 17, 18:—)	17	z_{29}	—
S. III *arizonae* (Ar. 12:17, 20:—)	17	z_{36}	—

Type	Somatic O-antigen	Flagellar H-antigen	
		Phase 1	Phase 2

Group O:18 (K)

Type	Somatic O-antigen	Phase 1	Phase 2
S. brazos	6, 14, 18	a	e, n, z_{15}
S. fluntern	6, 14, 18	b	1, 5
S. rawash	6, 14, 18	c	e, n, x
S. groenekan	18	d	1, 5
S. usumbura	18	d	1, 7
S. pontypridd	18	g, m	—
S. III arizonae (Ar. 7a, 7b:13, 14:—)	18	g, z_{51}	—
S. II	18	m, t	1, 5
S. langenhorn	18	m, t	—
S. memphis	18	k	1, 5
S. III arizonae (Ar. 7a, 7b:22:25)	18	(k)	z_{53}
S. III arizonae (Ar. 7a, 7b:22:34)	18	(k)	z_{54}
S. III arizonae (Ar. 7a, 7b:23:28)	18	l, v	e, n, x, z_{15}
S. orlando	18	l, v	e, n, z_{15}
S. III arizonae (Ar. 7a, 7b:23:21)	18	l, v	z
S. toulon	18	l, w	e, n, z_{15}
S. III arizonae (Ar. 7a, 7b:24:31)	18	r	z
S. II	18	y	e, n, x, z_{15}
S. cerro	6, 14, 18	z_4, z_{23}	[1, 5]
S. aarhus	18	z_4, z_{23}	z_{64}
S. II	18	z_4, z_{23}	—
S. III arizonae (Ar. 7a, 7b:1, 2, 5:—) (Ar. 7a, 7b:1, 2, 6:—)	18	z_4, z_{23}	—
S. blukwa	18	z_4, z_{24}	—
S. III arizonae (Ar. 7a, 7b:1, 7, 8:—)	18	z_4, z_{32}	—
S. carnac	18	z_{10}	z_6
S. II zeist	18	z_{10}	z_6
S. II beloha	18	z_{36}	—
S. IV	18	z_{36}, z_{38}	—
S. sinthia	18	z_{38}	—
S. cotia	18	—	1, 6

Type	Somatic O-antigen	Flagellar H-antigen Phase 1	Flagellar H-antigen Phase 2
Group O:21 (L)			
S. assen	21	a	[1, 5]
S. ghana	21	b	1, 6
S. minnesota[a]	21	b	e, n, x
S. hydra	21	c	1, 6
S. rhone	21	c	e, n, x
S. II	21	c	e, n, x
S. spartel	21	d	1, 5
S. magwa	21	d	e, n, x
S. madison	21	d	z_6
S. good	21	f, g	e, n, x
S. III *arizonae* (Ar. 22:13, 14:—)	21	g, z_{51}	—
S. diourbel	21	i	1, 2
S. III *arizonae* (Ar. 22:33:30)	21	i	1, 5, 7
S. III *arizonae* (Ar. 22:33:28)	21	i	e, n, x, z_{15}
S. III *arizonae* (Ar. 22:29:28)	21	k	e, n, x, z_{15}
S. III *arizonae* (Ar. 22:29:31)	21	k	z
S. III *arizonae* (Ar. 22:23:31)	21	l, v	z
S. III *arizonae* (Ar. 22:23:40_a, 40_c)	21	l, v	z_{57}
S. keve	21	l, w	—
S. ruiru	21	y	e, n, x
S. II	21	z	—
S. baguida	21	z_4, z_{23}	—
S. III *arizonae* (Ar. 22:1, 2, 6:—)	21	z_4, z_{23}	—
S. IV *soesterberg*	21	z_4, z_{23}	—
S. II *gwaai*	21	z_4, z_{24}	—
S. III *arizonae* (Ar. 22:1, 3, 11:—)	21	z_4, z_{24}	—
S. III *arizonae* (Ar. 22:27:28)	21	z_{10}	e, n, x, z_{15}
S. III *arizonae* (Ar. 22:27:31)	21	z_{10}	z
S. II *wandsbek*	21	z_{10}	z_6

[a] May possess a R-phase H-antigen: z_{33}; z_{49}.

Type	Somatic O-antigen	Flagellar H-antigen	
		Phase 1	Phase 2
S. III *arizonae* (Ar. 22:16, 17, 18:—)	21	z_{29}	—
S. *gambaga*	21	z_{35}	e, n, z_{15}
S. III *arizonae* (Ar. 22:32^b:28)	21	z_{65}	e, n, x, z_{15}

^{b} The antigenic factor described for this strain as Ar. 32a, 32c is very different from other factors H_{32} of *Arizona* 32a, 32b. Factor 32b is strongly related to *Salmonella* H factor c.

Group O:28 (M)			
S. *solna*	28	a	1, 5
S. *dakar*	28	a	1, 6
S. *bakau*	28	a	1, 7
S. *seattle*	28	a	e, n, x
S. *honelis*	28	a	e, n, z_{15}
S. *moero*	28	b	1, 5
S. *ashanti*	28	b	1, 6
S. *bokanjac*	28	b	1, 7
S. *langford*	28	b	e, n, z_{15}
S. II *kaltenhausen*	28	b	z_6
S. *hermannswerder*	28	c	1, 5
S. *eberswalde*	28	c	1, 6
S. *halle*	28	c	1, 7
S. *dresden*	28	c	e, n, x
S. *wedding*	28	c	e, n, z_{15}
S. *techimani*	28	c	z_6
S. *amoutive*	28	d	1, 5
S. *hatfield*	28	d	1, 6
S. *mundonobo*	28	d	1, 7
S. *mocamedes*	28	d	e, n, x
S. *patience*	28	d	e, n, z_{15}
S. *cullingworth*	28	d	l, w
S. *kpeme*	28	e, h	1, 7
S. II	28	e, n, x	1, 7
S. *friedrichsfelde*	28	f, g	—
S. *abadina*	28	g, m	[e, n, z_{15}]
S. II *llandudno*	28	g, (m), [s], t	1, 5
S. *croft*	28	g, m, s	—
S. II	28	g, m, t	e, n, x
S. II	28	g, s, t	e, n, x
S. *ona*	28	g, s, t	—
S. II	28	m, t	[e, n, x]
S. *vinohrady*	28	m, t	—

Type	Somatic O-antigen	Flagellar H-antigen Phase 1	Phase 2
S. doorn	28	i	1, 2
S. cotham	28	i	1, 5
S. volkmarsdorf	28	i	1, 6
S. dieuppeul	28	i	1, 7
S. warnemuende	28	i	e, n, x
S. kuessel	28	i	e, n, z_{15}
S. guildford	28	k	1, 2
S. ilala	28	k	1, 5
S. adamstown	28	k	1, 6
S. ikeja	28	k	1, 7
S. taunton	28	k	e, n, x
S. ank	28	k	e, n, z_{15}.
S. leoben	28	l, v	1, 5
S. vitkin	28	l, v	e, n, x
S. nashua	28	l, v	e, n, z_{15}
S. ramsey	28	l, w	1, 6
S. fajara	28	l, z_{28}	e, n, x
S. bassadji	28	r	1, 6
S. kibusi	28	r	e, n, x
S. II oevelgoenne	28	r	e, n, z_{15}
S. chicago	28	r, [i]	1, 5
S. banco	28	r, i	1, 7
S. sanktgeorg	28	r, [i]	e, n, z_{15}
S. oskarshamn	28	y	1, 2
S. nima	28	y	1, 5
S. pomona	28	y	1, 7
S. kitenge	28	y	e, n, x
S. telaviv	28	y	e, n, z_{15}
S. shomolu	28	y	l, w
S. selby	28	y	z_6
S. ezra	28	z	1, 7
S. brisbane	28	z	e, n, z_{15}
S. II ceres	28	z	z_{39}
S. teltow	28	z_4, z_{23}	1, 6
S. babelsberg	28	z_4, z_{23}	[e, n, z_{15}]
S. rogy	28	z_{10}	1, 2
S. farakan	28	z_{10}	1, 5
S. malaysia	28	z_{10}	1, 7
S. umbilo	28	z_{10}	e, n, x
S. luckenwalde	28	z_{10}	e, n, z_{15}
S. moroto	28	z_{10}	l, w
S. III arizonae (Ar. 35:27:[40a, 40c])	28	z_{10}	[z_{57}]
S. djermaia	28	z_{29}	—
S. babili	28	z_{35}	1, 7
S. aderike	28	z_{38}	e, n, z_{15}

Type	Somatic O-antigen	Flagellar H-antigen Phase 1	Phase 2
		Group O:30 (N)	
S. overvecht	30	a	1, 2
S. zehlendorf	30	a	1, 5
S. guarapiranga	30	a	e, n, x
S. doulassame	30	a	e, n, z_{15}
S. II *odijk*	30	a	z_{39}
S. louga	30	b	1, 2
S. aschersleben	30	b	1, 5
S. urbana	30	b	e, n, x
S. neudorf	30	b	e, n, z_{15}
S. II	30	b	z_{16}
S. zaire	30	c	1, 7
S. morningside	30	c	e, n, z_{15}
S. II	30	c	z_{39}
S. messina	30	d	1, 5
S. livulu	30	e, h	1, 2
S. II *slatograd*	30	f, g, t	—
S. godesberg	30	g, m	—
S. II	30	g, m, s	e, n, x
S. giessen	30	g, m, s	—
S. sternschanze[a]	30	g, s, t	—
S. wayne	30	g, z_{51}	—
S. landau	30	i	1, 2
S. morehead	20	i	1, 5
S. soerenga	30	i	l, w
S. hilversum	30	k	1, 2
S. ramatgan	30	k	1, 5
S. aqua	30	k	1, 6
S. angoda	30	k	e, n, x
S. odozi	30	k	e, n, [x], z_{15}
S. II	30	k	e, n, x, z_{15}
S. ligeo	30	l, v	1, 2
S. donna	30	l, v	1, 5
S. morocco	30	l, z_{13}, z_{28}	e, n, z_{15}
S. gege	30	r	1, 5
S. matopeni	30	y	1, 2
S. bietri	30	y	1, 5
S. steinplatz	30	y	1, 6
S. baguirmi	30	y	e, n, x
S. nijmegen	30	y	e, n, z_{15}
S. bodjonegoro	30	z_4, z_{24}	—
S. II	30	z_6	1, 6
S. sada	30	z_{10}	1, 2

[a] May possess a R-phase H-antigen: z_{58}.

Type	Somatic O-antigen	Flagellar H-antigen	
		Phase 1	Phase 2
S. kumasi	30	z_{10}	e, n, z_{15}
S. aragua	30	z_{29}	—
S. kokoli	30	z_{35}	1, 6
S. wuiti	30	z_{35}	e, n, z_{15}
S. ago	30	z_{38}	—
S. II	30	z_{39}	1, 7

Group O:35 (O)

S. umhlatazana	35	a	e, n, z_{15}
S. tchad	35	b	—
S. yolo	35	c	—
S. dembe[a]	35	d	l, w
S. gassi	35	e, h	z_6
S. adelaide[b]	35	f, g	—
S. II	35	f, g, t	1, 5
S. ealing	35	g, m, s	—
S. II	35	g, m, s, t	—
S. ebrie	35	g, m, t	—
S. anecho	35	g, s, t	—
S. II	35	g, t	z_{42}
S. agodi	35	g, t	—
S. III arizonae (Ar. 20:13, 14:—)	35	g, z_{51}	—
S. monschaui	35	m, t	—
S. III arizonae (Ar. 20:33:28)	35	i	e, n, x, z_{15}
S. gambia	35	i	e, n, z_{15}
S. bandia	35	i	l, w
S. III arizonae (Ar. 20:33:31)	35	i	z
S. III arizonae (Ar. 20:33:21)	35	i	z_{35}
S. III arizonae (Ar. 20:29:31)	35	k	z
S. III arizonae (Ar. 20:22:31)	35	(k)	z
S. III arizonae (Ar. 20:22:21)	35	(k)	z_{35}
S. III arizonae[c] (Ar. 20:29:25)	35	k	z_{53}

[a] May possess a R-phase H-antigen: z_{59}.
[b] May possess a R-phase H-antigen: z_{27}.
[c] May possess a R-phase H-antigen: z_{50}.

Type	Somatic O-antigen	Flagellar H-antigen	
		Phase 1	Phase 2
S. III *arizonae* (Ar. 20:23:30)	35	l, v	1, 5, 7
S. III *arizonae* (Ar. 20:23:21)	35	l, v	z_{35}
S. II	35	l, z_{28}	—
S. III *arizonae* (Ar. 20:24:28)	35	r	e, n, x, z_{15}
S. massakory	35	r	l, w
S. III *arizonae* (Ar. 20:24:21)	35	r	z_{35}
S. III *arizonae* (Ar. 20:24:41)	35	r	z_{61}
S. alachua[d]	35	z_4, z_{23}	—
S. III *arizonae* (Ar. 20:1, 2, 6:—)	35	z_4, z_{23}	—
S. westphalia	35	z_4, z_{24}	—
S. III *arizonae* (Ar. 20:1, 7, 8:—)	35	z_4, z_{32}	—
S. camberene	35	z_{10}	1, 5
S. enschede	35	z_{10}	l, w
S. ligna	35	z_{10}	z_6
S. III *arizonae* (Ar. 20:27:21)	35	z_{10}	z_{35}
S. II *utbremen*	35	z_{29}	e, n, x
S. widemarsh	35	z_{29}	—
S. III *arizonae* (Ar. 20:16, 17, 18:—)	35	z_{29}	—
S. III *arizonae* (Ar. 20:17, 20:—)	35	z_{36}	—
S. haga	35	z_{38}	—
S. III. *arizonae* (Ar. 20:26:30)	35	z_{52}	1, 5, 7
S. III *arizonae* (Ar. 20:26:28)	35	z_{52}	e, n, x, z_{15}
S. III *arizonae* (Ar. 20:26:31)	35	z_{52}	z
S. III *arizonae* (Ar. 20:26:21)	35	z_{52}	z_{35}

[d] May possess a R-phase H-antigen: z_{45}.

Type	Somatic O-antigen	Flagellar H-antigen	
		Phase 1	Phase 2

Group O:38 (P)

Type	Somatic O-antigen	Phase 1	Phase 2
S. II	38	b	1, 2
S. rittersbach	38	b	e, n, z_{15}
S. sheffield	38	c	1, 5
S. kidderminster	38	c	1, 6
S. II *carletonville*	38	d	[1, 5]
S. thiaroye	38	e, h	1, 2
S. kasenyi	38	e, h	1, 5
S. korovi	38	g, m, [s]	—
S. II *foulpointe*	38	g, t	—
S. III *arizonae* (Ar. 16:13, 14:–)	38	g, z_{51}	—
S. IV	38	g, z_{51}	—
S. mgulani	38	i	1, 2
S. lansing	38	i	1, 5
S. III *arizonae* (Ar. 16:33:25)	38	i	z_{53}
S. echa	38	k	1, 2
S. mango	38	k	1, 5
S. inverness	38	k	1, 6
S. njala	38	k	e, n, x
S. III *arizonae* (Ar. 16:29:31)	38	k	z
S. III *arizonae* (Ar. 16:29:25)	38	k	z_{53}
S. III *arizonae* (Ar. 16:22:30)	38	(k)	1, 5, 7
S. III *arizonae* (Ar. 16:22:31)	38	(k)	z
S. III *arizonae* (Ar. 16:22:21:[38])	38	(k)	$z_{35}:[z_{56}]$
S. III *arizonae* (Ar. 16:22:34)	38	(k)	z_{54}
S. III *arizonae* (Ar. 16:22:37)	38	(k)	z_{55}
S. alger	38	l, v	1, 2
S. kimberley	38	l, v	1, 5
S. roan	38	l, v	e, n, x
S. III *arizonae* (Ar. 16:23:31)	38	l, v	z
S. III *arizonae* (Ar. 16:23:21)	38	l, v	z_{35}
S. III *arizonae* (Ar. 16:23:25:[34]	38	l, v	$z_{53}:[z_{54}]$

Type	Somatic O-antigen	Flagellar H-antigen Phase 1	Flagellar H-antigen Phase 2
S. lindi	38	r	1, 5
S. III *arizonae* (Ar. 16:24:30)	38	r	1, 5, 7
S. emmastad	38	r	1, 6
S. III *arizonae* (Ar. 16:24:31:[40_a, 40_b])	38	r	z:[z_{57}]
S. III *arizonae* (Ar. 16:24:21)	38	r	z_{35}
S. freetown	38	y	1, 5
S. colombo	38	y	1, 6
S. perth	38	y	e, n, x
S. yoff	38	z_4, z_{23}	1, 2
S. IV	38	z_4, z_{23}	—
S. bangkok	38	z_4, z_{24}	—
S. III *arizonae* (Ar. 16:27:31)	38	z_{10}	z
S. III *arizonae* (Ar. 16:27:25)	38	z_{10}	z_{53}
S. klouto	38	z_{38}	—
S. III *arizonae* (Ar. 16:39:25)	38	z_{47}	z_{53}
S. III *arizonae* (Ar. 16:26:21)	38	z_{52}	z_{35}
S. III *arizonae* (Ar. 16:26:25)	38	z_{52}	z_{53}

Group O:39 (Q)

Type	Somatic O-antigen	Flagellar H-antigen Phase 1	Flagellar H-antigen Phase 2
S. II	39	a	z_{39}
S. wandsworth	39	b	1, 2
S. abidjan	39	b	1, w
S. II	39	c	e, n, x
S. logone	39	d	1, 5
S. mara	39	e, h	1, 5
S. hofit	39	i	1, 5
S. champaign	39	k	1, 5
S. kokomlemle	39	l, v	e, n, x
S. oerlikon	39	l, v	e, n, z_{15}
S. II *mondeor*	39	l, z_{28}	e, n, x
S. anfo	39	y	1, 2
S. windermere	39	y	1, 5

Type	Somatic O-antigen	Flagellar H-antigen	
		Phase 1	Phase 2

Group O:40 (R)

Type	Somatic O-antigen	Phase 1	Phase 2
S. shikmonah	40	a	1, 5
S. greiz	40	a	z_6
S. II	1, 40	a	z_6
S. II springs	40	a	z_{39}
S. riogrande	40	b	1, 5
S. saugus	40	b	1, 7
S. johannesburg	1, 40	b	e, n, x
S. duval	1, 40	b	e, n, z_{15}
S. benguella	40	b	z_6
S. II	40	b	—
S. II suarez	1, 40	c	e, n, x, z_{15}
S. II	1, 40	c	z_{39}
S. driffield	1, 40	d	1, 5
S. II ottershaw	40	d	—
S. tilene	1, 40	e, h	1, 2
S. II	1, 40	(f), g	e, n, x, z_{15}
S. bijlmer	1, 40	g, m	—
S. II boksburg	40	g, m, s, t	e, n, x
S. II alsterdorf	1, 40	g, m, t	1, 5
S. II	1, 40	g, t	1, 5
S. II	1, 40	g, t	e, n, x
S. II	1, 40	g, t	z_{42}
S. III arizonae (Ar. 10a, 10b:13, 14:28)	40	g, z_{51}	e, n, x, z_{15}
S. IV seminole	1, 40	g, z_{51}	—
S. II	40	m, t	z_{39}
S. II	1, 40	m, t	z_{42}
S. IV	40	m, t	—
S. III arizonae (Ar. 10a, 10b:33:30)	40	i	1, 5, 7
S. goulfey	1, 40	k	1, 5
S. allandale	1, 40	k	1, 6
S. hann	40	k	e, n, x
S. II sunnydale	1, 40	k	e, n, x, z_{15}
S. III arizonae (Ar. 10a, 10b:29:31:40a, 40c)	40	k	$z:z_{57}$
S. III arizonae (Ar. 10a, 10b:29:25)	40	k	z_{53}
S. millesi	1, 40	l, v	1, 2
S. III arizonae (Ar. 10a, 10b, (10c):23:31)	40	l, v	z

Type	Somatic O-antigen	Flagellar H-antigen Phase 1	Phase 2
S. III *arizonae* (Ar. 10a, 10b:23:25)	40	1, v	z_{53}
S. overchurch	40	1, w	—
S. bukavu	1, 40	1, z_{28}	1, 5
S. santhiaba	40	1, z_{28}	1, 6
S. II *bulawayo*	1, 40	z	1, 5
S. casamance	40	z	e, n, x
S. nowawes	40	z	z_6
S. II	1, 40	z	z_6
S. II	40	z	z_{39}
S. III *arizonae* (Ar. 10a, 10b:1, 2, 5:—) (Ar. 10a, 10b:1, 2, 6:—)	40	z_4, z_{23}	—
S. IV *sachsenwald*	1, 40	z_4, z_{23}	—
S. II *degania*	40	z_4, z_{24}	z_{39}
S. III *arizonae* (Ar. 10a, 10b:1, 3, 11:—)	40	z_4, z_{24}	—
S. IV	40	z_4, z_{24}	—
S. III *arizonae* (Ar. 10a, 10b:1, 7, 8:—)	40	z_4, z_{32}	—
S. IV	40	z_4, z_{32}	—
S. II	1, 40	z_6	1, 5
S. trotha	40	z_{10}	z_6
S. III *arizonae* (Ar. 10a, 10b:27:21)	40	z_{10}	z_{35}
S. omifisan	40	z_{29}	—
S. III *arizonae* (Ar. 10a, 10b:16, 17, 18:—)	40	z_{29}	—
S. II *fandran*	1, 40	z_{35}	e, n, x, z_{15}
S. III *arizonae* (Ar. 10a, 10b:17, 20:—)	40	z_{36}	—
S. II *grunty*	1, 40	z_{39}	1, 6
S. karamoja	1, 40	z_{41}	1, 2
S. II	1, 40	$[z_{42}]$	1, (5), 7

Group O:41 (S)

Type	Somatic O-antigen	Phase 1	Phase 2
S. II	41	b	[1, 5]
S. II	41	b	1, 7
S. vietnam	41	b	$[z_6]$
S. III *arizonae* (Ar. 13:32a, 32b:28)	41	c	e, n, x, z_{15}
S. II	41	c	z_6

Type	Somatic O-antigen	Flagellar H-antigen Phase 1	Flagellar H-antigen Phase 2
S. egusi	41	d	[1, 5]
S. II *hennepin*	41	d	z_6
S. II *lethe*	41	g, t	—
S. III *arizonae* (Ar. 13:13, 14:−)	41	g, z_{51}	—
S. leatherhead	41	m, t	1, 6
S. II	41	k	—
S. III *arizonae* (Ar. 13:22:21)	41	(k)	z_{35}
S. II	41	l, z_{13}, z_{28}	e, n, x, z_{15}
S. lubumbashi	41	r	1, 5
S. II *dubrovnik*	41	z	1, 5
S. waycross	41	z_4, z_{23}	—
S. III *arizonae* (Ar. 13:1, 2, 5:−) (Ar. 13:1, 2, 6:−)	41	z_4, z_{23}	—
S. IV	41	z_4, z_{23}	—
S. III *arizonae* (Ar. 13:1, 6, 7:−)	41	z_4, z_{23}, z_{32}	—
S. ipswich	41	z_4, z_{24}	[1, 5]
S. III *arizonae* (Ar. 13:1, 3, 11:−)	41	z_4, z_{24}	—
S. III *arizonae* (Ar. 13:1, 7, 8:−)	41	z_4, z_{32}	—
S. II *negev*	41	z_{10}	1, 2
S. leipzig	41	z_{10}	1, 5
S. landala	41	z_{10}	1, 6
S. inpraw	41	z_{10}	e, n, x
S. II *lurup*	41	z_{10}	e, n, x, z_{15}
S. II *lichtenberg*	41	z_{10}	z_6
S. lodz	41	z_{29}	—
S. III *arizonae* (Ar. 13:16, 17, 18:−)	41	z_{29}	—
S. III *arizonae* (Ar. 13:17, 20:−)	41	z_{36}	—
S. offa	41	z_{38}	—
S. II	41	—	1, 6

Group O:42 (T)

Type	Somatic O-antigen	Flagellar H-antigen Phase 1	Flagellar H-antigen Phase 2
S. faji	*1*, 42	a	e, n, z_{15}
S. II *chinovum*	42	b	1, 5
S. II *uphill*	42	b	e, n, x, z_{15}
S. tomegbe	*1*, 42	b	e, n, z_{15}

Type	Somatic O-antigen	Flagellar H-antigen	
		Phase 1	Phase 2
S. egusitoo	*1*, 42	b	z_6
S. antwerpen	*1*, 42	c	e, n, z_{15}
S. kampala	*1*, 42	c	z_6
S. II *fremantle*	42	(f), g, t	—
S. maricopa	*1*, 42	g, z_{51}	1, 5
S. III *arizonae*	42	g, z_{51}	—
(Ar. 15:13, 14:—)			
S. II	42	m, t	[e, n, x, z_{15}]
S. waral	*1*, 42	m, t	—
S. kaneshie	*1*, 42	i	l, w
S. middlesbrough	*1*, 42	i	z_6
S. haferbreite	42	k	1, 6
S. III *arizonae*	42	k	z
(Ar. 15:29:31)			
S. gwale	*1*, 42	k	z_6
S. III *arizonae*	42	(k)	z_{35}
(Ar. 15:22:21)			
S. III *arizonae*	42	k	—
(Ar. 15:29:—)			
S. III *arizonae*	42	l, v	1, 5, 7
(Ar. 15:23:30)			
S. II *portbech*	42	l, v	e, n, x, z_{15}
S. III *arizonae*	42	l, v	e, n, x, z_{15}
(Ar. 15:23:28)			
S. coogee	42	l, v	e, n, z_{15}
S. III	42	l, v	z
(Ar. 15:23:31)			
S. III *arizonae*	42	l, v	z_{53}
(Ar. 15:23:25)			
S. II	42	l, z_{13}, z_{28}	z_6
S. II	42	l, z_{28}	—
S. sipane	*1*, 42	r	e, n, z_{15}
S. brive	*1*, 42	r	l, w
S. III *arizonae*	42	r	z
(Ar. 15:24:31)			
S. III *arizonae*	42	r	z_{53}
(Ar. 15:24:25)			
S. II *nairobi*	42	r	—
S. III *arizonae*[a]	42	r	—
(Ar. 15:24:—)			
S. harvestehude	*1*, 42	y	z_6
S. II *detroit*	42	z	1, 5
S. ursenbach	42	z	1, 6

[a] May possess a R-phase H-antigen: z_{50}.

Type	Somatic O-antigen	Flagellar H-antigen Phase 1	Flagellar H-antigen Phase 2
S. II *rand*	42	z	e, n, x, z_{15}
S. II *nuernberg*	42	z	z_6
S. gera	*1*, 42	z_4, z_{23}	1, 6
S. III *arizonae*	42	z_4, z_{23}	—
(Ar. 15:1, 2, 5:—)			
(Ar. 15:1, 2, 6:—)			
S. toricada	*1*, 42	z_4, z_{24}	—
S. III *arizonae*	42	z_4, z_{24}	—
(Ar. 15:1, 3, 11:—)			
S. II	42	z_6	1, 6
S. II	42	z_{10}	e, n, x, z_{15}
S. III *arizonae*		z_{10}	e, n, x, z_{15}
(Ar. 15:27:28)			
S. III *arizonae*	42	z_{10}	z
(Ar. 15:27:31)			
S. loenga	*1*, 42	z_{10}	z_6
S. II	42	z_{10}	z_6
S. III *arizonae*	42	z_{10}	z_{35}
(Ar. 15:27:21)			
S. III *arizonae*	42	z_{10}	z_{56}
(Ar. 15:27:38)			
S. djama	*1*, 42	z_{29}	—
S. kahla	*1*, 42	z_{35}	1, 6
S. weslaco	42	z_{36}	—
S. IV	42	z_{36}	—
S. vogan	*1*, 42	z_{38}	z_6
S. taset	*1*, 42	z_{41}	—
S. III *arizonae*	42	z_{52}	z
(Ar. 15:26:31)			
S. II	42	—	1, 6

Group O:43 (U)

Type	Somatic O-antigen	Phase 1	Phase 2
S. graz	43	a	1, 2
S. berkeley	43	a	1, 5
S. II	43	a	z_6
S. II *kommetje*	43	b	z_{42}
S. montreal	43	c	1, 5
S. II	43	d	e, n, x, z_{15}
S. II	43	d	z_{39}
S. II	43	d	z_{42}
S. II	43	e, n, x, z_{15}	1, (5), 7
S. II	43	e, n, x, z_{15}	1, 6
S. milwaukee	43	f, g	—

Type	Somatic O-antigen	Flagellar H-antigen	
		Phase 1	Phase 2
S. II	43	f, g, t	1, 5
S. II *mosselbay*	43	g, m, [s], t	$[z_{42}]$
S. veddel	43	g, t	—
S. IV	43	g, z_{51}	—
S. II	43	g, z_{62}	e, n, x
S. mbao	43	i	1, 2
S. thetford	43	k	1, 2
S. ahuza	43	k	1, 5
S. III *arizonae* (Ar. 21:29:31)	43	k	z
S. III *arizonae* (Ar. 21:23:25)	43	l, v	z_{53}
S. III *arizonae* (Ar. 21:23:38)	43	l, v	z_{56}
S. III *arizonae* (Ar. 21:24:28)	43	r	e, n, x, z_{15}
S. III *arizonae* (Ar. 21:24:31)	43	r	z
S. III *arizonae* (Ar. 21:24:25)	43	r	z_{53}
S. farcha	43	y	1, 2
S. kingabwa	43	y	1, 5
S. ogbete	43	z	1, 5
S. II	43	z	1, 5
S. III *arizonae* (Ar. 21:1, 2, 5:−)	43	z_4, z_{23}	—
S. IV *houten*	43	z_4, z_{23}	—
S. III *arizonae* (Ar. 21:1, 3, 11:−)	43	z_4, z_{24}	—
S. IV	43	z_4, z_{24}	—
S. IV *tuindorp*	43	z_4, z_{32}	—
S. adana	43	z_{10}	1, 5
S. II	43	z_{29}	e, n, x
S. II	43	z_{29}	z_{42}
S. IV	43	z_{29}	—
S. ahepe	43	z_{35}	1, 6
S. III *arizonae* (Ar. 21:17, 20:−)	43	z_{36}	—
S. IV *volksdorf*	43	z_{36}, z_{38}	—
S. irigny	43	z_{38}	—
S. II *bunnik*	43	z_{42}	[1, 5, 7]
S. III *arizonae* (Ar. 21:26:25)	43	z_{52}	z_{53}

Type	Somatic O-antigen	Flagellar H-antigen	
		Phase 1	Phase 2
Group O:44 (V)			
S. niakhar	44	a	1, 5
S. tiergarten	44	a	e, n, x
S. niarembe	44	a	l, w
S. sedgwick	44	b	e, n, z_{15}
S. madigan	44	c	1, 5
S. quebec	44	c	e, n, z_{15}
S. bobo	44	d	1, 5
S. kermel	44	d	e, n, x
S. fischerstrasse	44	d	e, n, z_{15}
S. II	1, 44	e, n, x	1, 6
S. vleuten	44	f, g	—
S. gamaba	44	g, m, s	—
S. II	44	g, t	z_{42}
S. carswell	44	g, z_{51}	—
S. IV	44	g, z_{51}	—
S. muguga	44	m, t	—
S. lawra	44	k	e, n, z_{15}
S. malika	44	l, z_{28}	1, 5
S. brefet	44	r	e, n, z_{15}
S. V *camdeni*	44	r	—
S. uhlenhorst	44	z	l, w
S. kua	44	z_4, z_{23}	—
S. II	44	z_4, z_{23}	—
S. III *arizonae* (Ar. 1, 3:1, 2, 5:−) (Ar. 1, 3:1, 2, 6:−)	44	z_4, z_{23}	—
S. IV	44	z_4, z_{23}	—
S. III *arizonae* (Ar. 1, 3:1, 6, 7, 9:−) (Ar. 1, 3:1, 2, 10:−)	44	z_4, z_{23}, z_{32}	—
S. christiansborg	44	z_4, z_{24}	—
S. III *arizonae* (Ar. 1, 3:1, 3, 11:−)	44	z_4, z_{24}	—
S. IV	44	z_4, z_{24}	—
S. III *arizonae* (Ar. 1, 3:1, 7, 8:−)	44	z_4, z_{32}	—
S. IV *lohbruegge*	44	z_4, z_{32}	—
S. guinea	44	z_{10}	[1, 7]
S. IV	44	z_{36}, [z_{38}]	—
S. koketime	44	z_{38}	—
S. II *clovelly*	1, 44	z_{39}	[e, n, x, z_{15}]

		Flagellar H-antigen	
Type	Somatic O-antigen	Phase 1	Phase 2

Group O:45 (W)

Type	Somatic O-antigen	Phase 1	Phase 2
S. II *vrindaban*	45	a	e, n, x
S. *meekatharra*	45	a	e, n, z_{15}
S. II *ejeda*	45	a	z_{10}
S. *riverside*	45	b	1, 5
S. *fomeco*	45	b	e, n, z_{15}
S. *deversoir*	45	c	e, n, x
S. *dugbe*	45	d	1, 6
S. *karachi*	45	d	e, n, x
S. *suelldorf*	45	f, g	—
S. *tornow*	45	g, m, [s]	—
S. II *windhoek*	45	g, m, s, t	1, 5
S. II *bremen*	45	g, m, s, t	e, n, x
S. II *perinet*	45	g, m, t	e, n, x, z_{15}
S. *binningen*	45	g, s, t	—
S. III *arizonae* (Ar. 11:13, 14:—)	45	g, z_{51}	—
S. IV	45	g, z_{51}	—
S. II	45	m, t	1, 5
S. *apapa*	45	m, t	—
S. *casablanca*	45	k	1, 7
S. *cairns*	45	k	e, n, z_{15}
S. II *klapmuts*	45	z	z_{39}
S. IV	45	z_4, z_{23}	—
S. III *arizonae* (Ar. 11:1, 3, 11:—)	45	z_4, z_{24}	—
S. III *arizonae* (Ar. 11:1, 7, 8:—)	45	z_4, z_{32}	—
S. II	45	z_{29}	1, 5
S. II	45	z_{29}	z_{42}
S. *jodhpur*	45	z_{29}	—
S. III *arizonae* (Ar. 11:16, 17, 18:—)	45	z_{29}	—
S. *lattenkamp*	45	z_{35}	1, 5
S. *balcones*	45	z_{36}	—

Group O:47 (X)

Type	Somatic O-antigen	Phase 1	Phase 2
S. II *bilthoven*	47	a	[1, 5]
S. II	47	a	e, n, x, z_{15}
S. II *phoenix*	47	b	1, 5
S. II *khami*	47	b	[e, n, x, z_{15}]

Type	Somatic O-antigen	Flagellar H-antigen Phase 1	Phase 2
S. saka	47	b	—
S. sya	47	b	z_6
S. III arizonae (Ar. 28:43:—)	47	b	—
S. III arizonae (Ar. 28:32:30)	47	c	1, 5, 7
S. III arizonae (Ar. 23:32:28) (Ar. 28:32:28:[40_a, 40_c])	47	c	e, n, x, z_{15}:[z_{57}]
S. III arizonae (Ar. 28:32:31)	47	c	z
S. III arizonae (Ar. 28:32:21)	47	c	z_{35}
S. kodjovi	47	c	—
S. stellingen	47	d	e, n, x
S. II quimbamba	47	d	z_{39}
S. sljeme	1, 47	f, g	—
S. luke	1, 47	g, m	—
S. anie	47	(g), m, t	—
S. II	47	g, t	e, n, x
S. mesbit	47	m, t	e, n, z_{15}
S. III arizonae (Ar. 23:33:28)	47	i	e, n, x, z_{15}
S. bergen	47	i	e, n, z_{15}
S. III arizonae (Ar. 28:33:31)	47	i	z
S. III arizonae (Ar. 23:33:21) (Ar. 28:33:21)	47	i	z_{35}
S. III arizonae (Ar. 23:33:25) (Ar. 28:33:25:[40_a, 40_c])	47	i	z_{53}:[z_{57}]
S. staoueli	47	k	1, 2
S. bootle	47	k	1, 5
S. III arizonae (Ar. 28:29:30)	47	k	1, 5, 7
S. dahomey[a]	47	k	1, 6
S. III arizonae (Ar. 28:29:28)	47	k	e, n, x, z_{15}
S. lyon	47	k	e, n, z_{15}
S. III arizonae (Ar. 28:29:31)	47	k	z
S. III arizonae (Ar. 23:29:21)	47	k	z_{35}

[a] May possess a R-phase H-antigen: z_{58}.

Type	Somatic O-antigen	Flagellar H-antigen Phase 1	Phase 2
S. III *arizonae* (Ar. 23:29:25)	47	k	z_{53}
S. III *arizonae*[b] (Ar. 23:23:30)	47	l, v	1, 5, (7)
S. III *arizonae* (Ar. 28:23:28)	47	l, v	e, n, x, z_{15}
S. III *arizonae* (Ar. 28:23:21)	47	l, v	z_{35}
S. III *arizonae* (Ar. 28:23:25)	47	l, v	z_{53}
S. III *arizonae* (Ar. 28:23:40a, 40c)	47	l, v	z_{57}
S. teshie	1, 47	l, z_{13}, z_{28}	e, n, z_{15}
S. dapango	47	r	1, 2
S. III *arizonae* (Ar. 23:24:30)	47	r	1, 5, 7
S. III *arizonae* (Ar. 23:24:31)	47	r	z
S. III *arizonae* (Ar. 23:24:21)	47	r	z_{35}
S. III *arizonae* (Ar. 23:24:—)	47	r	—
S. III *arizonae*[c] (Ar. 23:24:25[44]	47	r	$z_{53}[z_{60}]$
S. moualine	47	y	1, 6
S. blitta	47	y	e, n, x
S. mountpleasant	47	z	1, 5
S. kaolack	47	z	1, 6
S. II	47	z	e, n, x, z_{15}
S. II *chersina*	47	z	z_6
S. tabligbo	47	z_4, z_{23}	e, n, z_{15}
S. bere[d]	47	z_4, z_{23}	z_6
S. tamberma	47	z_4, z_{24}	—
S. II	47	z_6	1, 6
S. III *arizonae* (Ar. 28:27:30)	47	z_{10}	1, 5, 7
S. III *arizonae* (Ar. 28:27:31)	47	z_{10}	z
S. III *arizonae* (Ar. 28:27:21)	47	z_{10}	z_{35}
S. ekpoui	47	z_{29}	—
S. III *arizonae* (Ar. 28:16, 17, 18:—)	47	z_{29}	—

[b] May possess a R-phase H-antigen: z_{50}.
[c] May possess a R-phase H-antigen: z_{50}.
[d] May possess a R-phase H-antigen: z_{45}.

Type	Somatic O-antigen	Flagellar H-antigen	
		Phase 1	Phase 2
S. bingerville	47	z_{35}	e, n, z_{15}
S. alexanderplatz	47	z_{38}	—
S. quinhon	47	z_{44}	—
S. III arizonae (Ar. 28:26:30)	47	z_{52}	1, 5, 7
S. III arizonae (Ar. 28:26:28)	47	z_{52}	e, n, x, z_{15}
S. III arizonae (Ar. 28:26:31)	47	z_{52}	z
S. III arizonae (Ar. 28:26:21)	47	z_{52}	z_{35}

Group O:48 (Y)

Type	Somatic O-antigen	Flagellar H-antigen	
S. hisingen	48	a	1, 5, 7
S. II	48	a	z_6
S. III arizonae (Ar. 5:35:[21])	48	a	$[z_{35}]$
S. II	48	a	z_{39}
S. II	48	b	z_6
S. III arizonae (Ar. 5, 29:32:31)	48	c	z
S. II hagenbeck	48	d	z_6
S. fitzroy	48	e, h	1, 5
S. II hammonia	48	e, n, x, z_{15}	z_6
S. II erlangen	48	g, m, t	—
S. III arizonae (Ar. 5:13, 14:—)	48	g, z_{51}	—
S. IV marina	48	g, z_{51}	—
S. III arizonae (Ar. 5, 29:33:31)	48	i	z
S. III arizonae (Ar. 29:33:21:[40])	48	i	z_{35}:$[z_{57}]$
S. III arizonae (Ar. 5:33:25)	48	i	z_{53}
S. III arizonae (Ar. 5:29:30)	48	k	1, 5, (7)
S. II	48	k	e, n, x, z_{15}
S. III arizonae (Ar. 5:29:28)	48	k	e, n, x, z_{15}
S. dahlem	48	k	e, n, z_{15}
S. III arizonae (Ar. 5, 29:29:31)	48	k	z
S. III arizonae (Ar. 5:29:21)	48	k	z_{35}

Type	Somatic O-antigen	Flagellar H-antigen	
		Phase 1	Phase 2
S. II *sakaraha*	48	k	z_{39}
S. III *arizonae* (Ar. 5, 29:29:25)	48	k	z_{53}
S. III *arizonae* (Ar. 5:22:25)	48	(k)	z_{53}
S. III *arizonae*[a] (Ar. 5:23:30)	48	l, v	1, 5, (7)
S. III *arizonae* (Ar. 5, 29:23:31)	48	l, v	z
S. III *arizonae* (Ar. 5:24:28)	48	r	e, n, x, z_{15}
S. III *arizonae* (Ar. 5, 29:24:31)	48	r	z
S. *toucra*[b]	48	z	1, 5
S. III *arizonae* (Ar. 5:1, 2, 5:−) (Ar. 5:1, 2, 5, 6;−) (Ar. 5:1, 6:−)	48	z_4, z_{23}	—
S. III *arizonae* (Ar. 5:1, 6, 7:−)	48	z_4, z_{23}, z_{32}	—
S. *djakarta*	48	z_4, z_{24}	—
S. III *arizonae* (Ar. 5:1, 3, 11;−)	48	z_4, z_{24}	—
S. III *arizonae* (Ar. 5:1, 2, 10:−) (Ar. 5:1, 7, 8:−)	48	z_4, z_{32}	—
S. IV	48	z_4, z_{32}	—
S. II *ngozi*	48	z_{10}	[1, 5]
S. *isaszeg*	48	z_{10}	e, n, x
S. III *arizonae* (Ar. 5:27:28)	48	z_{10}	e, n, x, z_{15}
S. III *arizonae* (Ar. 5, 29:27:31)	48	z_{10}	z
S. II	48	z_{29}	—
S. V *bongor*	48	z_{35}	—
S. III *arizonae* (Ar. 5, 29:17, 20:−)	48	z_{36}	—
S. IV	48	z_{36}, z_{38}	—
S. V *balboa*	48	z_{41}	—
S. III *arizonae* (Ar. 5, 29:26:28)	48	z_{52}	e, n, x, z_{15}
S. III *arizonae* (Ar. 5:26:31)	48	z_{52}	z

[a] May possess a R-phase H-antigen: z_{47}; z_{50}.
[b] May possess a R-phase H-antigen: z_{58}.

| | | Flagellar H-antigen | |
Type	Somatic O-antigen	Phase 1	Phase 2
Group O:50 (Z)			
S. rochdale	50	b	e, n, x
S. II	50	b	z_6
S. II *krugersdorp*	50	e, n, x	1, 7
S. II *namib*	50	g, m, s, t	1, 5
S. IV *wassenaar*	50	g, z_{51}	—
S. II *atra*	50	m, t	z_6:z_{42}
S. III *arizonae* (Ar. 9a, 9c:33:30)	50	i	1, 5, 7
S. III *arizonae* (Ar. 9a, 9c:33:28)	50	i	e, n, x, z_{15}
S. III *arizonae* (Ar. 9a, 9c:33:31)	50	i	z
S. III *arizonae* (Ar. 9a, 9c:29:30)	50	k	1, 5, 7
S. III *arizonae* (Ar. 9a, 9c:29:28)	50	k	e, n, x, z_{15}
S. III *arizonae* (Ar. 9a, 9b:29:31:—)[a] (Ar. 9a, 9c:29:31)	50	k	z
S. III *arizonae* (Ar. 9a, 9b:22:31)	50	(k)	z
S. II *seaforth*	50	k	z_6
S. III *arizonae* (Ar. 9a, 9b:29:21)	50	k	z_{35}
S. III *arizonae* (Ar. 9a, 9b:22:21)	50	(k)	z_{35}
S. III *arizonae* (Ar. 9a, 9c:29:25)	50	k	z_{53}
S. fass	50	l, v	1, 2
S. III *arizonae* (Ar. 9a, 9b:23:28)	50	l, v	e, n, x, z_{15}
S. III *arizonae* (Ar. 9a, 9c:23:31)	50	l, v	z
S. III *arizonae* (Ar. 9a, 9c:23:21)	50	l, v	z_{35}
S. II	50	l, w	e, n, x, z_{15}:z_{42}
S. II	50	l, z_{28}	z_{42}
S. III *arizonae* (Ar. 9a, 9b:24:30)	50	r	1, 5, (7)
S. III *arizonae* (Ar. 9a, 9c:24:28)	50	r	e, n, x, z_{15}

[a] May possess a R-phase H-antigen: z_{50}.

Type	Somatic O-antigen	Flagellar H-antigen	
		Phase 1	Phase 2
S. III *arizonae* (Ar. 9a, 9b:24:31) (Ar. 9a, 9c:24:31)	50	r	z
S. III *arizonae* (Ar. 9a, 9b:24:21)	50	r	z_{35}
S. III *arizonae* (Ar. 9a, 9b:24:25)	50	r	z_{53}
S. *dougi*	50	y	1, 6
S. II *greenside*	50	z	e, n, x
S. III *arizonae* (Ar. 9a, 9b:1, 2, 5:−) (Ar. 9a, 9b:1, 2, 6:−)	50	z_4, z_{23}	—
S. IV *flint*	50	z_4, z_{23}	—
S. III *arizonae* (Ar. 9a, 9b:1, 6, 7:−)	50	z_4, z_{23}, z_{32}	—
S. III *arizonae* (Ar. 9a, 9b:1, 3, 11:−)	50	z_4, z_{24}	—
S. IV	50	z_4, z_{24}	—
S. III *arizonae* (Ar. 9a, 9b:1, 2, 10:−) (Ar. 9a, 9b: 1, 7, 8:−)	50	z_4, z_{32}	—
S. IV *bonaire*	50	z_4, z_{32}	—
S. III *arizonae* (Ar. 9a, 9c: 27:31:[38])	50	z_{10}	$z:[z_{56}]$
S. II *hooggraven*	50	z_{10}	$z_6:z_{42}$
S. III *arizonae* (Ar. 9a, 9c:27:25)	50	z_{10}	z_{53}
S. III *arizonae* (Ar. 9a, 9b:16, 17, 18:−)	50	z_{29}	—
S. III *arizonae* (Ar. 9a, 9b: 17, 20:−)	50	z_{36}	—
S. II *faure*	50	z_{42}	1, 7
S. III *arizonae* (Ar. 9a, 9b:26:30) (Ar. 9a, 9c:26:30)	50	z_{52}	1, 5, 7
S. III *arizonae* (Ar. 9a, 9b: 26:31) (Ar. 9a, 9c: 26:31)	50	z_{52}	z
S. III *arizonae* (Ar. 9a, 9b:26:21) (Ar. 9a, 9c:26:21)	50	z_{52}	z_{35}
S. III *arizonae* (Ar. 9a, 9b:26:25) (Ar. 9a, 9c:26:25)	50	z_{52}	z_{53}

Type	Somatic O-antigen	Flagellar H-antigen Phase 1	Phase 2
Group O:51			
S. tione	51	a	e, n, x
S. II	51	c	—
S. gokul	1, 51	d	[1, 5]
S. meskin	51	e, h	1, 2
S. III arizonae (Ar. 1, 2:13, 14:−)	51	g, z_{51}	—
S. kabete	51	i	1, 5
S. dan	51	k	e, n, z_{15}
S. III arizonae (Ar. 1, 2:29:21)	51	k	z_{35}
S. overschie	51	l, v	1, 5
S. dadzie	51	l, v	e, n, x
S. III arizonae (Ar. 1, 2:23:31)	51	l, v	z
S. II askraal	51	l, z_{28}	[z_6]
S. antsalova	51	z	1, 5
S. treforest	1, 51	z	1, 6
S. lechler	51	z	e, n, z_{15}
S. III arizonae (Ar. 1, 2:1, 2, 5:−) (Ar. 1, 2:1, 2, 6:−)	51	z_4, z_{23}	—
S. IV harmelen	51	z_4, z_{23}	—
S. III arizonae (Ar. 1, 2:1, 3, 11:−)	51	z_4, z_{24}	—
S. II	51	z_{29}	e, n, x, z_{15}
S. II roggeveld	51	—	1, 7
Group O:52			
S. uithof	52	a	1, 5
S. ord	52	a	e, n, z_{15}
S. molesey	52	b	1, 5
S. flottbek	52	b	[e, n, x]
S. II	52	c	k
S. utrecht	52	d	1, 5
S. II	52	d	e, n, x, z_{15}
S. butare	52	e, h	1, 6
S. derkle	52	e, h	1, 7
S. saintemarie	52	g, t	—
S. II	52	g, t	—

Type	Somatic O-antigen	Flagellar H-antigen	
		Phase 1	Phase 2
S. III *arizonae* (Ar. 31:29:21)	52	k	z_{35}
S. III *arizonae* (Ar. 31:29:25)	52	k	z_{53}
S. III *arizonae* (Ar. 31:23:25)	52	l, v	z_{53}
S. II *lobatsi*	52	z_{44}	1, 5, 7
S. III *arizonae* (Ar. 31:26:31)	52	z_{52}	z

Group O:53

Type	Somatic O-antigen	Phase 1	Phase 2
S. II	53	d	1, 5
S. II	1, 53	d	z_{39}
S. II	53	d	z_{42}
S. III *arizonae* (Ar. 1, 4:13, 14:−)	53	g, z_{51}	—
S. IV	1, 53	g, z_{51}	—
S. III *arizonae* (Ar. 1, 4:33:31)	53	i	z
S. III *arizonae* (Ar. 1, 4:29:28)	53	k	e, n, x, z_{15}
S. III *arizonae* (Ar. 1, 4:29:31)	53	k	z
S. III *arizonae* (Ar. 1, 4:22:31)	53	(k)	z
S. III *arizonae* (Ar. 1, 4:22:21)	53	(k)	z_{35}
S. III *arizonae* (Ar. 1, 4:23:28)	53	l, v	e, n, x, z_{15}
S. III *arizonae* (Ar. 1, 4:23:21)	53	l, v	z_{35}
S. II *midhurst*	53	l, z_{28}	z_{39}
S. III *arizonae* (Ar. 1, 4:24:31)	53	r	z
S. III *arizonae* (Ar. 1, 4:24:21)	53	r	z_{35}
S. III *arizonae* (Ar. 1, 4:24:38)	53	r	z_{56}
S. II	53	z	1, 5
S. III *arizonae* (Ar. 1, 4:31:30)	53	z	1, 5, (7)
S. II	53	z	z_{6}

| Type | Somatic O-antigen | Flagellar H-antigen | |
		Phase 1	Phase 2
S. III *arizonae* (Ar. 1, 4:1, 2, 5:−) (Ar. 1, 4:1, 2, 6:−)	53	z_4, z_{23}	—
S. IV	53	z_4, z_{23}	—
S. III *arizonae* (Ar. 1, 4:1, 6, 7:−) (Ar. 1, 4:1, 6, 7, 9:−)	53	z_4, z_{23}, z_{32}	—
S. II *humber*	53	z_4, z_{24}	—
S. III *arizonae* (Ar. 1, 4:1, 3, 11:−)	53	z_4, z_{24}	—
S. III *arizonae* (Ar. 1, 4, :27:21)	53	z_{10}	z_{35}
S. III *arizonae* (Ar. 1, 4:16, 17, 18:−)	53	z_{29}	—
S. IV *bockenheim*	1, 53	z_{36}, z_{38}	—
S. III *arizonae* (Ar. 1, 4:26:21)	53	z_{52}	z_{35}
S. III *arizonae* (Ar. 1, 4:26:25)	53	z_{52}	z_{53}

Group O:54[a]

S. tonev	21, 54	b	e, n, x
S. winnipeg	54	e, h	1, 5
S. rossleben	54	e, h	1, 6
S. borreze	54	f, g, s	—
S. uccle	3, 54	g, s, t	—
S. poeseldorf	8, 20, 54	i	z_6
S. ochsenwerder	6, 7, 54	k	1, 5
S. czernyring	54	r	1, 5
S. steinwerder	3, 15, 54	y	1, 5
S. yerba	54	z_4, z_{23}	—
S. canton	54	z_{10}	e, n, x

[a] This group is not homogeneous and certain serotypes possess factors other than 54. Moreover, factor O:54 (which has some antigenic resemblance to O:42) can be lost by certain serotypes: *S. tonev*, which possesses factor 21, then becomes similar to *S. minnesota*, *S. uccle* retains factor 3 on this segregation. *S. poeseldorf*, which possesses factors 8, 20, becomes similar to *S. kentucky*, and *S. ochsenwerder*, which possesses factors 6_1, 6_2, 7, becomes similar to *S. thompson*. *S. steinwerder* can, moreover, be converted by phage ε_{34} and acquire factors 34 and 12_2.

Group O:55

S. II *tranoroa*	55	k	z_{39}

Type	Somatic O-antigen	Flagellar H-antigen	
		Phase 1	Phase 2

Group O:56

Type	Somatic O-antigen	Phase 1	Phase 2
S. II *artis*	56	b	—
S. II	56	d	—
S. II	56	e, n, x	1, 7
S. II	56	l, z_{28}	—
S. III *arizonae* (Ar. 14:1, 2, 5:—) (Ar. 14:1, 2, 6:—)	56	z_4, z_{23}	—
S. III *arizonae* (Ar. 14:1, 6, 7, 9:—)	56	z_4, z_{23}, z_{32}	—
S. II	56	z_{10}	e, n, x
S. III *arizonae* (Ar. 14:16, 18:—)	56	z_{29}	—

Group O:57

Type	Somatic O-antigen	Phase 1	Phase 2
S. antonio	57	a	z_6
S. maryland	57	b	1, 7
S. III *arizonae* (Ar. 34:32:31:44)	57	c	$z:z_{60}$
S. II	57	d	1, 5
S. II	57	g, m, s, t	z_{42}
S. II	57	g, t	—
S. III *arizonae* (Ar. 34:33:28)	57	i	e, n, x, z_{15}
S. III *arizonae* (Ar. 34:33:31)	57	i	z
S. IV	57	z_4, z_{23}	—
S. II *locarno*	57	z_{29}	z_{42}
S. II *manombo*	57	z_{34}	e, n, x, z_{15}
S. II *tokai*	57	z_{42}	1, 6:z_{53}

Group O:58

Type	Somatic O-antigen	Phase 1	Phase 2
S. II	58	a	[z_6]
S. II	58	b	1, 5
S. II	58	c	z_6
S. II	58	d	z_6
S. III *arizonae* (Ar. 1, 33:33:28)	58	i	e, n, x, z_{15}
S. III *arizonae* (Ar. 1, 33:23:28)	58	l, v	e, n, x, z_{15}

| Type | Somatic O-antigen | Flagellar H-antigen | |
		Phase 1	Phase 2
S. III arizonae (Ar. 1, 33:23:21)	58	1, v	z_{35}
S. II basel	58	1, z_{13}, z_{28}	1, 5
S. III arizonae (Ar. 1, 33:24:28)	58	r	e, n, x, z_{15}
S. III arizonae (Ar. 1, 33:24:31)	58	r	z
S. III arizonae[a] (Ar. 1, 33:24:25: [40_a, 40_c])	58	r	z_{53}:[z_{57}]
S. II	58	z_{10}	1, 6
S. II	58	z_{10}	z_6
S. III arizonae (Ar. 1, 33:26:31)	58	z_{52}	z
S. III arizonae (Ar. 1, 33:26:21)	58	z_{52}	z_{35}

[a] May possess a R-phase H antigen: z_{47}.

Group O:59			
S. III arizonae (Ar. 19:32:28)	59	c	e, n, x, z_{15}
S. III arizonae (Ar. 19:33:31)	59	i	z
S. III arizonae (Ar. 19:33:21)	59	i	z_{35}
S. III arizonae (Ar. 19:22:28)	59	(k)	e, n, x, z_{15}
S. II betioky	59	k	(z)
S. III arizonae (Ar. 19:22:31)	59	(k)	z
S. III arizonae (Ar. 19:22:21)	59	(k)	z_{35}
S. III arizonae (Ar. 19:29:25)	59	k	z_{53}
S. III arizonae (Ar. 19:23:31)	59	1, v	z
S. III arizonae (Ar. 19:23:25)	59	1, v	z_{53}
S. III arizonae (Ar. 19:1, 2, 5:—) (Ar. 19:1, 2, 6:—)	59	z_4, z_{23}	—
S. III arizonae (Ar. 19:27:25)	59	z_{10}	z_{53}

Type	Somatic O-antigen	Flagellar H-antigen	
		Phase 1	Phase 2
S. III *arizonae* (Ar. 19:27:40$_a$, 40$_c$)	59	z_{10}	z_{57}
S. III *arizonae* (Ar. 19:16, 17, 18:−)	59	z_{29}	—
S. III *arizonae* (Ar. 19:17,20:−)	59	z_{36}	—
S. III *arizonae* (Ar. 19:26:−)	59	z_{52}	—

Group O:60

Type	Somatic O-antigen	Flagellar H-antigen Phase 1	Phase 2
S. II *setubal*	60	g, m, t	z_6
S. III *arizonae* (Ar. 24:33:28)	60	i	e, n, x, z_{15}
S. III *arizonae* (Ar. 24:33:21)	60	i	z_{35}
S. III *arizonae* (Ar. 24:29:31)	60	k	z
S. III *arizonae* (Ar. 24:29:21)	60	k	z_{35}
S. III *arizonae* (Ar. 24:22:25)	60	(k)	z_{53}
S. III *arizonae* (Ar. 24:23:31)	60	l, v	z
S. III *arizonae* (Ar. 24:24:28)	60	r	e, n, x, z_{15}
S. III *arizonae* (Ar. 24:24:31)	60	r	z
S. III *arizonae* (Ar. 24:24:21)	60	r	z_{35}
S. III *arizonae* (Ar. 24:24:25)	60	r	z_{53}
S. II *luton*	60	z	e, n, x
S. III *arizonae* (Ar. 24:27:31)	60	z_{10}	z
S. III *arizonae* (Ar. 24:27:21)	60	z_{10}	z_{35}
S. III *arizonae* (Ar. 24:26:30)	60	z_{52}	1, 5, 7
S. III *arizonae* (Ar. 24:26:31)	60	z_{52}	z
S. III *arizonae* (Ar. 24:26:21)	60	z_{52}	z_{35}
S. III *arizonae* (Ar. 24:26:25)	60	z_{52}	z_{53}

| Type | Somatic O-antigen | Flagellar H-antigen | |
		Phase 1	Phase 2
Group O:61			
S. III arizonae (Ar. 26:32:30)	61	c	1, 5, (7)
S. III arizonae (Ar. 26:32:21)	61	c	z_{35}
S. III arizonae (Ar. 26:33:28)	61	i	e, n, x, z_{15}
S. III arizonae (Ar. 26:33:31)	61	i	z
S. III arizonae (Ar. 26:33:21)	61	i	z_{35}
S. III arizonae (Ar. 26:33:25)	61	i	z_{53}
S. III arizonae (Ar. 26:29:30)	61	k	1, 5, (7)
S. III arizonae (Ar. 26:22:25)	61	(k)	z_{53}
S. III arizonae (Ar. 26:23:30: $[40_a, 40_b]$)	61	l, v	1, 5, 7:$[z_{57}]$
S. III arizonae (Ar. 26:23:31)	61	l, v	z
S. III arizonae (Ar. 26:23:21)	61	l, v	z_{35}
S. III arizonae (Ar. 26:24:30)	61	r	1, 5, 7
S. III arizonae (Ar. 26:24:21)	61	r	z_{35}
S. III arizonae (Ar. 26:24:25)	61	r	z_{53}
S. III arizonae (Ar. 26:27:21)	61	z_{10}	z_{35}
S. III arizonae (Ar. 26:26:30)	61	z_{52}	1, 5, 7
S. III arizonae (Ar. 26:26:31)	61	z_{52}	z
S. III arizonae (Ar. 26:26:21)	61	z_{52}	z_{35}
S. III arizonae (Ar. 26:26:25)	61	z_{52}	z_{53}

Type	Somatic O-antigen	Flagellar H-antigen	
		Phase 1	Phase 2

Group O:62

Type	Somatic O-antigen	Phase 1	Phase 2
S. III *arizonae* (Ar. 6:13, 14:−)	62	g, z_{51}	—
S. III *arizonae* (Ar. 6:1, 2, 5:−)	62	z_4, z_{23}	—
S. III *arizonae* (Ar. 6:1, 7, 8:−)	62	z_4, z_{32}	—

Group O:63

Type	Somatic O-antigen	Phase 1	Phase 2
S. III *arizonae* (Ar. 8:13, 14:−)	63	g, z_{51}	—
S. III *arizonae* (Ar. 8:1, 2, 5)	63	z_4, z_{23}	—
S. III *arizonae* (Ar. 8:1, 7, 8:−)	63	z_4, z_{32}	—
S. III *arizonae* (Ar. 8:17, 20:−)	63	z_{36}	—

Group O:65[a]

Type	Somatic O-antigen	Phase 1	Phase 2
S. III *arizonae* (Ar. 30:32:30)	65	c	1, 5, 7
S. III *arizonae* (Ar. 30:32:31)	65	c	z
S. III *arizonae* (Ar. 30:32:25)	65	c	z_{53}
S. II	65	(f), g, t	—
S. III *arizonae* (Ar. 30:33:28)	65	i	e, n, x, z_{15}
S. III *arizonae* (Ar. 30:22:31)	65	(k)	z
S. III *arizonae* (Ar. 30:22:21)	65	(k)	z_{35}
S. III *arizonae* (Ar. 30:22:25)	65	(k)	z_{53}
S. III *arizonae* (Ar. 30:23:28)	65	l, v	e, n, x, z_{15}

[a] Group O:64 is combined with Group O:48.
Winkle, I., *Ann. Microbiol. (Paris)* 1976, **127B**, 463–472.

Type	Somatic O-antigen	Flagellar H-antigen	
		Phase 1	Phase 2
S. III *arizonae* (Ar. 30:23:31)	65	1, v	z
S. III *arizonae* (Ar. 30:23:21)	65	1, v	z_{35}
S. III *arizonae* (Ar. 30:23:25)	65	1, v	z_{53}
S. III *arizonae* (Ar. 30:27:28)	65	z_{10}	e, n, x, z_{15}
S. III *arizonae* (Ar. 30:27:31)	65	z_{10}	z
S. III *arizonae* (Ar. 30:26:31)	65	z_{52}	z
S. III *arizonae* (Ar. 30:26:21)	65	z_{52}	z_{35}
S. III *arizonae* (Ar. 30:26:25)	65	z_{52}	z_{53}
S. II	65	—	1, 6

Group O:66

S. V *maregrosso*	66	z_{35}	—
S. V *brookfield*	66	z_{41}	—
S. V *malawi*	66	z_{65}	—

Group O:67

S. *crossness*	67	r	1, 2

Appendix II
Antigenic formulae of the *Salmonella*

Species	O-antigen group	Species	O-antigen group
Subgenus I		S. alabama	D_1
S. aarnus	K	S. alachua	O
S. aba	C_2	S. alagbon	C_3
S. abadina	M	S. alamo	C_1
S. abaetuba	F	S. albany	C_3
S. aberdeen	F	S. albert	B
S. abidjan	Q	S. albuquerque	H
S. ablogame	I	S. alexanderplatz	X
S. abobo	I	S. alexanderpolder	C_3
S. abony	B	S. alfort	E_1
S. abortusbovis	B	S. alger	P
S. abortuscanis	B	S. allandale	R
4, 5, 12:b:z_5 (phase R)		S. allerton	E_1
S. abortusequi	B	S. alminko	C_3
S. abortusovis	B	S. altendorf	B
S. accra	E_4	S. altona	C_3
S. adabraka	E_1	S. amager	E_1
S. adamstown	M	S. amba	F
S. adamstua	F	S. amersfoort	C_1
S. adana	U	S. amierstiana	C_3
S. adelaide	O	S. amina	I
S. adeoyo	I	S. aminatu	E_1
S. aderike	M	S. amounderness	E_1
S. adime	C_1	S. amoutive	M
S. adjame	G_2	S. amsterdam	E_1
S. aesch	C_2	S. amunigun	I
S. aequatoria	C_1	S. anatum	E_1
S. aflao	H	S. anderlecht	E_1
S. africana	B	S. anecho	O
S. afula	C_1	S. anfo	Q
S. agama	B	S. angers	C_3
S. agbeni	G_2	S. angoda	N
S. agege	E_1	S. angouleme	I
S. ago	N	S. anie	X
S. agodi	O	S. ank	M
S. agona	B	S. anna	G_2
S. agoueve	G_1	S. annedal	I
S. ahanou	J	S. antarctica	D_1
S. ahepe	U	S. antonio	57
S. ahmadi	E_4	S. antsalova	51
S. ahuza	U	S. antwerpen	T
S. ajiobo	G_2	S. apapa	W
S. akanji	C_2	S. apeyeme	C_3
S. akuafo	I	S. aqua	N

Species	O-antigen group	Species	O-antigen group
S. aragua	N	*S. bargny*	C_3
S. ardwick	C_4	*S. barmbek*	I
S. arechavaleta	B	*S. barranquilla*	I
S. arkansas	E_3	*S. basingstoke*	D_2
S. aschersleben	N	*S. bassa*	C_2
S. ashanti	M	*S. bassadji*	M
S. assen	L	"*S. batavia*" = *S. lexington*	E_1
S. assinie	E_1	*S. battle*	I
S. atakpame	C_3	*S. bazenheid*	C_3
S. atento	F	*S. be*	C_3
"*S. atherton*" = *S. waycross*	S	*S. beaudesert*	H
S. athinai	C_1	*S. bedford*	E_4
S. atlanta (combined with		*S. belem*	C_2
S. mississippi)	G_2	*S. belfast*	C_2
S. augustenborg	C_1	*S. benfica*	E_1
S. austin	C_1	*S. benguella*	R
S. avignon	I	*S. benin*	D_2
S. avonmouth	E_4	*S. bere*	X
S. ayinde	B	*S. bergedorf*	D_2
S. ayton	B	*S. bergen*	X
S. azteca	B	*S. berkeley*	U
S. babelsberg	M	*S. berlin*	J
S. babili	M	*S. berta*	D_1
S. baguida	L	*S. bessi*	E_1
S. baguirmi	N	*S. biafra*	E_1
S. bahati	G_1	*S. bietri*	N
S. bahrenfeld	H	*S. bignona*	J
S. baiboukoum	C_1	*S. bijlmer*	R
S. baildon	D_2	*S. bilu*	E_4
S. bakau	M	*S. bingerville*	X
S. balboa	Y	*S. binningen*	W
S. balcones	W	*S. binza*	E_2
S. ball	B	*S. birkenhead*	C_1
S. bama	J	*S. birmingham*	E_1
S. bambesa (combined with		*S. bispebjerg*	B
S. miami)	D_1	*S. blegdam*	D_1
S. bamboye	D_2	*S. blijdorp*	H
S. bambylor	D_2	*S. blitta*	X
S. banalia	C_2	*S. blockley*	C_2
S. banana	B	*S. blukwa*	K
S. banco	M	*S. bobo*	V
S. bandia	O	*S. bochum*	B
S. bangkok	P	*S. bodjonegoro*	N
S. banjul	H	*S. boecker*	H
("*S. bantam*" = *S. meleagridis*	E_1	*S. bokanjac*	M
S. bardo	C_3	*S. bolombo*	E_1
S. bareilly	C_1	*S. bolton*	E_1

Species	O-antigen group	Species	O-antigen group
S. bonames	J	*S. buzu*	H
S. bonariensis	C_2	*S. caen*	I
S. bongor	Y	*S. cairina*	E_1
S. bonn	C_1	*S. cairns*	W
S. bootle	X	*S. cairo* (combined with	
S. borbeck	G_1	*S. stanley*)	B
S. bornum	C_4	*S. calabar*	E_4
S. borreze	54	*S. california*	B
S. bournemouth	D_1	*S. camberene*	O
S. bousso	H	*S. camberwell*	D_1
S. bovismorbificans	C_2	*S. cambridge*	E_2
S. bracknell	G_2	*S. camdeni*	V
S. bradford	B	*S. campinense*	D_1
S. braenderup	C_1	*S. canada*	B
S. brandenburg	B	*S. cannonhill*	E_4
S. brancaster	B	*S. cannstatt*	E_4
S. brazil	I	*S. canoga*	E_3
S. brazos	K	*S. canton*	54
S. brazzaville	C_1	*S. caracas*	H
S. bredeney	B	"*S.* cardiff" 6, 7,:k:1,10	
S. brefet	V	(phase R)	C_1
S. breukelen	C_2	*S. carmel*	J
S. brevik	I	*S. carnac*	K
S. brezany	B	*S. carno*	E_4
S. brijbhumi	F	*S. carrau*	H
S. brikama	C_3	*S. carswell*	V
S. brisbane	M	*S. casablanca*	W
S. bristol	G_1	*S. casamance*	R
S. brive	T	*S. catanzaro*	H
S. bron	G_1	*S. cayar*	C_1
S. bronx	C_2	*S. cerro*	K
S. brookfield	66	*S. ceyco*	D_2
S. broughton	E_4	*S. chagoua*	G_2
"*S.* broxbourne" = *S. wien*	B	*S. chailey*	C_2
S. bruck	C_1	*S. champaign*	Q
S. brunei	C_3	*S. chandans*	F
S. budapest	B	*S. charity*	H
"*S.* buenos-aires" = *S. bonariensis*	C_2	*S. charlottenburg*	C_2
S. bukavu	R	*S. chester*	B
S. bukuru	C_2	*S. chicago*	M
S. bulgaria	C_2	*S. chichiri*	H
S. bullbay	F	*S. chincol*	C_2
S. burgas	I	*S. chingola*	F
S. bury	B	*S. chiredzi*	F
S. businga	C_1	*S. chittagong*	E_4
S. butantan	E_1	*S. choleraesuis*	C_1
S. butare	52	*S. chomedey*	C_3

Species	O-antigen group	Species	O-antigen group
S. christiansborg	V	S. derby	B
S. clackamas	B	S. derkle	52
S. claibornei	D_1	S. dessau	E_4
S. clerkenwell	E_1	S. deversoir	W
S. cleveland	C_2	S. dieuppeul	M
S. clichy	E_2	S. diguel	G_1
S. cochin	D_2	S. diogoye	C_3
S. cocody	C_3	S. diourbel	L
S. coeln	B	S. djakarta	Y
S. coleypark	C_1	S. djama	T
S. colindale	C_1	S. djelfa	C_3
S. colobane	F	S. djermaia	M
S. colombo	P	S. djibouti	J
S. colorado	C_1	S. djugu	C_1
S. concord	C_1	S. doba	D_2
S. congo	G_2	S. doncaster	C_2
S. coogee	T	S. donna	N
"S. cook" 39:z_{48}:1, 5 (phase R)	Q	S. doorn	M
S. coquilhatville	E_1	S. dougi	Z
S. corvallis	C_3	S. doulassame	N
S. cotham	M	S. dresden	M
S. cotia	K	S. driffield	R
S. cremieu	C_2	S. drogana	B
S. croft	M	S. drypool	E_2
S. crossness	67	S. dublin	D_1
S. cubana	G_2	S. duesseldorf	C_2
S. cuckmere	E_1	S. dugbe	W
S. cullingworth	M	S. duisburg	B
S. curacao	C_2	S. dumfries	E_1
S. cyprus	C_2	S. durban	D_1
S. dabou	C_3	S. durham	G_2
S. dadzie	51	S. duval	R
S. dahlem	Y	S. ealing	O
S. dahomey	X	S. eastbourne	D_1
S. dakar	M	S. eberswalde	D_1
S. dakota	I	S. eboko	C_2
S. dalat (combined with S. ball)	B	S. ebrie	O
		S. echa	P
S. dallgow	E_4	S. edinburg	C_1
S. dan	51	S. edmonton	C_2
S. dapango	X	S. egusi	S
S. daytona	C_1	S. egusitoo	T
S. decatur (combined with S. choleraesuis)	C_1	S. eimsbuettel	C_4
		S. eingedi	C_1
S. dembe	O	S. eko	B
S. demerara	G_2	S. ekotedo	D_2
S. denver	C_1	S. ekpoui	X

Species	O-antigen group	Species	O-antigen group
S. elisabethville	E_1	*S. freetown*	P
S. elokate	D_1	*S. freiburg*	E_1
S. elomrane	D_1	*S. fresno*	D_2
S. emek	C_3	*S. friedenau*	G_1
S. emmastad	P	*S. friedrichsfelde*	M
S. encino	H	*S. frintrop*	D_1
S. enschede	O	*S. fufu*	E_1
S. entebbe	B	*S. fulica*	B
S. enteritidis	D_1	*S. fyris*	B
S. enugu	I	*S. gabon*	C_1
S. epicrates	E_1	*S. gafsa*	I
S. epinay	F	*S. galiema*	C_1
S. eppendorf	B	*S. galil*	E_1
S. escanaba	C_1	*S. gallen*	F
S. eschersheim	E_2	*S. gallinarum*	D_1
S. eschweiler	C_1	*S. gamaba*	V
S. essen	B	*S. gambaga*	L
S. etterbeek	F	*S. gambia*	O
S. ezra	M	*S. gaminara*	I
S. fajara	M	*S. garba*	H
S. faji	T	*S. garoli*	C_1
S. falkensee	E_1	*S. gassi*	O
S. fallowfield	E_1	*S. gateshead*	D_2
S. fann	F	*S. gatineau*	E_4
S. fanti	G_2	*S. gatow*	C_1
S. farakan	M	*S. gatuni*	C_2
S. farcha	U	*S. gbadago*	E_1
S. fareham	E_4	*S. gdansk*	C_1
S. farmsen	G_2	*S. gege*	N
S. fass	Z	*S. gelsenkirchen*	C_4
S. fayed	C_2	*S. georgia*	C_1
S. ferlac	H	*S. gera*	T
S. ferruch	C_3	*S. geraldton*	D_2
S. findorff	F	*S. ghana*	L
S. finkenwerder	H	*S. giessen*	N
S. fischerhuette	I	*S. give*	E_1
S. fischerkietz	H	*S. giza*	C_3
S. fischerstrasse	V	*S. glasgow*	I
S. fitzroy	Y	*S. glidji*	F
S. florian	E_1	*S. glostrup*	C_2
S. florida	H	*S. gloucester*	B
S. flottbek	52	*S. gnesta*	E_4
S. fluntern	K	*S. godesberg*	N
S. fomeco	W	*S. goelzau*	E_1
S. fortlamy	I	*S. goerlitz*	E_2
S. fortune	B	*S. goeteborg*	D_1
S. frankfurt	I	*S. goettingen*	D_1

Species	O-antigen group	Species	O-antigen group
S. gokul	51	*S. heron*	I
S. goldcoast	C_2	*S. herston*	C_2
S. goma	C_1	*S. herzliya*	F
S. gombe	C_1	*S. hessarek*	B
S. good	L	*S. heves*	H
S. gori	J	*S. hidalgo*	C_2
S. goulfey	R	*S. hiduddify*	C_2
S. goverdhan	D	*S. hillegersberg*	D_2
S. grampian	C_1	*S. hillingdon*	D_2
S. granlo	J	*S. hillsborough*	C_1
S. graz	U	*S. hilversum*	N
S. greiz	R	*S. hindmarsh*	C_3
S. groenekan	K	*S. hisingen*	Y
S. grumpensis	G_2	*S. hissar*	C_4
S. guarapiranga	N	*S. hithergreen*	I
S. guerin	D_2	*S. hofit*	Q
S. guildford	M	*S. hoghton*	E_1
S. guinea	V	*S. holcomb*	C_2
S. gustavia	F	*S. homosassa*	H
S. gwale	T	*S. honelis*	M
S. gwoza	E_4	*S. horsham*	H
S. haardt	C_3	*S. huddinge*	E_1
S. hadar	C_2	*S. hull*	I
S. haelsingborg	C_1	*S. huvudsta*	E_1
S. haferbreite	T	*S. hvittingfoss*	I
S. haga	O	*S. hydra*	L
S. haïfa	B	*S. ibadan*	G_1
S. halle	M	*S. idikan*	G_2
S. hallfold	B	*S. ikayi*	E_1
S. halmstad	E_2	*S. ikeja*	M
"S. hamilton" 3, *15*:e, h:1, 2:Z_{27}	E_2	*S. ilala*	M
(phase R) (combined with		*S. illinois*	E_3
S. goerlitz)		*S. ilugun*	E_4
S. handen	G_2	*S. inchpark*	C_2
S. hann	R	*S. india*	D_2
S. hannover	I	*S. indiana*	B
S. haouaria	G_1	*S. infantis*	C_1
S. harburg	H	*S. inganda*	C_1
S. harrisonburg	E_3	*S. inglis*	D_2
S. hartford	C_1	*S. inpraw*	S
S. harvestehude	T	*S. inverness*	P
S. hatfield	M	*S. ipeko*	D_1
S. hato	B	*S. ipswich*	S
S. havana	G_2	*S. irenea*	J
S. heerlen	F	*S. irigny*	U
S. heidelberg	B	*S. irumu*	C_1
S. hermannswerder	M	*S. isangi*	C_1

Species	O-antigen group	Species	O-antigen group
S. isaszeg	Y	*S. kaolack*	X
S. israel	D_1	*S. kapemba*	D_1
S. istanbul	C_3	*S. kaposvar* (combined with	
S. isuge	G_2	*S. reading*)	B
"*S. italiana*" 9, 12:1, v:1, 11		*S. karachi*	W
(phase R)	D_1	*S. karamoja*	R
S. itami	D_1	*S. kasenyi*	P
S. ituri	B	*S. kassberg*	H
S. itutaba	D_2	*S. kedougou*	G_2
"*S. iwojima*" = *S. kentucky*	C_3	*S. kentucky*	C_3
S. jaffna	D_1	*S. kenya*	C_1
S. jaja (combined with		*S. kermel*	V
S. stanleyville)	B	*S. keve*	L
S. jalisco	F	*S. khartoum*	E_3
S. jamaica	D_1	*S. kiambu*	B
S. jangwani	J	*S. kibi*	I
S. java (combined with		*S. kibusi*	M
S. paratyphi B)	B	*S. kidderminster*	P
S. javiana	D_1	*S. kiel*	A
S. jedburgh	E_1	*S. kikoma*	I
S. jericho	B	*S. kimberley*	P
S. jerusalem	C_4	*S. kimpese*	D_2
S. joal	E_1	*S. kimuenza*	B
S. jodhpur	W	*S. kingabwa*	U
S. joenkoeping (combined with		*S. kingston*	B
S. kingston)	B	*S. kinondoni*	J
S. johannesburg	R	*S. kinshasa*	E_2
S. jos	B	*S. kintambo*	G_2
S. juba	E_4	*S. kirkee*	J
S. jubilee	J	*S. kisangani*	B
S. jukestown	G_2	*S. kisarawe*	F
S. kaapstad	B	*S. kisii*	C_1
S. kabete	51	*S. kitenge*	M
S. kaduna	C_4	*S. kivu*	C_1
S. kahla	T	*S. klouto*	P
S. kaitaan	H	*S. kodjovi*	X
S. kalamu	B	*S. koenigstuhl*	B
S. kalina	E_1	*S. koketime*	V
S. kalumburu	C_2	*S. kokoli*	N
S. kambole	C_1	*S. kokomlemle*	Q
S. kamoru	B	*S. konstanz*	C_3
S. kampala	T	*S. korbol*	C_3
"*S. kanda*" = *S. meleagridis*	E_1	*S. korlebu*	E_4
S. kande	E_4	*S. korovi*	P
S. kandla	J	*S. kortrijk*	C_1
S. kaneshi	T	*S. kottbus*	C_2
S. kano	B	*S. kotte*	C_1

Species	O-antigen group	Species	O-antigen group
S. kouka	E_4	S. lindern	H
S. koumra	C_1	S. lindi	P
S. kpeme	M	S. linguere	D_2
S. kralingen	C_3	S. lingwala	I
S. krefeld	E_4	S. linton	G_2
S. kristianstad	E_1	S. lisboa	I
S. kua	V	S. lishabi	D_2
S. kubacha	B	S. litchfield	C_2
S. kuessel	M	S. liverpool	E_4
S. kumasi	N	S. livingstone	C_1
S. kunduchi	B	S. livulu	N
S. kuru	C_2	S. ljubljana	B
S. labadi	C_3	S. llandoff	E_4
S. lagos	B	S. loanda	C_2
S. lamin	E_1	S. lockleaze	C_4
S. landala	S	S. lode	J
S. landau	N	S. lodz	S
S. landwasser	E_1	S. loenga	T
S. langenhorn	K	S. logone	Q
S. langensalza	E_1	S. lokstedt	E_4
S. langford	M	S. loma-linda	D_1
S. lanka	E_2	S. lome	D_1
S. lansing	P	S. lomita	C_1
S. larochelle	C_1	S. lomnava	I
S. lattenkamp	W	S. london	E_1
S. lawndale	D_1	S. losangeles	I
S. lawra	V	S. louga	N
S. leatherhead	S	S. louisiana	D_2
S. lechler	51	S. lovelace	G_1
S. leeuwarden	F	S. lubumbashi	S
S. legon	B	S. luciana	F
S. leiden	G_1	S. luckenwalde	M
S. leipzig	S	S. luke	X
S. leith	C_2	S. lyon	X
S. lekke	E_1	S. maastricht	F
S. lene	F	S. macallen	E_1
S. leoben	M	S. machaga	E_4
S. leopoldville	C_1	S. madelia	H
S. lerum	E_4	S. madiago	E_4
S. lexington	E_1	S. madigan	V
S. lezennes	C_2	S. madison	L
S. ligeo	N	S. madjorio	E_1
S. ligna	O	S. magumeri	H
S. lika	C_1	S. magwa	L
S. lille	C_1	S. maiduguri	E_4
S. limete	B	S. makiso	C_1
S. lindenburg	C_2	S. malakal	I

Species	O-antigen group	Species	O-antigen group
S. malawi	66	*S. midway*	H
S. malysia	M	*S. mikawasima*	C_1
S. malika	V	*S. millesi*	R
S. malmoe	C_2	*S. milwaukee*	U
S. malstatt	I	*S. mim*	G_1
S. mampeza	H	*S. minna*	H
S. mampong	G_1	*S. minneapolis*	E_3
S. manchester	C_2	*S. minnesota*	L
S. mandera	I	*S. mishmarhaemek*	G_2
S. mango	P	*S. mission* (combined with	
S. manhattan	C_2	*S. isangi*)	C_1
S. manila	E_2	*S. mississippi*	G_2
S. mapo	C_2	*S. miyazaki*	D_1
S. mara	Q	*S. mkamba*	C_1
S. maracaibo	F	*S. mocamedes*	M
S. maricopa	T	*S. moero*	M
S. marienthal	E_1	*S. mokola*	E_1
S. maron	E_1	*S. molade*	C_3
S. marseille	F	*S. molesey*	52
S. marshall	G_1	*S. mono*	B
S. maryland	57	*S. mons*	B
S. marylebone	D_2	*S. monschaui*	O
S. masembe	E_1	*S. montevideo*	C_1
S. massakory	O	*S. montreal*	U
S. massenya	B	*S. morehead*	N
S. matadi	J	*S. morningside*	N
S. mathura	D_2	*S. mornington*	H
S. matopeni	N	*S. morocco*	N
S. mayday	D_2	*S. morotai*	J
S. mbandaka	C_1	*S. moroto*	M
S. mbao	U	*S. moscow*	D_1
S. meekatharra	W	*S. moualine*	X
S. meleagridis	E_1	*S. mountpleasant*	X
S. memphis	K	*S. moussoro*	H
S. menden	C_1	*S. mowanjum*	C_2
S. mendoza	D_1	*S. mpouto*	I
S. menhaden	E_3	*S. muenchen*	C_2
S. menston	C_1	*S. muenster*	E_1
S. mesbit	X	*S. muguga*	V
S. meskin	51	*S. mundonobo*	M
S. messina	N	*S. mura*	B
S. mexicana (combined with		*S. naestved*	D_1
S. muenchen)	C_2	*S. nagoya*	C_2
S. mgulani	P	*S. nakuru*	B
S. miami	D_1	*S. nancy*	E_2
S. michigan	J	*S. nanergou*	C_2
S. middlesbrough	T	*S. nanga*	G_2

Species	O-antigen group	Species	O-antigen group
S. napoli	D_1	S. obogu	C_1
S. narashino	C_2	S. ochsenwerder	54
S. nashua	M	S. odozi	N
S. naware	I	S. oerlikon	Q
S. nchanga	E_1	S. oevelgoenne	M
S. ndjamena	H	S. offa	S
S. ndolo	D_1	S. ogbote	U
S. neftenbach	B	S. ohio	C_1
S. nessa	H	S. ohlstedt	E_1
S. nessziona	C_1	S. okatie	G_2
S. neudorf	N	S. okefoko	E_1
S. neukoelln	C_1	S. okerara	E_1
S. neumuenster	B	S. oldenburg	I
S. newbrunswick	E_2	S. olten	D_2
S. newhaw	E_2	S. omderman	C_4
S. newington	E_2	S. omifisan	R
S. newlands	E_1	S. ona	M
S. newmexico	D_1	S. onarimon	D_1
S. newport	C_2	S. onderstepoort	H
S. newrochelle	E_1	S. onireke	E_1
S. newyork	G_1	S. ontario	D_2
S. ngili	C_1	S. oranienburg	C_1
S. ngor	E_4	S. ord	52
S. niakhar	V	S. ordonez	G_2
S. niamey	J	S. oregon (combined with	
S. niarembe	V	S. muenchen)	C_2
S. nienstedten	C_4	S. orientalis	I
S. nieukerk	C_4	S. orion	E_1
S. nigeria	C_1	S. oritamerin	C_1
S. nijmegen	N	S. orlando	K
S. nikolaifleet	I	S. os	D_1
S. niloese	E_4	S. oskarshamn	M
S. nima	M	S. oslo	C_1
S. nimes	G_1	S. osnabrueck	F
S. nissii	C_4	S. othmarschen	C_1
S. nitra	A	S. ottawa	D_1
S. njala	P	S. ouakam	D_2
S. nordufer	C_2	S. oudwijk	G_1
S. norton	C_1	S. overchurch	R
S. norwich	C_1	S. overshie	51
S. nottingham	I	S. overvecht	N
S. nowawes	R	S. oxford	E_1
S. nuatja	I	S. oyonnax	C_1
S. nyanza	F	S. pakistan	C_3
S. nyborg	E_1	S. palime	C_1
S. nyeko	I	S. panama	D_1
S. oakland	C_1	S. pankow	E_2

Species	O-antigen group	Species	O-antigen group
S. papuana	C_1	S. redlands	I
S. paratyphi A	A	S. regent	E_1
S. paratyphi B	B	S. reinickendorf	B
S. paratyphi C	C_1	S. remete	F
S. paris	C_3	S. remo	B
S. parkroyal	E_4	S. reubeuss	C_3
S. pasing	B	S. rhone	L
S. patience	M	S. rhydyfelin	I
S. penarth	D_1	S. richmond	C_1
S. pensacola	D_1	S. rideau	E_4
S. perth	P	S. ridge	D_1
S. pharr	F	S. ried	G_1
S. pikine (combined with		S. riggil	C_1
S. altona)	C_3	S. riogrande	R
S. pisa	I	S. rissen	C_1
S. plymouth	D_2	S. rittersbach	P
S. poano	H	S. riverside	W
S. poeseldorf	54	S. roan	P
S. pomona	M	S. rochdale	Z
S. pontypridd	K	S. rogy	M
S. poona	G_1	S. romanby	G_2
S. portland	D_1	S. roodepoort	G_1
S. portsmouth	E_2	S. rosenthal	E_2
S. potosi	H	S. rossleben	54
S. potsdam	C_1	S. rostock	D_1
S. potto	D_2	S. rottnest	G_1
S. pramiso	E_1	S. rovaniemi	I
S. praha	C_2	S. rubislaw	F
S. presov	C_2	S. ruiru	L
S. preston	B	S. ruki (combined with S. ball)	B
S. pretoria	F	S. rumford	C_1
S. pueris (combined with		S. runby	H
S. newport)	C_2	"S. rutgers" 3, 10:1, z_{40}:1, 7	
S. pullorum	D_1	(phase R)	E_1
S. putten	G_2	S. ruzizi	E_1
S. quebec	V	S. saarbruecken	D_1
S. quentin	D_2	S. saboya	I
S. quinhon	X	S. sada	N
S. quiniela	C_2	S. saintemarie	52
S. ramatgan	N	S. saintpaul	B
S. ramsey	M	S. saka	X
S. raus	G_1	"S. sakai" = S. postdam	C_1
S. rawash	K	S. salford	I
S. reading	B	S. salinatis	B
S. rechovot	C_3	S. saloniki	I
S. redba	C_1	S. sambre	E_4
S. redhill	F	S. san-diego	B

Species	O-antigen group	Species	O-antigen group
S. sandow	C_2	*S. singapore*	C_1
S. sanga	C_3	*S. sinstorf*	E_1
S. sangalkam	D_2	*S. sinthia*	K
S. sangera	I	*S. sipane*	T
S. san-juan	C_1	*S. skansen*	C_2
S. sanktgeorg	M	*S. sladun* (combined with	
S. sanktmarx	E_4	*S. abony*)	B
S. santhiaba	R	*S. sljeme*	X
S. santiago	C_3	*S. sloterdijk*	B
S. sao	E_4	*S. soahanina*	H
S. saphra	I	*S. soerenga*	N
S. sara	H	*S. sokode*	D_2
S. sarajane	B	*S. solna*	M
S. saugus	R	*S. solt*	F
S. schalkwijk	H	*S. somone*	C_1
S. schleissheim	B	*S. southampton*	B
S. schoeneberg	E_4	*S. southbank*	E_1
"*S. schottmuelleri*" =		*S. souza*	E_1
S. paratyphi B	B	*S. spartel*	L
S. schwarzengrund	B	*S. stanley*	B
S. schwerin	C_2	*S. stanleyville*	B
S. seattle	M	*S. staoueli*	X
S. sedgwick	V	*S. steinplatz*	N
S. seegefeld	E_1	*S. steinwerder*	54
S. sekondi	E_1	*S. stelligen*	X
S. selandia	E_2	*S. stendal*	F
S. selby	M	*S. sternschanze*	N
S. sendai	D_1	*S. sterrenbos*	C_2
S. senegal	F	*S. stockholm*	E_1
S. senftenberg	E_4	*S. stormont*	E_1
S. seremban	D_1	*S. stourbridge*	C_2
S. shamba	I	*S. straengnaes*	F
S. shangai	I	*S. strasbourg*	D_2
S. shangani	E_1	*S. stratford*	E_4
S. shannon	E_1	*S. stuivenberg*	E_4
S. sharon	F	*S. suberu*	E_1
S. sheffield	P	*S. suelldorf*	W
S. sherbrooke	I	"*S. suez*" = *S. shubra*	B
S. shikmonah	R	"*S. suipestifer*" =	
S. shipley	C_3	*S. cholerae-suis*	C_1
S. shomolu	M	*S. sundsvall*	H
S. shoreditch	D_2	*S. sunnycove*	C_3
S. shubra	B	*S. surat*	H
S. simi	E_1	*S. sya*	X
"*S. simsbury*" 1, 3, 19 :z_{27}:—		*S. szentes*	I
(phase R)	E_4	*S. tabligbo*	X
S. sinchew	E_1	*S. tado*	C_3

Species	O-antigen group	Species	O-antigen group
S. tafo	B	*S. tomelilla*	E_4
"S. taihoku" = *S. meleagridis*	E_1	*S. tonev*	54
S. takoradi	C_2	*S. toowong*	F
S. taksony	E_4	*S. toricada*	T
S. tallahassee	C_2	*S. tornow*	W
S. tamale	C_3	*S. toronto*	D_2
S. tambacounda	E_4	*S. toucra*	Y
S. tamberma	X	*S. toulon*	K
S. tamilnadu	C_1	*S. tounouma*	C_3
S. tananarive	C_2	*S. tournai*	E_2
S. tanger	G_1	*S. trachau*	B
S. tanzania	G_1	*S. travis*	B
S. tarshyne	D_1	*S. treforest*	51
S. taset	T	*S. trimdon*	D_2
S. taunton	M	*S. trotha*	R
S. tchad	O	*S. truro*	E_1
S. tchamba	J	*S. tschangu*	G_2
S. techimani	M	*S. tsevie*	B
S. teddington	B	*S. tshiongwe*	C_2
S. tees	I	*S. tucson*	H
S. tejas	B	*S. tudu*	B
S. teko	H	*S. tuebingen*	E_2
S. telaviv	M	*S. tunis*	G_2
S. telelkebir	G_2	*S. typhi*	D_1
S. telhashomer	F	*S. typhimurium*	B
S. teltow	M	*S. typhisuis*	C_1
S. tennessee	C_1	*S. tyresoe*	B
S. tennyson	B	*S. uccle*	54
S. teshie	X	*S. uganda*	E_1
S. texas	B	*S. ughelli*	E_1
S. thaygen	B	*S. uhlenhorst*	V
S. thetford	U	*S. uithof*	52
S. thiaroye	P	*S. ullevi*	G_2
S. thielallee	C_4	*S. umbilo*	M
S. thomasville	E_3	*S. umhlali*	C_1
S. thompson	C_1	*S. umhlatazana*	O
S. tiergarten	V	*S. uno*	C_2
S. tilburg	E_4	*S. uppsala*	B
S. tilene	R	*S. urbana*	N
S. tim (combined with		*S. ursenbach*	T
S. newington)	E_2	*S. usumbura*	K
S. tinda	B	*S. utah*	C_2
S. tione	51	*S. utrecht*	52
S. togba	I	*S. uzaramo*	H
S. togo	B	*S. vaertan*	G_1
S. tokoin	B	*S. vancouver*	I
S. tomegbe	T	*S. vejle*	E_1

Species	O-antigen group	Species	O-antigen group
S. vellore	B	S. wilhelmsburg	B
S. veneziana	F	S. willemstad	G_1
S. venusberg (combined with		S. wimborne	E_1
S. nchanga)	E_1	S. windermere	Q
S. victoria	D_1	S. wingrove	C_2
S. victoriaborg	J	S. winnipeg	54
S. vietnam	S	S. winterthur	E_4
S. vilvoorde	E_4	S. wippra	C_2
S. vinohrady	M	S. wisbech	I
S. virchow	C_1	S. wohlen	F
S. virginia	C_3	S. womba (combined with	
S. visby	E_4	S. altendorf)	B
S. vitkin	M	S. worb	D_2
S. vleuten	V	S. worthington	G_2
S. vogan	T	S. wuerzburg (combined with	
S. volksmarsdorf	M	S. miami	D_1
S. volta	F	S. wuiti	N
S. vom	B	S. wuppertal	D_2
S. wagenia	B	S. wyldegreen	G_2
S. wandsworth	Q	S. yaba	E_1
S. wangata	D_1	S. yalding	E_4
S. waral	T	S. yaounde	B
S. warengo	J	S. yarm	C_2
S. warnemuende	M	S. yarrabah	G_2
S. warnow	C_2	S. yeerongpilly	E_1
S. warragul	H	S. yehuda	F
S. washington	G_1	S. yerba	54
S. waycross	S	S. yoff	P
S. wayne	N	S. yokoe	C_3
S. wedding	M	S. yolo	O
S. welikade	I	S. yovokome	C_3
S. weltevreden	E_1	S. yundum	E_1
S. wentworth	F	S. zadar	D_2
S. wernigerode	D_2	S. zagreb (combined with	
S. weslaco	T	S. saintpaul)	B
S. westeinde	I	S. zaire	N
S. westerstede	E_4	S. zanzibar	E_1
S. westhampton	E_1	S. zega	D_1
S. westminster	E_2	S. zehlendorf	N
S. weston	I	S. zerifin	C_2
S. westphalia	O	S. zongo	E_1
S. weybridge	E_1	S. zuilen	E_4
S. wichita	G_2	S. zwickau	I
S. widemarsh	O		
S. wien	B	**Subgenus II**	
S. wil	C_1	S. II acres	G_2
S. wildwood	E_3	S. II alexander	E_1

Species	O-antigen group	Species	O-antigen group
S. II *alsterdorf*	R	*S.* II *etosha*	Y
S. II *angola*	D_1	*S.* II *fandran*	R
S. II *artis*	56	*S.* II *faure*	Z
S. II *askraal*	51	*S.* II *finchley*	E_1
S. II *atra*	Z	*S.* II *foulpointe*	P
S. II *bacongo*	C_1	*S.* II *fremantle*	T
S. II *baragwanath*	C_2	*S.* II *fuhlsbuettel*	E_1
S. II *basel*	58	*S.* II *germiston*	C_2
S. II *bechuana*	B	*S.* II *gilbert*	C_1
S. II *bellville*	I	*S.* II *glencairn*	F
S. II *beloha*	K	*S.* II *gojenberg*	G_2
S. II *betioky*	59	*S.* II *goodwood*	G_1
S. II *bilthoven*	X	*S.* II *grabouw*	F
S. II *blankenese*	D_1	*S.* II *greenside*	Z
S. II *bleadon*	J	*S.* II *grunty*	R
S. II *bloemfontein*	C_1	*S.* II *gwaai*	L
S. II *boksburg*	R	*S.* II *haarlem*	D_2
S. II *bornheim*	H	*S.* II *haddon*	I
S. II *boulders*	G_2	*S.* II *hagenbeck*	Y
S. II *bremen*	W	*S.* II *hamburg*	D_1
S. II *bulawayo*	R	(combined with *S.* II 1, 9, 12:	
S. II *bunnik*	U	g, m, [s], t:[1, 5]:[z_{42}])	
S. II *caledon*	B	*S.* II *hammonia*	Y
S. II *calvinia*	C_1	*S.* II *heilbron*	C_1
S. II *canastel*	D_1	*S.* II *helsinki*	B
S. II *cape*	C_1	*S.* II *hennepin*	S
S. II *carletonville*	P	*S.* II *hillbrow*	J
S. II *ceres*	M	*S.* II *hooggraven*	Z
S. II *chersina*	X	*S.* II *hueningen*	D_1
S. II *chinovum*	T	*S.* II *huila*	F
S. II *chudleigh*	E_1	*S.* II *humber*	53
S. II *clifton*	G_1	*S.* II *islington*	E_1
S. II *clovelly*	V	*S.* II *jacksonville*	I
S. II *constantia*	J	*S.* II *kaltenhausen*	M
S. II *daressalaaem*	D_1	*S.* II *katesgrove*	G_2
S. II *degania*	R	*S.* II *khami*	X
S. II *detroit*	T	*S.* II *kilwa*	B
S. II *dubrovnik*	S	*S.* II *klapmuts*	W
S. II *duivenhoks*	D_2	*S.* II *klutjenfelde*	B
S. II *durbanville*	B	*S.* II *kommetje*	U
S. II *eilbek* (combined with	61	*S.* II *kraaifontein* (combined	G_2
S. III *arizonae* 61 :i :z)		*S.* II *luanshya*)	
S. II *ejeda*	W	*S.* II *krugersdorp*	Z
S. II *elsiesrivier*	I	*S.* II *kuilsrivier*	D_1
S. II *emmerich*	H	*S.* II *lethe*	S
S. II *epping*	G_2	*S.* II *lichtenberg*	S
S. II *erlangen*	Y	*S.* II *limbe*	G_1

Species	O-antigen group	Species	O-antigen group
S. II lincoln	F	S. II rhodesiense	D_1
S. II lindrick	D_1	S. II roggeveld	51
S. II llandudno	M	S. II rooikrantz	H
S. II lobatsi	52	S. II rotterdam	G_1
S. II locarno	57	S. II rowbarton	I
S. II louwbester	I	S. II sakaraha	Y
S. II luanshya	G_2	S. II sarepta	I
S. II lundby	D_2	S. II seaforth	Z
S. II lurup	S	S. II setubal	60
S. II luton	60	S. II shomron	K
S. II maarssen	D_2	(combined with S. III arizonae	
S. II makoma	B	$18:z_4, z_{32}:-$)	
S. II makumira	B	S. II simonstown	H
S. II manica (combined with S.	D_1	S. II slangkop	H
II 1, 9, 12:g, m, [s], t:[1, 5]:		S. II slatograd	N
[z_{42}])		S. II sofia	B
S. II manombo	57	S. II soutpan	F
S. II matroosfontein	E_1	S. II springs	R
S. II merseyside	I	S. II srinagar	F
S. II midhurst	53	S. II stellenbosch	D_1
S. II mjimwema	D_1	S. II stevenage	G_2
S. II mobeni	I	S. II stikland	E_1
S. II mondeor	Q	S. II suarez	R
S. II montgomery	F	S. II suederelbe	D_1
S. II mosselbay	U	S. II sullivan	C_1
S. II mpila	E_1	S. II sunnydale	R
S. II muizenbert (combined with S.	D_1	S. II sydney (combined with S.	Y
II 1, 9, 12:g, m, [s], t:[1, 5]:		S. III arizonae 48:i:z)	
[z_{42}])		S. II tafelbaai	
S. II nachshonim	G_2	S. II tokai	57
S. II nairobi	T	S. II tosamanga	C_1
S. II namib	Z	S. II tranoroa	55
S. II neasden	D_1	S. II tulear	C_2
S. II negev	S	S. II tygerberg	G_2
S. II ngozi	Y	S. II uphill	T
S. II noordhoek	I	S. II utbremen	O
S. II nordenham	B	S. II veddel	U
S. II nuernberg	T	S. II verity	J
S. II odijk	N	S. II vredelust	G_2
S. II ottershaw	R	S. II vrindaban	W
S. II oysterbeds	C_1	S. II wandsbek	L
S. II parow	E_2	S. II westpark	E_1
S. II perinet	W	S. II wilhemstrasse (combined	52
S. II phoenix	X	S. II lobatsi)	
S. II portbech	T	S. II winchester	E_1
S. II quimbamba	X	S. II windhoek	W
S. II rand	T	S. II woerden	J

Species	O-antigen group	Species	O-antigen group
S. II *woodstock*	I	*S.* IV *mundsburg*	F
S. II *worcester*	G_2	*S.* IV *ochsenzoll*	I
S. II *wynberg*	D_1	*S.* IV *parera*	F
S. II *zeist*	K	*S.* IV *roterberg*	C_1
S. II *zuerich*	D_3	*S.* IV *sachsenwald*	R
		S. IV *seminole*	R
Subgenus IV		*S.* IV *soesterberg*	L
S. IV *argentina*	C_1	*S.* IV *tuindorp*	U
S. (IV) *bern*	R	*S.* IV *volksdorf*	U
(combined with S. IV 40:z_4, z_{32})		*S.* IV *wassenaar*	Z
S. IV *bockenheim*	53		
S. IV *bonaire*	Z	**Subgenus V**	
S. IV *chameleon*	I	*S.* V *balboa*	Y
S. IV *flint*	Z	*S.* V *bongor*	Y
S. IV *harmelen*	51	*S.* V *brookfield*	66
S. IV *houten*	U	*S.* V *camdeni*	V
S. IV *kralendyk*	C_1	*S.* V *malawi*	66
S. IV *lohbruegge*	V	*S.* V *maregrosso*	66
S. IV *marina*	Y		

Addendum

It has been recently shown by DNA relatedness and numerical taxonomy studies that the genus *Salmonella* consists of a single species having six subspecies corresponding to (1) the former "subgenus" I, (2) the former "subgenus" II, (3) the monophasic serotype of the former "subgenus" III (*Arizona*), (4) the diphasic serotype of the former "subgenus" III (*Arizona*), (5) the former "subgenus" IV, (6) a new subspecies composed of strains that are positive for dulcitol, ONPG and KCN. [Le Minor L., Véron M. and Popoff M. (1982) *Ann. Microbiol.* (*Inst. Pasteur*) **133B**, 223–243 and 245–254.]

2

Biochemical and Serological Characterization of *Citrobacter*

B. LÁNYI

National Institute of Hygiene, Budapest, Hungary

I. Definition of taxon

The genus *Citrobacter* consists of Gram-negative peritrichous rods, which conforms with the definition of the family Enterobacteriaceae. The majority of the isolates ferment lactose promptly or slowly, and practically all give a positive β-galactosidase (*o*-nitro-phenyl β-D-galactopyranoside, ONPG) test. The methyl red test is invariably positive, acetylmethylcarbinol is not produced and rapid utilization of citrate as a sole source of carbon is characteristic. Citrobacters do not produce phenylalanine deaminase, do not liquefy gelatin and hydrolyse urea slowly or not at all. Their inability to decarboxylate lysine distinguishes them from *Salmonella*.

METHODS IN MICROBIOLOGY
VOLUME 15 ISBN 0–12–521515–0

The genus is subdivided into two species: (i) *Citrobacter freundii* (Braak) Werkman and Gillen, 1932 includes mostly H_2S-positive, indole-negative and adonitol-negative cultures. The neotype strain for the species is ATCC 8090. (ii) *Citrobacter diversus* (Burkey) Werkman and Gillen, 1932 is composed of H_2S-negative, indole-positive and adonitol-positive cultures. The neotype strain designated for this species is ATCC 27156.

Members of the genus are divided into serogroups by O-antigens and the serogroups are subdivided into serotypes by H-antigens. A separate antigenic scheme has been elaborated for each species.

Citrobacters appear to be normal inhabitants of the intestinal tract of man and animals. When ingested in large doses by susceptible individuals, they may give rise to diarrhoeal disease. Similarly to other members of the Enterobacteriaceae family, they are associated with a variety of extra-intestinal pathological conditions.

II. History of the taxonomy and nomenclature of *Citrobacter*

In 1932 Werkman and Gillen described a group of bacteria termed coli-aerogenes intermediates. They proposed the name *Citrobacter* as a new genus and *Citrobacter freundii* Braak as the type species. Tittsler and Sandholzer (1935) and Carpenter and Fulton (1937) recommended that such intermediate bacteria, which utilized citrate and failed to produce acetylmethylcarbinol, should be included in the *Escherichia* genus. Borman *et al.* (1944) suggested the name *Colobactrum freundii*. In "Bergey's Manual of Determinative Bacteriology" (Yale, 1948, 1957) and in Kauffmann (1954) the name *Escherichia freundii* was used.

An organism with properties similar to the above group was first named *Salmonella ballerup* (Kauffmann and Møller, 1940) but was later removed from *Salmonella* (Harhoff, 1949; Bruner *et al.*, 1949) and it was designated as the test strain for the new Ballerup group. *Salmonella hormaechei* described by Monteverde (1944) was also included in this group (Bruner *et al.*, 1949). Related bacteria described by Stuart *et al.* (1943), Barnes and Cherry (1946) and Edwards *et al.* (1948) were classified together and termed the Bethesda group. West and Edwards (1954) combined the two groups into one group (Bethesda-Ballerup) and called attention to the close biochemical relationship between these bacteria and *Escherichia freundii*. Combination of the Bethesda and Ballerup groups and their classification in *Escherichia freundii* was suggested also by Kauffmann (1954).

The revival of the name *Citrobacter* was recommended first by Kauffmann (1956) and then by Sedlák (1957). Kauffmann (1956) reclassified *Citrobacter* in a main group containing klebsiellae and related bacteria. Ewing and

Edwards (1960, 1962) and Ewing (1963, 1967) recognized that citrobacters were more closely related to *Salmonella* than to *Escherichia* or to *Klebsiella*, and classified them accordingly in the tribe Salmonelleae. They also recommended the revival of the term *Citrobacter freundii*, which had priority over the other names. This has been generally accepted for H_2S-positive and indole-negative members of the genus (Macierewicz, 1966; Young *et al.*, 1971; Cowan, 1974; Sedlák, 1974).

Bacteria showing the main biochemical features of *Citrobacter* but aberrant in the H_2S and indole reactions have been regarded by many workers as belonging to a taxon separate from *C. freundii*.

The species named *Citrobacter diversum* by Werkman and Gillen (1932) consisted of indole-producing bacteria that fermented adonitol. Vaughn and Levine (1942) suggested that intermediate bacteria that failed to produce H_2S should be classified as *Escherichia intermedium* (Werkman and Gillen) comb. nov. which was recognized in the 6th edition of "Bergey's Manual of Determinative Bacteriology" (Yale, 1948). In the 7th edition of "Bergey's Manual of Determinative Bacteriology", this name was changed to *Escherichia intermedia* (Yale, 1957).

A group of H_2S-negative and adonitol-negative bacteria resembling *C. freundii* but producing indole was described by Macierewicz (1966). She proposed the generic name *Padlewskia* in memory of the Polish bacteriologist L. Padlewski.

Frederiksen (1970) proposed the name *Citrobacter koseri* for H_2S-negative, adonitol-positive and indole-positive bacteria resembling *C. freundii* in other features. A similar group, *Citrobacter intermedius* characterized by Sedlák *et al.* (1971), consisted of H_2S-negative, indole-positive, KCN-positive and malonate-positive organisms.

Young *et al.* (1971) proposed a new genus, *Levinea*, named in honour of M. Levine. The significant characteristics of this group of bacteria which distinguished it from typical members of *C. freundii*, were the positive indole reaction and the lack of H_2S production. Young *et al.* (1971) divided the proposed genus into two species: *L. malonatica* and *L. amalonatica*. The two species differed in the utilization of sodium malonate and the fermentation of adonitol. Young *et al.* (1971) pointed out the close relationship between *L. amalonatica* and *Padlewskia*.

Booth and McDonald (1971) referred to H_2S-negative, indole-positive and malonate-positive bacteria as a new species of *Citrobacter*, without proposing a specific name.

The classification of Young *et al.* (1971) was essentially accepted for the 8th edition of "Bergey's Manual of Determinative Bacteriology" (Sedlák, 1974) but the two taxa were united into one species, *Citrobacter intermedius*, taking into consideration the earlier name *Escherichia intermedia*. According to this classification *L. amalonatica* corresponds to *C. intermedius* biotype a and *L. malonatica* corresponds to *C. intermedius* biotype b.

Ewing and Davis (1972) proposed a revival of the name *Citrobacter diversum* of Werkman and Gillen (1932) as *Citrobacter diversus*, changing the specific epithet in accordance with current rules of nomenclature. According to Ewing and Davis (1972) *C. diversus* consists of H_2S-negative, indole-positive, adonitol-positive and KCN-negative organisms; strains that grow in KCN and fail to ferment adonitol should be classified as *C. freundii*, irrespective of their indole and H_2S reaction. Consequently, *Padlewskia* and *L. amalonatica* constitute a biogroup of *C. freundii*.

Soon after the publication of Ewing and Davis (1972), the name *C. diversus* became popular in America (Jones *et al.*, 1973; Smith *et al.*, 1973; Zwadik, 1976). In England Gross and Rowe (1974) and Rowe *et al.* (1975) employed the name *C. koseri* "for convenience only, pending an agreement on the nomenclature of these organisms". They have emphasized that *C. koseri, C. diversus* and *L. malonatica* should be regarded as members of a single species, whereas *L. amalonatica* constitutes a different entity. The identity of *C. koseri* and *C. diversus* has also been recognized by Cowan (1974).

Summarizing the literature it may be concluded that classification of the genus *Citrobacter* into two species is justified on the basis of well-defined biochemical characteristics tested on a sufficient number of isolates. Hydrogen sulphide-negative and/or indole-positive strains that grow in KCN and fail to ferment adonitol may be distinguished from other isolates as a separate biogroup of *C. freundii*. As regards the other species of the genus, *C. diversus* may be recommended as a validly published and legitimate term, taking priority over other names. The biochemical characteristics of *Citrobacter*, as summarized from the literature, are presented in Table I. Synonyms are shown in the footnotes to Table I.

III. Methods of identification and biogrouping of *Citrobacter*

A. Cultural properties

At 37°C citrobacters grow readily on ordinary media. On nutrient agar, 18- to 24-h colonies are relatively large with an average diameter of 2–3 mm. They are circular, moderately convex with an entire edge and a smooth surface. Rough or dwarf colonies occur infrequently in freshly isolated cultures. Colonies on MacConkey's neutral red bile salt lactose agar or on eosin methylene blue lactose agar are nearly as large as those on nutrient agar. On MacConkey's agar, isolated colonies of strains fermenting lactose rapidly are brick red in colour, whereas those of the late lactose fermenters are not coloured. On eosin methylene blue agar, the former organisms produce dark

TABLE I
Biochemical reactions of *Citrobacter*

	Citrobacter freundii		*Citrobacter diversus*[c]
	Biogroup a[a]	Biogroup b[b]	
Motility	+	+	+
Oxidase	−	−	−
Nitrite from nitrate	+	+	+
H$_2$S	+	−	−
Indole	−	+	+
Urease	(d)	(d)	(d)
Methyl red	+	+	+
Voges–Proskauer	−	−	−
Ammonium citrate	+	+	+
Sodium malonate	d	d	+
KCN	+	+	−
Gelatin liquefaction	−	−	−
Phenylalanine deaminase	−	−	−
Arginine dihydrolase	(+)	(+)	(+)
Lysine decarboxylase	−	−	−
Ornithine decarboxylase	−	d	+
Glucose	+ +	+ +	+ +
Adonitol	−	−	+
Arabinose	+	+	+
Dulcitol	d	d	d
Inositol	−	−	−
Lactose	d	d	+ or (+)
Mannitol	+	+	+
Raffinose	d	d	−
Rhamnose	+	+	+
Salicin	(d)	(d)	(d)
Sorbitol	+	+	+
Sucrose	d	d	d
Xylose	+	+	+
Aesculin hydrolysis	−	(d)	(+)
ONPG	+	+	+
Cephalothin 5 μg ml^{-1}	R	R	S

Symbols: + +, fermentation of glucose with acid and gas; +, 90–100% strains positive; d, 10–90% strains positive; −, 0–10% strains positive; () delayed reaction; R, resistant; S, sensitive.

[a] First described as *Citrobacter freundii* by Werkman and Gillen (1932). Synonyms: *Colobactrum freundii* (Borman *et al.*, 1944), *Salmonella ballerup* (Kauffmann and Moller, 1940), *Salmonella hormaèchei* (Monteverde, 1944), *Escherichia freundii* (Yale, 1948, 1957), Bethesda–Ballerup group (West and Edwards, 1954).

[b] Synonyms: *Padlewskia* (Macierewicz, 1966), *Citrobacter intermedius* (Sedlák *et al.*, 1971), *Levinea amalonatica* (Young *et al.*, 1971), *Citrobacter freundii* hydrogen sulphide-negative, indole-positive biotype (Ewing and Davis, 1972), *Citrobacter intermedius* biotype a (Sedlák, 1974).

[c] First described as *Citrobacter diversum* by Werkman and Gillen (1932) and corrected to *Citrobacter diversus* by Ewing and Davis (1972). Synonyms: *Citrobacter koseri* (Frederiksen, 1970), *Levinea malonatica* (Young *et al.*, 1971), *Citrobacter* sp. (Booth and McDonald, 1971), *Citrobacter intermedius* biotype b (Sedlák, 1974).

violet colonies and the latter grow in colourless colonies sometimes with a bluish centre.

On deoxycholate citrate agar, citrobacters are inhibited; they may produce raised opaque red colonies (early lactose fermenters) or yellowish colonies (late lactose fermenters). Bismuth sulphite agar usually inhibits their growth; occasional strains form small, slightly raised greenish or yellowish brown colonies. On brilliant green agar, lactose-fermenting strains may produce coloured colonies resembling *E. coli*, whilst late lactose fermenters frequently grow in colourless colonies similar to those of salmonellae.

In triple sugar iron (TSI) medium, strains which ferment lactose promptly produce acid slant and acid butt with gas, usually obscured by the black colour as a result of H_2S production. Late lactose fermenters grow in TSI usually as salmonellae; they produce alkaline slant and acid butt with gas and H_2S. *C. diversus* strains usually show the characteristics of late lactose-fermenting *E. coli* (acid butt with gas, no H_2S production).

B. Biochemical differentiation

Biochemical reactions differentiating subgeneric entities of *Citrobacter* from one another and from other Enterobacteriaceae are presented in Table I.

H_2S-positive *C. freundii* strains that fail to ferment lactose or do so slowly may be confused with salmonellae, especially in laboratories not using bismuth sulphite agar to cultivate faecal specimens. The lack of lysine decarboxylase and the growth in KCN medium is characteristic of *C. freundii*, whereas salmonellae are the opposite. Most *Citrobacter* cultures give a positive ONPG test within one day (Pickett and Goodman, 1966); in contrast, salmonellae (except *S. arizonae*) produce no β-galactosidase.

C. freundii biogroup b and *C. diversus* strains can be distinguished most readily by adonitol and KCN. It is also characteristic that *C. diversus* strains are sensitive to first-generation cephalosporins, whilst *C. freundii* isolates (both producers and non-producers of H_2S) are usually resistant to these antibiotics (Booth and McDonald, 1971; Puppel and Mannheim, 1972; Jones *et al.*, 1973; Smith *et al.*, 1973).

H_2S-negative citrobacters may be confused with other Enterobacteriaceae. The main features differentiating between *E. coli* and *C. freundii* biogroup b are the citrate and KCN tests, and between *E. coli* and *C. diversus* the citrate, malonate and adonitol tests. *C. freundii* biogroup b and *C. diversus* can be distinguished from *Enterobacter* by the indole, gelatin, methyl red and Voges–Proskauer reactions. To differentiate between *Enterobacter* and *C. diversus*, the KCN test is also of value.

In epidemiological typing of *Citrobacter* some biochemical reactions may have a limited value. *C. freundii* biogroups a and b and *C. diversus* can be subdivided into biotypes (or biovars) on the basis of their carbohydrate fermentation pattern. The most useful substrates giving fairly consistent results for *C. freundii* strains having a single ancestral origin are dulcitol, raffinose and sucrose. *C. diversus* can be subdivided by the fermentation of dulcitol and sucrose.

IV. Antigenic structure of *Citrobacter*

A. History of *Citrobacter* serology

The first *Citrobacter* strain analysed for serological properties contained the Vi-antigen of *Salmonella typhi* and part of the O-antigen of *Salmonella senftenberg*. Kauffmann and Møller (1940) called this organism *Salmonella ballerup*. Monteverde (1944) described a similar organism: *Salmonella hormaechei*. The serology of such isolates, termed later as Ballerup bacteria, was studied by Harhoff (1949) and Bruner *et al.* (1949).

The antigenic structure of strains belonging to another related group, the Bethesda group, was first investigated by Edwards *et al.* (1948) and then by Moran and Bruner (1949). West and Edwards (1954) recognizing the close biochemical and antigenic relationship between these bacteria, joined them into one group (the Bethesda–Ballerup group). Using strains isolated mostly in the United States, they established an antigenic scheme which became the basis for the serological classification of *Citrobacter*. This scheme (West and Edwards, 1954; reprinted in Kauffmann, 1954) distinguished 32 O-antigen groups and 167 serological types. According to this scheme West and Edwards (1954) examined *C. freundii* strains that fermented lactose rapidly, and concluded that the time of lactose utilization and serological characterization were apparently independent of each other. These data supported views that the Bethesda–Ballerup group should be placed in the species *C. freundii*.

Using the West–Edwards scheme different authors were able to determine the O-antigen group of *Citrobacter* strains in 63–81% and the serotype of the same isolates in 39–57% (Edwards and Ewing, 1972; Sedlák and Šlajsová, 1966a).

Sedlák and Šlajsová (1966a,b) added ten O-antigen groups (O:33–O:42) to the West–Edwards scheme. Reference strains for the new O:33, 34, 35 and 36 groups were originally described as representing *E. coli* O:67, 72, 94 and 122 groups respectively, but were later identified as belonging to *Citrobacter* (Kauffmann, 1954). The test strain for *Citrobacter* O:37 was described by Kauffman (1941) as *Salmonella coli* 1. Strains for *C. freundii* O:38 and O:39 were supplied by E. van Oye, Brussels, and those for O:40, O:41 and O:42 by B. Lányi, Budapest. Using the extended scheme Sedlák and Šlajsová (1966a,b) classified more than 90% of almost 5000 *Citrobacter* isolates according to their O-antigens.

An independent antigenic scheme has been elaborated for *C. diversus*. Gross and Rowe (1974) examined serological cross-reactions of 12 strains described by different authors as *C. koseri, L. amalonatica* and *L. malonatica*. These strains fell into seven O-antigen groups. Of the 28 clinical isolates identified by Gross and Rowe (1974) as *C. koseri*, 22 were identified according to the proposed scheme, three were rough and three were ungroupable. The proposed neotype strain of *C. diversus* (Ewing and Davis, 1972) belonged to the O:1-antigen group.

In a subsequent study Gross and Rowe (1975) extended their scheme by examining a further 165 strains from clinical sources. They proposed that the specific O-antigen components of the 14 test strains should form the basis of a serotyping scheme for *C. koseri* (*C. diversus*). Strains described as *L. amalonatica* (Young et al., 1971) were not related serologically to the above strains and were not included in the proposed scheme.

B. Characteristics of *Citrobacter* antigens

1. Somatic antigens

Similarly to other enterobacteria, *Citrobacter* has a mosaic-like antigenic structure. Certain partial O-antigens are frequently present in different serogroups and are responsible for numerous cross-reactions. In serological characteristics the O-antigens of *Citrobacter* are similar to *Salmonella* O-antigens. Their immunogenicity, agglutinability and agglutinin-binding capacity remain unchanged after treatment with 50% ethanol, N-hydrochloric acid, saturated sodium chloride and after heating at 60–100°C (Kauffmann, 1954; Edwards and Ewing, 1972; Sedlák, 1968; Lányi and Czirók, 1979).

Living cultures of *Citrobacter*, as a rule, agglutinate readily in homologous O-sera. Those containing Vi-antigen, similar to the situation in *Salmonella typhi* Vi^+-strains, may react less readily in homologous O-serum. The serological specificity of the Vi-antigen in *Citrobacter* strains is identical to, or closely related to, the Vi-antigen specificity of *S. typhi* and *S. paratyphi-C*. It is destroyed by heating at 100°C and by exposure to 1 N hydrochloric acid. Ethanol (50% v/v) does not destroy the immunogenic and antibody-binding capacity of the Vi-antigen, but has the ability to cause inagglutination of the cells in Vi-serum (Kauffmann, 1954). *Citrobacter* strains with Vi-antigen frequently split off Vi^--colonies, which are characterized by their translucent appearance as opposed to the opaque colonies of Vi^+-bacteria (Kauffmann and Møller, 1940).

2. *H-antigens*

Citrobacter H-antigens are more complex than O-antigens and most of the isolates have two or three different partial H-antigen factors. The *Citrobacter* H-antigens exhibit features similar to those of Enterobacteriaceae H-antigens; they are destroyed by hydrochloric acid, ethanol and by heating at 100°C, but not by formaldehyde (0.2–0.5%). True phase variation, characteristic of the majority of salmonellae does not occur in *Citrobacter*. However, the H-antigens of certain strains may segregate into different partial factors which apparently are stable (Edwards and Ewing, 1972).

3. *Fimbriae*

In addition to flagella, *Citrobacter* may produce other antigenic filamentous appendages known as fimbriae or pili. Baturo *et al.* (1971) demonstrated the presence of fimbriae in 20 out of 91 strains of *Citrobacter*. They observed fimbriate strains in serogroups O:2, 3, 4, 5, 7, 8, 12, 13, 26, 28 and 29.

4. *Alpha antigen*

Some *Citrobacter* strains possess the alpha antigen, which is a common factor found in different Enterobacteriaceae species (Kauffmann, 1954; Edwards and Ewing, 1972). Such cultures should be avoided when selecting strains for serum production.

C. Antigenic schemes for *Citrobacter*

Table II shows the *C. freundii* antigenic scheme as described by West and Edwards (1954) and extended by Sedlák and Šlajsová (1966a,b). Later Sedlák *et al.* (1971) distinguished further partial antigens: *C. freundii* O:22a,22b,22c (= *Salmonella* O:4,5,12) and *C. freundii* O:38a,38b (= *Salmonella* O:8,20).

The independent scheme for *C. diversus* O-antigens is derived from the paper by Gross and Rowe (1975). In view of the numerous cross-reactions among the test strains, the majority of *C. diversus* O-sera should be absorbed by the cultures indicated (Table III).

152 B. LÁNYI

TABLE II
Antigenic scheme for *Citrobacter freundii* West and Edwards
(1954), supplemented by Sedlák and Slajsová (1966a, b)

O-antigen group	O-antigen	H-antigen
1	1a, 1b, 1c	1, 2
	1a, 1b, 1c	14, 15, 16
	1a, 1b, 1c	(21), 25, 26
	1a, 1b, 1c	(25), 21, 27
	1a, 1b, 1c	39
	1a, 1b, 1c	(9), . . .
2	2a, 1b	1, 2
	2a, 1b	5, 6
	2a, 1b	7, (8), 10
	2a, 1b	8, 10, 11
	2a, 1b	14, 15, 16
	2a, 1b	(13), 17
	2a, 1b	21, 22
	2a, 1b	(23), 28
	2a, 1b	32, 34
	2a, 1b	35, 37
	2a, 1b	39
3	3a, 3b, 1c	4, 5
	3a, 3b, 1c	5, 6
	3a, 3b, 1c	7, (8), 10
	3a, 3b, 1c	8
	3a, 3b, 1c	8, 9
	3a, 3b, 1c	(9), 13, 14
	3a, 3b, 1c	14, 15, 16
	3a, 3b, 1c	(13), 17
	3a, 3b, 1c	21, 22
	3a, 3b, 1c	21, 23
	3a, 3b, 1c	21, 24
	3a, 3b, 1c	(21), 25, 27
	3a, 3b, 1c	(9), 29, 30
	3a, 3b, 1c	(9), 29, 31
	3a, 3b, 1c	32, 33
	3a, 3b, 1c	32, 34
	3a, 3b, 1c	39
	3a, 3b, 1c	47
	3a, 3b, 1c	(4), (33), . . .
	3a, 3b, 1c	—
4	4a, 4b	4, 5
	4a, 4b	5, 6
	4a, 4b	7, (8), 10
	4a, 4b	(9), 13, 14
	4a, 4b	(13), 17
	4a, 4b	(13), 18, 19
	4a, 4b	19, 20

TABLE II— *continued*

O-antigen group	O-antigen	H-antigen
	4a, 4b	(9), 29, 30
	4a, 4b	(9), 29, 31
	4a, 4b	32, 34
	4a, 4b	44, 45
	4a, 4b	44, 46
	4a, 4b	(31), (45), . . .
5	5a, 5b, 4b	53, 54
	5a, 5b, 4b	63
	5a, 5b, 4b, (Vi)	73
	5a, 5b, 4b	—
6	6, 4b, 5b	72
	6, 4b, 5b	(54), . . .
7	7, 3b, 1c	4, 5
	7, 3b, 1c	7, (8), 10
	7, 3b, 1c	8, 9
	7, 3b, 1c	(9), 13, 14
	7, 3b, 1c	(13), 17
	7, 3b, 1c	21, 22
	7, 3b, 1c	21, 23
	7, 3b, 1c	(21), 25, 27
	7, 3b, 1c	39
	7, 3b, 1c	68
	7, 3b, 1c	(4), (33), . . .
8	8a, 1c	1, 2
	8a, 1c	5, 6
	8a, 1c	(9), 13, 14
	8a, 1c	(9), 13, 15
	8a, 1c	21, 22
	8a, 1c	(21), 25, 27
	8a, 1c	(9), 29, 30
	8a, 1c	32, 33
	8a, 1c	39
	8a, 1c	53, 55
	8a, 1c	67
	8a, 8b	1, 2
	8a, 8b	8, 12
	8a, 8b	(9), 13, 14
	8a, 8b	(21), 25, 26
	8a, 8b	(21), 25, 27
	8a, 8b	35, 37
	8a, 8c	5, 6
	8a, 8c	(9), 13, 14
	8a, 8c	(13), 17
	8a, 8c	(21), 25, 27

B. LÁNYI

TABLE II — *continued*

O-antigen group	O-antigen	H-antigen
	8a, 8c	32, 33
	8a, 8c	35, 37
9	9a, 9b	1, 2
	9a, 9b	2, 3
	9a, 9b	4, 5
	9a, 9b	8, 9
	9a, 9b	(13), 17
	9a, 9b	(21), 25, 26
	9a, 9b	32, 33
	9a, 9b	32, 34
	9a, 9b	39
	9a, 9b	48
	9a, 9b	—
10	10, 9b	(13), 17
	10, 9b	(9), 29, 30
	10, 9b	—
11	11	(9), 13, 14
	11	14, 15, 16
	11	32, 33
	11	35, 37
	11	35, (36), 38
	11	77, 78
	11	83
12	12a, 12b	5, 6
	12a, 12b	(13), 17
	12a, 12b	35, 36
	12a, 12b	57
	12a, 12c	57
	12a, 12c	62
13	13	5, 6
	13	59
	13	65
	13	66
	13	69
14	14	40, 41
	14	61
15	15	(13), 18, 19
	15	21, 22
	15	32, 34
16	16	21, 24
	16	58
17	17	21, 24
	17	44, 45
	17	75
18	18	56
19	19	14, 15, 16
	19	87

TABLE II — *continued*

O-antigen group	O-antigen	H-antigen
20	20	40, 41, 43
21	21a, 21b	60
22	22a, 22b, 22c	64
23	23	52
24	24	49, 51
25	25	35, (36), 38
26	26	49, 50
	26	59
27	27	40, 41
28	28, 1c	70
29	29	8, 10, 11
	29	40, 42
	29, (Vi)	73
	29, (Vi)	74
	29, (Vi)	74, 75
	29, (Vi)	75
	29, (Vi)	77
	29	77, 78
	29, (Vi)	77, 79
	29	77, 80
	29	77, 81
	29	77, 88
	29, (Vi)	82
	29, (Vi)	83
	29	84
	29	85
	29, (Vi)	86
	29	87
	29	—
30	30, 21b	76
31	31	71
32	32	(23), 28
33	33	87
34	34	68
35	35	.
36	36	35, 36
	36	44, 45
37	37a, 37b, 37c	.
38	38a, 38b	.
39	39	61
40	40a, 40b	76
41	41	.
42	42	63

Figures in parentheses indicate antigens which may be absent; ..., further antigens may be present; ., H-antigen not investigated; —, non-motile.

156 B. LÁNYI

TABLE III
Antigenic scheme for *Citrobacter diversus*[a] Gross and Rowe (1975)

O-group	Reference strain	Reaction of corresponding serum with other O-antigens	Absorptions required to prepare group-specific serum
1	NCTC 10786	O:2, O:3, O:4, O:8	O:2 and O:4
2	E.1597/71	O:1, O:3, O:4	O:1
3	NCTC 10768	O:1, O:4	O:1 and O:2
4	E.1558/72	O:1, O:3	O:1
5	NCTC 10769	O:13	O:13
6	NCTC 10770	O:8	O:8
7	E.1599/71	O:9, O:11	O:9
8	E.2641/72	O:2, O:6	O:2 and O:6
9	E.2567/72	O:7, O:11	O:11
10	E.2576/72	—	
11	E.2569/72	O:7, O:9	O:9
12	E.2577/72		—
13	E.508/74	O:5, O:10	O:5
14	E.1020/74		—

[a] Synonyms: *Citrobacter koseri, Levinea malonatica*.

D. Antigenic relationships between *Citrobacter* and other organisms

1. O-antigenic relationships

The inter-generic O-antigenic relationships of *C. freundii* are summarized in Table IV on the basis of data described by West and Edwards (1954), Sedlák and Šlajsová (1966a,b), Sedlák *et al.* (1971), Edwards and Ewing (1972), Lányi *et al.* (1972) and Winkle *et al.* (1972). The relationships are diagnostically important: the cross-reactions may be expected in routine slide agglutination tests and may cause confusion in the identification of taxonomically related bacteria (e.g. *Citrobacter* and *Salmonella*).

C. *diversus* serogroups are related to other Enterobacteriaceae antigens by minor unilateral cross-reactions: O:2 to *E. coli* O:79 and *Salmonella* O:1, O:5 to *Shigella boydii* O:7 and O:6 to *E. coli* O:20 and O:32 (Gross and Rowe, 1974).

2. H-antigenic relationships

West and Edwards (1954) observed no relationship between the H-antigens of *Citrobacter* and *Salmonella*. Later it was shown that some *Citrobacter* isolates possessing H-antigens other than those included in the *Citrobacter* scheme

TABLE IV
O-antigenic relationships of *Citrobacter freundii*

C. freundii O-antigen	Related O-antigen of other species
1a, 1b, 1c	*E. coli* 9
3a, 3b, 1c	*Salmonella* 41
3a, 3b, 3c	*Salmonella* (3, 10)
7, 3b, 1c	*Salmonella* 6, 14; *E. coli* 99
8a, 8b	*E. coli* 93; *P. aeruginosa* (6)
9a, 9b	*Salmonella* 30; *E. coli* 7; *V. cholerae*
10, 9b	*E. coli* 71
11	*Salmonella* (1, 2, 12); (6, 14, 24); 40; *E. coli* 6
12a, 12b	*Salmonella* 44; (1, 51); 57; *E. coli* 23
12a, 12c	*Salmonella* 57; *E. coli* 38
14	*Salmonella* 38; *E. coli* 21
15	*E. coli* 15; 57
16	*P. aeruginosa* (4)
17	*E. coli* 101
18	*E. coli* 15; 100
19	*Salmonella* 1, 6, 14, 25; 28; *E. coli* 77
20	*Salmonella* 17
21a, 21b	*Salmonella* 6, 14, 24; *E. coli* 73
22a, 22b, 22c	*Salmonella* 4, 5, 12; *P. aeruginosa* (4)
23	*Salmonella* 18
26	*Salmonella* 21; *E. coli* 76
28, 1c	*E. coli* 9; 44; 73
29	*E. coli* 53; 62; 78
31	*E. coli* 52
32	*P. aeruginosa* (2)
33	*P. mirabilis* (29)
35	*P. aeruginosa* (2)
37a, 37b, 37c	*Salmonella* 48
38a, 38b	*Salmonella* 8, 20
39	*Salmonella* 3, 10; *P. aeruginosa* (12)
40a, 40b	*Salmonella* 57
41	*Salmonella* 55
42	*Salmonella* 54
43	*Salmonella* 28

[a] Figures in parentheses indicate unilateral or slight bilateral relationships. *P. aeruginosa* O-antigens are designated as recommended for international use by Lányi and Bergan (1978).

cross-reacted in *Salmonella* H:z_{27}-, z_{36}- or z_{46}-sera (Edwards and Ewing, 1972).

E. Immunoelectrophoretic grouping of *Citrobacter* O-antigens

Tsvetkova *et al.* (1978) divided 13 *Citrobacter* O-antigen reference strains into two groups on the basis of the immunoelectrophoretic patterns of their antigens. Representatives of Group 2 were characterized by the lack of precipitation arcs towards the cathode.

F. Immunochemistry of *Citrobacter* O-antigens

As in other members of the Enterobacteriaceae, the O-antigens of *Citrobacter* are composed of polysaccharide, lipid and protein, which corresponds to the endotoxin. The polysaccharide contains a central core which exhibits R specificity and is common to all serogroups. Attached to the core are side chains with sugar constituents that determine O-antigen specificity.

The first studies on *Citrobacter* immunochemistry were performed with strains related in antigenic structure to salmonellae (Westphal *et al.*, 1960; Lüderitz *et al.*, 1966; Sedlák *et al.*, 1971). Keleti *et al.* (1971) classified 45 *Citrobacter* serogroups into 20 chemotypes; to this scheme Tsvetkova *et al.* (1978) added another chemotype. As shown in Table V *Citrobacter* lipopolysaccharides have five constituents in common (2-keto-deoxyoctonic acid, heptose, glucose, galactose and glucosamine). Other carbohydrates confer specificity of the O-antigens. In Table V roman numerals designate those chemotypes which also occur in *Salmonella* (S) and/or *E. coli* (E). Chemotypes not found in these genera are designated as *Citrobacter* chemotypes (e.g. CC-A, CC-B).

Table VI links *Citrobacter* chemotypes and O-antigen groups. Several unrelated O-antigens fall in the same chemotype, whereas cross-reacting O-antigen groups may belong to different chemotypes. Members of serogroups O:29 and O:36 are each associated with two different chemotypes (I and II, and CC-D and CC-G respectively), otherwise each serogroup is associated with only one chemotype.

Citrobacter and *Salmonella* serogroups that are related serologically usually exhibit the same or a similar chemotype pattern. Out of the 16 serologically related *Citrobacter* and *Salmonella* pairs, six were identical and ten were different in the sugar composition of their lipopolysaccharides (Westphal *et al.*, 1960; Lüderitz *et al.*, 1966; Sedlák *et al.*, 1971; Keleti *et al.*, 1971).

The lipopolysaccharides of certain *Citrobacter* strains may be even more complex. Barry *et al.* (1963) demonstrated the presence of *N*-acetylneuraminic acid and 4-oxonorleucine in *C. freundii* O:5 and in the serologically related

TABLE V

Chemotypes of *Citrobacter freundii* Keleti *et al.* (1971)

Designation of chemotype	Common constituents	3-Aminofucose	3-Aminoquinovose	2-D-Fucosamine	Mannosamine	Galactosamine	D-Xylose	D-Mannose	Fucose	Rhamnose	6-Deoxytalose	4-Deoxy-D-idose	Abequose	Sialic acid	Unidentified sugar
I (S, E)	+														
II (S, E)	+					+									
III (S, E)	+						+								
IV (S, E)	+					+	+								
V (S, E)	+							+							
VI (S, E)	+					+		+							
VIII (S, E)	+					+				+					
XIII (S)	+								+	+					
XIV (S)	+								+	+		+			
XXXII (S)	+				?										
CC-A	+					+			+	+					
CC-B	+			+		+	+								
CC-C	+		+			+					+				
CC-D	+	+									+				
CC-E	+					+	+				+				
CC-F	+							+	+		+				
CC-G	+					+							+		
CC-H	+					+		+					+		
CC-J	+				(+)						?		(+)		
CC-K	+									+	+				+
CC-L[a]	+							+			+				+

[a] Supplemented by Tsvetkova *et al.* (1978).

Common constituents: 2-keto-deoxyoctonic acid, heptose, glucose, galactose and glucosamine; +, present; (+), present in low amount; ?, presence doubtful.

TABLE VI
Correlation between chemotype and O-antigens of *Citrobacter freundii* Keleti *et al.*
(1971)

Designation of chemotype	O-antigens
I (S, E)	9a, 9b; 10a, 9b; 13; 14; 29a; 33
II (S, E)	5a, 5b, 4b; 12a, 12c; 16; 17; 29b; 30, 21b; 42
III (S, E)	21a, 21b
IV (S, E)	2a, 1b; 28, 1c
VI (S, E)	6, 4b, 5b
VIII (S, E)	1a, 1b, 1c; 7, 3b, 1c; 12a, 12b; 15; 18
XIII (S)	3a, 3b, 1c; 8a, 1c
XIV (S)	22; 38
XXXII (S)	20
CC-A	26
CC-B	11
CC-C	19
CC-D	36b; 41
CC-E	8a, 8b
CC-F	25, 32
CC-G	4a, 4b; 27; 36a
CC-H	23
CC-J	37a, 37b, 37c
CC-K	35
CC-L[a]	8a, 8c

[a] Supplemented by Tsvetkova *et al.* (1978).

Salmonella dahlem. In *Citrobacter* ATCC 8090 Raff and Wheat (1968) showed, in addition to the known common components, 2-acetamido-2-deoxy-D-glucose and 3-acetamido-3,6-dideoxy-D-glucose, O-acetyl and trace amino acids.

V. Methods of examining *Citrobacter* antigens

A. Determination of O-antigens

1. *Preparation of O-antisera*

(a) *Preparation of antigens.* Colonies showing specific O-agglutination are selected and subcultured overnight on nutrient agar slants. The growth from the slants is washed off and used for the inoculation of nutrient agar plates, poured preferably in Roux flasks. After incubation at 37°C for 18–20 h, the growth is suspended in saline. To inactivate heat-labile antigens, the suspension is heated at 100°C for 2–2.5 h. After heating the suspension should be homogeneous. Auto-agglutinability upon heating indicates O-antigen deficiency and such suspensions should not be used for immunization. The density of the suspension is adjusted to about 5×10^9 cells per millilitre.

(b) *Immunization and preservation of sera.* Rabbits are injected at four-day intervals with graded intravenous doses of 0.5, 1.0, 2.0, and 4.0 ml of the antigen suspension. The rabbits are bled on the 4th to 6th day after the last injection. The antisera are preserved with phenol (9 vol. serum + 1 vol. 5% aqueous phenol solution) or with merthiolate (1 in 10 000 w/v) or by the addition of an equal volume of glycerol. The sera are stored in the refrigerator at 4°C. The titre against the homologous strain should be 1:2560 or more.

(c) *Determination of tube titre and working dilution.* Bacterial suspensions heated at 100°C for 2–2.5 h are used to determine the tube agglutination titre. The antigen suspension is adjusted to contain approximately 5×10^9 bacteria per millilitre. To 0.5 ml amounts of twofold serial dilutions of the serum, 0.05 ml bacterial suspension is added. The tubes are incubated at 37°C or at 50°C overnight.

Slide agglutination is performed with sera diluted so that the homologous culture agglutinates within 1–4 s (+ + + + reaction), and major antigenic relationships appear as distinct, but less prompt agglutination. In checking *Citrobacter* sera it must be remembered that only the antigens of major diagnostic significance are expressed in the current schemes. Since cross-reactions due to minor antigens appear rather frequently, it is emphasized that each serum should be tested against all antigen reference strains. If unexpected cross-reactions are distinct at the working dilution of the serum, it should be discarded and another batch tested. If agglutinins causing confusing cross-reactions are present in two or more different batches, the serum is absorbed with the corresponding culture. The serum is examined for the presence of alpha agglutinins: if agglutination occurs with living *Proteus inconstans* strain Wakefield, the serum is to be discarded.

(d) *Absorption of O-sera.* For the differentiation of serogroups with prominent related O-antigens, absorbed sera are used. The choice of serum and absorbent culture is evident from the antigenic scheme; e.g. for the determination of O:1a-antigen, serum O:1a,1b,1c is absorbed by cultures O:2a,1b + O:3a,3b,1c. For absorption, the bacteria are cultured on agar plates in Roux flasks at 37°C for 18–24 h. The growth is suspended in saline, steamed at 100°C for 1–2 h and centrifuged. The deposit is resuspended in diluted serum. For tube agglutination, the serum to be absorbed should be diluted 1:10–1:20, for slide agglutination the serum should be 3–4 times more concentrated than the working dilution of the unabsorbed serum. After incubation at 37°C for 2 h then overnight at 4°C, the bacteria are removed by centrifugation. The supernatant serum is preserved with phenol or merthiolate as described above.

(e) *Preparation of pooled O-sera. C. freundii* pooled sera may be prepared as recommended by Sedlák (1968). The amount of single serum components (Table VII) is chosen so that the pool gives distinct (+ + + +) slide agglutination with each of the homologous antigens, but does not react with strains belonging to other antigenic groups.

2. *Serogrouping of isolates*

Slide agglutination with living cultures in adequately checked sera is a satisfactory routine method. Bacteria grown on nutrient agar plates or on agar slants incubated at 37°C for 18–24 h are examined. It is advisable to test the strain under investigation first in pooled O-sera (Table VII) and then in unabsorbed O-sera. If the culture belongs to a serogroup of complex antigenic structure, it is examined further in the corresponding absorbed sera. The Vi-antigen of *Citrobacter* is determined in a Vi-serum prepared with *S. typhi.*

Strains that fail to agglutinate in any of the O-sera may be tested by tube agglutination after heating their suspensions at 100°C for 2 h. Heat-labile factors inhibiting the agglutination of live cultures occur infrequently in *Citrobacter*. Strains inagglutinable in the O-sera usually contain O-antigens not defined in the antigenic scheme.

TABLE VII

Ingredients of *Citrobacter freundii* pooled O-sera (Sedlák, 1968)

Designation	Ingredients, O-sera prepared with antigens
OA	1a, 1b, 1c; 2a, 1b; 3a, 3b, 1c; 4a, 4b; 5a, 5b, 4b; 6, 4b, 5b; 7, 3b, 1c
OB	8a, 1c; 8a, 8b; 9a, 9b; 10, 9b; 11; 12a, 12b; 12a, 12c; 13
OC	14; 16; 17; 18; 19; 20
OD	21a, 21b; 22; 23; 24; 25; 26
OE	27; 28, 1c; 29, Vi; 30, 21b; 31; 32
OF	33; 34; 35; 36; 37a, 37b, 37c
OG	38; 39; 40a, 40b; 41; 42

B. Determination of H-antigens

1. *Preparation of H-antisera*

(a) *Preparation of antigens.* It is essential that only very actively motile bacteria be used as antigens. The culture is inoculated into a U-tube containing semi-solid nutrient agar and incubated until the bacteria spread throughout the medium. The growth in the opposite branch of the U-tube is then further subcultured in U-tubes until the bacteria are able to grow through the tubes in 16–18 h. Then nutrient agar plates which are soft, moist and thick are spot inoculated and incubated overnight. Incubation at 37°C is usually satisfactory, but some strains develop their H-antigen best at 30–32°C. The swarming growth from the plates is suspended in saline and the bacteria are killed with formaldehyde (0.2%). The suspension is adjusted to contain about 5×10^9 cells per millilitre. Alternatively, broth cultures inoculated, incubated and treated with formalin as above may be used for immunization (Edwards and Ewing, 1972).

(b) *Immunization and preservation of sera.* The immunization schedule is the same as when preparing O-sera. Some workers prefer to give six instead of five doses (Edwards and Ewing, 1972). The formalin-treated H-antigen suspensions are used throughout one course of immunization, but they must not be employed after this period of time (stored suspensions may become heterogeneous). The H-sera are preserved and stored as described for the O-sera.

(c) *Determination of the tube titre and the working dilution.* Tube agglutination is carried out using formalin-treated suspensions prepared in the same manner as antigens for immunization. The suspension is adjusted to approximately 5×10^9 bacteria per millilitre and used as described for determining of O-antigen titres.

The working dilution for slide agglutination of H-sera is determined on the same principle as described for O-sera. The sera are tested with reference strains freshly grown on swarm agar plates. Alternatively, dense suspensions washed off swarm plates are treated with formalin and stored at room temperature; in one drop of serum dilution a double loopful of the preserved H-antigen is mixed.

Unabsorbed *Citrobacter* H-sera necessarily contain O-antibodies. Although the H-antigen titre in slide testing invariably exceeds the O-antigen titre, the O-agglutinins may cause confusing results in certain batches of sera. It is, therefore, essential to test each H-serum with a strain having an homologous O-antigen but an heterologous H-antigen compared to the immunizing culture. For example, unabsorbed serum H:5,6 (O:3a,3b,1c) at a working dilution should give the following pattern of agglutination: strain O:3a,3b,1c;H:5,6 + + + + (floccular type); strain O:4a,4b;H:5,6 + + + + (floccular type); strain O:3a,3b,1c;H:8,9 + (weak granular type agglutination). If O-agglutinins are present in titres that cannot be diluted out, they should be absorbed from the H-serum.

(d) *Absorption of H-sera.* Regarding their complex antigenic structure, many *Citrobacter* H-antigens are determined in absorbed sera. The bacterial suspensions should be chosen as indicated by the antigenic formula (Table II). For example, H:1-serum is prepared by absorbing serum H:1,2 (O:9a,9b) with strain O:9a,9b;H:2,3 – in this manner not only H:2-, but also O:9a,9b-agglutinins will be removed from the serum.

Absorption of H-agglutinins is carried out in the same manner as the absorption of O-agglutinins, except that the Roux flasks are inoculated with actively motile bacteria and the harvested culture is killed with formaldehyde.

(e) *Preparation of pooled O;H-serum.* The pooled serum presented in Table VIII may be recommended. The ingredients have been selected so that agglutination occurs with antigens most frequently encountered in *C. freundii.*

TABLE VIII
Ingredients of *Citrobacter freundii* pooled OH-serum (Edwards and Ewing, 1972)

Ingredients, OH-sera prepared with strains	
O-antigen	H-antigen
1a, 1b, 1c	1, 2
2a, 1b	7, (8), 10
2a, 1b	21, 22
3a, 3b, 1c	8, 9
3a, 3b, 1c	(13), 17
4a, 4b	(9), 29, 31
7, 3b, 1c	(9), 13, 14
7, 3b, 1c	(21), 25, 27
8a, 8c	5, 6
8a, 8c	32, 33
9a, 9b	4, 5
9a, 9b	39
29	74, 75

If the working dilution is set so that the serum reacts not only with H- but also with O-antigens, the serum is suitable for screening purposes of about 90% of *C. freundii* isolates (Edwards and Ewing, 1972). If the serum is employed at higher dilution and O-agglutinins still giving disturbing agglutinations are absorbed, it will react only with H-antigens. This pooled H-serum is used to detect the presence of the most frequent H-antigens of *C. freundii*.

2. *Determination of H-antigens*

Slide agglutination test in properly diluted H-sera is the procedure recommended. Isolates grown on nutrient agar or on TSI medium may be tested, but optimal results are obtained with bacteria swarming on soft agar plates. Sometimes it may be necessary to develop the flagella by subculturing the isolate in semi-solid agar.

VI. Application of biotyping and serotyping of *Citrobacter*

A. Indications

1. *Biotyping*

All clinical and public health laboratories should be capable of performing a limited number of tests in order to rapidly identify *C. freundii* (indole, urease, H_2S, ammonium citrate, methyl red, lysine decarboxylase, fermentation of glucose, lactose and some other carbohydrates). Rapid identification of *C. diversus* is more difficult and requires additional tests (Table I). In fact, *C. diversus* is frequently misidentified in the clinical laboratory as *E. coli*. To follow the nosocomial spreading of *Citrobacter*, biochemical tests described in Section III.A may be of some value.

2. *Serotyping*

Serotyping *Citrobacter* gives reproducible results and provides a large number of distinct epidemiological entities. Owing to the existence of an internationally recognized antigenic scheme, the results of different authors are comparable. The preparation of antisera and the typing of the isolates require special experience in bacteriological serology. Accordingly, a central reference laboratory should be appointed to perform serotyping for smaller laboratories. Serotyping is indicated when a nosocomial spread of *Citrobacter* is observed or when an outbreak of enteritis may be associated with this organism.

B. Frequency of biochemical and serological entities

1. *Incidence of* Citrobacter freundii *in human faecal specimens*

C. freundii is an inhabitant of the intestinal tract and is, accordingly, present in sewage, surface waters and in food contaminated with faecal material. In about 40 000 specimens of this kind, Sedlák (1968) reported a 7% incidence of *Citrobacter*.

A total of 6137 *Citrobacter* strains isolated from a wide variety of sources in Czechoslovakia during the years 1954 to 1966 belonged to 39 different serogroups (Sedlák, 1968). The most common serogroups were O:3 (10.2% of all strains), O:4 (6.8%), O:6 (7.0%), O:7 (10.4%), O:8 (13.9%), O:12 (5.7%) and O:22 (9.1%).

In association with *Citrobacter*, Sedlák (1968) described five outbreaks of febrile gastro-enteritis. The first one, involving 14 persons on a farm, was conveyed by milk; the aetiological role of *Citrobacter* serotype O:3a,3b,1c;H:8 was indicated by a high titre of specific agglutinins in all patients. Earlier Sedlák (1957) reported on a seroconversion in patients who had passed through enteritis associated with *Citrobacter*. Out of 213 patients involved in outbreaks, 70.7%, of 278 patients with sporadic enteritis 55.6 had a significant titre of antibodies against the corresponding *Citrobacter* serogroup. Two different outbreaks of enteritis among infants were associated with *Citrobacter* serotypes O:8a,8b;H:8,12. For two major food-borne infections serotypes O:8a,10;H and O:22;H:64 have been responsible (Sedlák, 1968). In view of these findings, Sedlák (1968) has concluded that at least *Citrobacter* serogroups O:8, O:22, O:7 and O:3 should be regarded as pathogenic. It is noticeable that *Citrobacter* O:22 is closely related antigenically to *Salmonella* O:4,5,12.

Kleinmaier and Schäfer (1956) isolated *Citrobacter* serotype O:8a,1c;H:(13),18,19 from stool specimens of seven patients with diarrhoea; the same serotype, however, occurred in the faeces of 22 persons free from enteric symptoms. From faeces of adults with sporadic enteritis Kahlich and Webershinke (1963) isolated *Citrobacter* in 7.1%, whereas in healthy persons it occurred in only 1.6%; the most frequent serogroups were O:4, O:12 and O:8. They concluded that *Citrobacter* may be associated with mild enteritis. Isolated parts of rabbit gut loops showed ulcero-phlegmonous and necrotic changes after the application of live broth cultures of *Citrobacter*. These changes cannot be considered as a sign of enterotoxin production, but are similar to those induced by *E. coli* associated with infantile diarrhoea.

Popovici *et al.* (1964) isolated *C. freundii* serotypes O:4a,4b;H:44,45 and O:7,3b,1c;H:68 in association with two sporadic cases of food poisoning. In their subsequent paper Popovici *et al.* (1967) described that the incidence of

Citrobacter was 15.1% in patients with enteritis and 12.5% in healthy children. *Citrobacter* strains isolated from the patients belonged in 20.7% to serogroup O:28; other frequent serogroups were O:1, O:21, O:9 and O:19. In the healthy control group of children, serogroups O:12, O:4, O:28, O:21, O:20 and O:7 were shown. In mouse pathogenicity there was no difference between strains isolated from the patients and from controls.

Ivanova (1966) by biochemical and serological characteristics classified 115 *C. freundii* strains isolated from the faeces of 2049 patients with enteritis, of 627 patients with extra-intestinal disease and of 1214 healthy persons. In contrast to other authors, she concluded that *Citrobacter* had no significant role in the aetiology of acute intestinal infections.

In the opinion of Kaganovskaya *et al.* (1973) *Citrobacter* may be associated with intestinal diseases. The most frequent serogroups in their material were O:3a,3b,1c, O:4a,4b, O:5a,5b,4b, O:8 and O:22.

Finn (1978) isolated *Citrobacter* strains more frequently from patients with enteric diseases (42.8–22.7%) than from healthy persons. The most frequent serogroups were O:3, O:9, O:23, O:1, and O:8. Serogroups O:1, O:5, O:8, O:17 and O:23 were frequently isolated from patients with aetiologically obscure dysentery; the incidence of these serogroups showed a seasonal increase during spring and summer.

Summarizing results available for the enteropathogenicity of *C. freundii*, it may be concluded that under certain conditions the organism is capable of causing primary intestinal disease. When ingested in large doses by susceptible individuals, it may even cause outbreaks of enteritis. The existence of serogroups or serotypes of special enteric pathogenicity is still in doubt.

2. Incidence of Citrobacter diversus in human faecal specimens

C. diversus, similarly to *C. freundii*, is an inhabitant of the human intestine. The isolates studied by Rowe *et al.* (1975) occurred in the faeces of persons mostly free from enteric symptoms. In an outbreak of diarrhoeal disease, however, the role of *C. diversus* serogroup O:5 was suspected. Rowe *et al.* (1975) have concluded that further study is needed to establish the role of this organism in diarrhoea.

3. Incidence of Citrobacter freundii in extra-intestinal pathological conditions

In miscellaneous human pathological specimens *C. freundii* occurs less frequently than any of the major Gram-negative pathogens such as *E. coli*, *Klebsiella*, *Enterobacter*, *Proteus* and *P. aeruginosa*. According to reports of the Hungarian Public Health Laboratory Service during the years 1975 to

1979 from a total of 565 742 urine samples examined in 27 laboratories, 10 511 strains of *C. freundii* were isolated (1.8%). From 873 778 other non-faecal clinical specimens, 3598 *C. freundii* strains were cultured (0.4%). In all non-faecal pathological specimens *C. freundii* comprised 2.9% of all Enterobacteriaceae isolates (14 109 from 488 235).

Sedlák (1968) has pointed out the association of citrobacters with urinary tract infections, otitis and cholecystitis. Serogroups O:8 strains were most commonly associated with these diseases.

Puppel and Mannheim (1972) reported on the significance of *Citrobacter* as a secondary invader of the human respiratory tract. Citrobacters were usually associated with chronic bronchitis and occurred in sputum and bronchial specimens in 4.2%. The incidence of H_2S-negative strains for all *Citrobacter* isolations was 12.5%, but indole-producing strains were not encountered among them. In the material of Puppel and Mannheim (1972) the incidence of *Citrobacter* was 1.1% in throat swabs and 0.8% in urine.

There are few data for the pathogenicity of *Citrobacter* to animals. Nestorescu *et al.* (1964) described three *Citrobacter* strains (serotype O:30,21b;H:76), the culture filtrates of which exerted a cytotoxic and lethal effect. Further studies (Szégli *et al.*, 1966) revealed that these organisms produced three kinds of antigenic extracellular products (i) a glycoprotein with lethal activity; (ii) a heat-labile protein with haemolytic and cytopathogenic properties (citrolysin) and (iii) two proteolytic enzymes.

4. *Incidence of* Citrobacter diversus *in extra-intestinal pathological conditions*

The frequency of *C. diversus* in clinical specimens is difficult to assess on the basis of laboratory reports because of the tendency of many laboratories to lump this organism into *E. coli* or non-identified Enterobacteriaceae.

C. diversus has been isolated from a variety of extra-intestinal infections as a primary or secondary invader (Frederiksen, 1970; Booth and McDonald, 1971; Jones *et al.*, 1973; Smith *et al.*, 1973). In diseases of the urinary tract, infected burns, otitis and so on the organism usually occurs sporadically and has not been shown as a major cause of hospital-associated infections. It may have a special role in meningitis and septicaemia in infants and should, in this respect, be regarded as potentially dangerous when present in hospital units (Rowe *et al.*, 1975).

The serological distribution of *C. diversus* isolates has been described by Rowe *et al.* (1975). The most common serogroups from all sources have been O:1 and O:2 (representing 42.4 and 21.2% of all typable strains).

References

Barnes, L. A. and Cherry, W. B. (1946). *Am. J. Public Health* **36**, 481–483.
Barry, G. T., Chen, F. and Roark, E. (1963). *J. Gen. Microbiol.* **33**, 97–116.
Baturo, A. P., Kalyaev, A. V. and Raginskaya, V. P. (1971). *Zh. Mikrobiol. Epidemiol. Immunobiol.* No. 1, 46–48.
Booth, E. V. and McDonald, S. (1971). *J. Med. Microbiol.* **4**, 329–336.
Borman, E. K., Stuart, C. A. and Wheeler, K. (1944). *J. Bacteriol.* **48**, 351–367.
Bruner, D. W., Edwards, P. R. and Hopson, A. S. (1949). *J. Infect. Dis.* **85**, 290–294.
Carpenter, P. L. and Fulton, M. (1937). *Am. J. Public Health* **27**, 822–827.
Cowan, S. T. (1974). "Manual for the Identification of Medical Bacteria", 2nd edn. Cambridge University Press, London.
Edwards, P. R. and Ewing, W. H. (1972). "Identification of Enterobacteriaceae", 3rd edn. Burgess, Minneapolis, Minnesota.
Edwards, P. R., West, M. G. and Bruner, D. W. (1948). *J. Bacteriol.* **55**, 711–719.
Ewing, W. H. (1963). *Int. Bull. Bacteriol. Nomencl. Taxon.* **13**, 95–110.
Ewing, W. H. (1967). "Revised Definitions for the Family Enterobacteriaceae, its Tribes and Genera". Center for Disease Control, Atlanta, Georgia.
Ewing, W. H. and Davis, B. R. (1972). *Int. J. Syst. Bacteriol.* **22**, 12–18.
Ewing, W. H. and Edwards, P. R. (1960). *Int. Bull. Bacteriol. Nomencl. Taxon.* **10**, 1–12.
Ewing, W. H. and Edwards, P. R. (1962). "The Principal Divisions and Groups of Enterobacteriaceae and their Differentiation". National Communicable Disease Center, Atlanta, Georgia.
Finn, V. G. (1978). *J. Hyg. Epidemiol. Microbiol. Immunol.* **22**, 338–343.
Frederiksen, W. (1970). *Public Fac. Sci. Univ. J.E. Purkyne, Brno* **47**, 89–94.
Gross, R. J. and Rowe, B. (1974). *J. Med. Microbiol.* **7**, 155–161.
Gross, R. J. and Rowe, B. (1975). *J. Hyg.* **75**, 121–127.
Harhoff, N. (1949). *Acta Pathol. Microbiol. Scand.* **26**, 167–174.
Ivanova, A. S. (1966). *Zh. Mikrobiol. Epidemiol. Immunobiol.* No. 12, 91–95.
Jones, S. R., Ragsdale, A. R., Kutscher, E. and Sanford, J. P. (1973). *J. Infect. Dis.* **128**, 563–565.
Kaganovskaya, S. N., Raginskaya, V. P., Lifshits, M. B. and Kazachkova, E. L. (1973). *Zh. Mikrobiol. Epidemiol. Immunobiol.* No. 3, 32–36.
Kahlich, R. and Webershinke, J. (1963). *Cs. Epidemiol. Mikrobiol. Immunol.* **12**, 55–64.
Kauffmann, F. (1941). "Die Bakteriologie der *Salmonella* Gruppe". Munksgaard, Copenhagen.
Kauffmann, F. (1954). "Enterobacteriaceae". Munksgaard, Copenhagen.
Kauffmann, F. (1956). *Zentralbl. Bakteriol. Parasitenkd. Infektionskr. Abt. Orig.* **165**, 344–353.
Kauffmann, F. and Møller, E. (1940). *J. Hyg.* **40**, 246–251.
Keleti, J., Lüderitz, O., Mlynarčik, D. and Sedlák, J. (1971). *Eur. J. Biochem.* **20**, 237–244.
Kleinmaier, H. and Schäfer, E. (1956). *Zentralbl. Bakteriol. Parasitenkd. Infektionskr. Abt. Orig.* **165**, 97–107.
Lányi, B. and Bergan, T. (1978). *In* "Methods in Microbiology" (T. Bergan and J. R. Norris, Eds). Vol. 10, pp.93–168. Academic Press, London and New York.
Lányi, B. and Czirók, E. (1979). *Zentralbl. Bakteriol. Parasitenkd. Infektionskr. Hyg. Abt. Orig. A* **243**, 308–320.

170   B. LÁNYI

Lányi, B., Adám, M. M. and Vörös, S. (1972). *Acta Microbiol. Acad. Sci. Hung.* **19**, 259–265.
Lüderitz, O., Staub, A. M. and Westphal, O. (1966). *Bacteriol. Rev.* **30**, 192–255.
Macierewicz, M. (1966). *Med. Dosw. Mikrobiol.* **18**, 333–339.
Monteverde, J. (1944). *Nature (London)* **154**, 676.
Moran, A. B. and Bruner, D. W. (1949). *J. Bacteriol.* **58**, 695–700.
Nestorescu, N., Popovici, M., Szégli, L., Negut, A., Negut, M. and Barbulescu, E. (1964). *Zentralbl. Bakteriol. Parasitenkd. Infektionskr. Abt. Orig.* **194**, 443–450.
Pickett, M. J. and Goodman, R. E. (1966). *Appl. Microbiol.* **14**, 178–182.
Popovici, M., Szégli, L., Soare, L., Negut, A., Dimitru, N. and Stanciu, V. (1964). *Arch. Roum. Pathol. Exp. Microbiol.* **23**, 1005–1010.
Popovici, M., Szégli, L., Racovitza, C., Badulescu, E., Florescu, D., Negut, M., Negut, A., Thomas, E. and Masek, S. (1967). *Zentralbl. Bakteriol. Parasitenkd. Infektionskr. Abt. Orig.* **204**, 112–121.
Puppel, H. and Mannheim, W. (1972). *Zentralbl. Bakteriol. Parasitenkd. Infektionskr. Hyg. Abt. Orig. Reihe A* **221**, 38–47.
Raff, R. A. and Wheat, R. W. (1968). *J. Bacteriol.* **95**, 2035–2043.
Rowe, B., Gross, R. J. and Allen, H. A. (1975). *J. Hyg.* **75**, 129–134.
Sedlák, J. (1957). *Zentralbl. Bakteriol. Parasitenkd. Infektionskr. Abt. Orig.* **168**, 10–20.
Sedlák, J. (1968). *In* "Enterobacteriaceae-Infektionen" (J. Sedlák and H. Rische, Eds), 2nd edn, pp. 521–530. Thieme, Stuttgart.
Sedlák, J. (1974). *In* "Bergey's Manual of Determinative Bacteriology" (R. E. Buchanan and N. E. Gibbons, Eds), 8th edn, pp. 296–298. Williams & Wilkins, Baltimore, Maryland.
Sedlák, J. and Šlajsová, M. (1966a). *Zentralbl. Bakteriol. Parasitenkd. Infektionskr. Abt. Orig.* **200**, 369–374.
Sedlák, J. and Šlajsová, M. (1966b). *J. Gen. Microbiol.* **43**, 151–158.
Sedlák, J., Puchmayerová-Šlajsová, M., Keleti, J. and Lüderitz, O. (1971). *J. Hyg. Epidemiol. Microbiol. Immunol.* **15**, 366–374.
Smith, R. F., Dayton, S. L. and Chipps, D. D. (1973). *Appl. Microbiol.* **25**, 157–158.
Stuart, C. A., Wheeler, K. M., Rustigian, R. and Zimmerman, A. (1943). *J. Bacteriol.* **45**, 101–119.
Szégli, L., Szégli, G., Popovici, M. and Nestorescu. N. (1966). *Zentralbl. Bakteriol. Parasitenkd. Infektionskr. Abt. Orig.* **200**, 497–509.
Tittsler, R. P. and Sandholzer, L. A. (1935). *J. Bacteriol.* **29**, 349–361.
Tsvetkova, N. V., Raginskaya, V. P. and Vaneeva, N. P. (1978). *Zh. Mikrobiol Epidemiol. Immunobiol.* No. 11, 33–38.
Vaughn, R. H. and Levine, M. (1942). *J. Bacteriol.* **44**, 487–505.
Werkman, C. H. and Gillen, G. F. (1932). *J. Bacteriol.* **23**, 167–182.
West, M. G. and Edwards, P. R. (1954). "The Bethesda–Ballerup Group of Paracolon Bacteria". Public Health Service Mongraphs No. 22 U. S. Government Printing Office, Washington, D.C.
Westphal, O., Kauffmann, F., Lüderitz, O. and Stierlin, H. (1960). *Zentralbl. Bakteriol. Parasitenkd. Infektionskr.* **179**, 336–342.
Winkle, S., Refai, M. and Rohde, R. (1972). *Ann. Inst. Pasteur* **123**, 775–881.
Yale, M. W. (1948). *In* "Bergey's Manual of Determinative Bacteriology" (R. S. Breed, E. G. D. Murray and A. P. Hitchens, Eds), 6th edn, pp. 444–453. Williams & Wilkins, Baltimore, Maryland.
Yale, M. W. (1957). *In* "Bergey's Manual of Determinative Bacteriology" (R. S. Breed,

E. G. D. Murray and N. R. Smith, Eds), 7th edn, pp. 334–341. Williams & Wilkins, Baltimore, Maryland.

Young, V. M., Kenton, D. M., Hobbs, B. J. and Moody, M. R. (1971). *Int. J. Syst. Bacteriol.* **21**, 58–63.

Zwadik, P. (1976). *In* "Zinsser Microbiology" (W. K. Joklik and H. Willett, Eds), 16th edn, pp. 566-574. Appleton, New York.

Note added in proof

For strains termed in Table I *Citrobacter freundii* biogroup b, the name *Citrobacter amalonaticus* has been recommended [Sonnenwirth, A. C. (1980). *In* "Gradwohl's Clinical Laboratory Methods and Diagnosis" (A. C. Sonnenwirth and L. Jarett, Eds), 8th edn, Vol. 2, pp. 1731–1852. Mosby, St. Louis, Missouri]. This name has been validated as *Citrobacter amalonaticus* comb. nov. (Young *et al.*) [Brenner and Farmer (1981) List No. 8 (1982). *Int. J. Syst. Bacteriol.* **32**, 266–268]. It should be noted that about 14% of H_2S-negative, indole-positive, KCN-positive and adonitol-negative strains do utilize sodium malonate; this single aberrant feature, however, is not a substantial reason for excluding such cultures from *C. amalonaticus*. The type strain for *C. amalonaticus* is ATTC 25405 (= NCTC 10805).

3
Serological Typing of *Serratia marcescens*

T. L. PITT AND Y. J. ERDMAN

Central Public Health Laboratory, Division of Hospital Infection,
London and Department of Microbiology, Middlesex Hospital Medical School,
London, England

I. Introduction

In the last 15 years, *Serratia marcescens* has emerged as an important cause of hospital-acquired infection in many countries. It has frequently been reported as a pathogen in urinary tract infection (Ball *et al.*, 1977) and in septicaemia (Dodson, 1968; Schaberg *et al.*, 1976) as well as an opportunist organism colonizing the upper respiratory tract (Ringrose *et al.*, 1968; Negut *et al.*, 1975).

METHODS IN MICROBIOLOGY
VOLUME 15 ISBN 0–12–521515–0

Infections are most common in the chronically ill compromised patient (Brooks et al., 1979; Young et al., 1980a), and appear to be associated with genito-urinary tract manipulations especially catheterization, respiratory ventilation and prior exposure to broad-spectrum antibiotics which are largely inactive against this species (Schaberg et al., 1976; Meers et al., 1978).

Common-source outbreaks of sepsis due to Gram-negative bacteria, such as S. marcescens, arise because of the persistence and multiplication of the organism in items of equipment, e.g. respirators or in solutions used in hospitals. Examples of common-source outbreaks are cited by Farmer et al. (1976). Gastro-intestinal carriage is rare (Rose and Schreier, 1968) but the urinary tract of asymptomatic and symptomatic patients have been implicated as a possible reservoir for nosocomial infection (Maki et al., 1973). Transmission of the organism by contaminated hands of attending medical and nursing personnel has been suggested by Schaberg et al. (1976) and Meers et al. (1978).

II. Identification of *Serratia marcescens*

The genus *Serratia* is relatively easy to detect in primary cultures if all non-lactose-fermenting Gram-negative bacilli are screened for the production of DNAase. In addition most strains are resistant to ampicillin, and this feature may serve to indicate its presence. Other characteristic features of *Serratia* which may serve as an aid to identification are rapid liquefaction of gelatin after two days and production of lipase as shown by hydrolysis of Tween 80. Brooks et al. (1979) found all strains to be DNAase-positive, oxidase-negative, fermentative in Hugh and Leifson's medium and did not produce phenylpyruvic acid from phenylalanine. *S. marcescens* may be distinguished from *S. liquefaciens* and *S. rubidaea* quite simply by its failure to produce acid from arabinose and raffinose in peptone water incubated aerobically.

Most clinical isolates are non-pigmented and appear as typical "coliforms" but grow more luxuriantly at 32°C than at 37°C. In our experience pigmented strains are usually isolated from the environment as opposed to patient sites. The biochemical and cultural characteristics of the genus have been comprehensively reviewed by Grimont and Grimont (1978).

III. Antigenic structure of *Serratia marcescens*

A. History

The first published account of the agglutination and biological relationships in the prodigiosus group (*Chromobacterium prodigosum* now *Serratia marcescens*) was that of Hefferan (1906). A variety of red pigment-producing bacilli were examined and rabbits or guinea-pigs were injected intra-peritoneally with live vaccines at intervals of two days to one week. The sera were tested by tube agglutination and some distinction between agglutinogens (probably O and H) was made by tests at different temperatures. Higher serum dilutions gave agglutination at the high temperatures and this was probably H-agglutination. Some doubt however must be cast on the relevance of this work to our present knowledge because the taxonomic position of the strains used was not clear, although some of the strains were probably *S. marcescens* as we recognize it today.

Apart from the limited studies of Bruynoghe (1936) and Jan (1939) no investigations of the serology of the species were made until the work of Davis and Woodward (1957). They distinguished six O-antigen groups amongst 16 strains of *S. marcescens* with sera prepared against boiled broth cultures. Ewing *et al.* (1959) extended the number of O-antigen groups to nine and subsequently to 15 in 1962. These O-antigen groups were recognized internationally and formed the basis of the O-antigen scheme used for the type identification of *S. marcescens* from hospital sources throughout the world. Le Minor and Pigache (1978) found a further five O-serogroups, which brings the present total to 20 groups.

Thirteen H flagellar types were described by Ewing *et al.* (1959) and seven additional types were added by Le Minor and Pigache (1977) and Traub and Fukushima (1979a).

To the authors' knowledge no typing schemes based on K-antigens or fimbriae (pili) of serratia have been reported in the literature.

IV. Heat-stable antigens

A. Characteristics of O-antigens

In the Enterobacteriaceae the structure of the lipopolysaccharide (LPS) determines the O-(somatic) antigen character, reistance to phagocytosis and to the bactericidal effect of normal human serum. The O-antigens of many enterobacteria have been studied in detail, and comprehensive reviews on their structure and function have been prepared by Freer and Salton (1971) and Nikaido (1973).

O-antigens are heat-, acid- and alcohol-stable and reside in the outer membrane above the rigid cell wall of the bacterium. The LPS consists of three distinct functional components. These are lipid A, an oligosaccharide complex or core region and an O-specific polysaccharide repeat unit. In

Salmonella the O-specific chain has a different composition in each of the O-antigen groups, but the core structure is similar in all O-antigen groups. The core region consists of a short oligosaccharide chain attached via ketodeoxyoctonate (KDO) to lipid A, a glucosamine, disaccharide containing long fatty acid chains. Lipid A is responsible for the endotoxic properties of LPS.

Relatively few publications dealing with the chemical structure of *S. marcescens* O-antigens have been reported. However, the studies of Tsang *et al.* (1974) and Wang *et al.* (1974) deserve mention as this group of workers dealt with the demonstration of O-specific substances in LPS preparations of the species.

Tsang *et al.* (1974) produced antisera by intra-peritoneal injection of animals with whole cells or purified whole endotoxin (PWE) prepared by trichloracetic acid extraction and suspended in Freund's adjuvant. The immunoelectrophoretic pattern given by PWE revealed three precipitin arcs. The line of precipitation given by an antigen migrating towards the anode was identified as the free protein moiety of the endotoxin. The antigen that migrated rapidly towards the cathode was the O-specific side chain and the line of precipitation formed nearest the antigen well was the intact endotoxin. Thus trichloracetic acid extraction caused some cleavage of the protein moiety from the O-specific side chain.

After purification by gel-filtration chromatography on Sepharose 4B, PWE gave a single line of precipitation in immunoelectrophoresis close to the antigen well (whole endotoxin). The extract had a high content of reducing sugars and D-glucosamine and relatively the same proportion of fatty acids and amino acids. Heptose accounted for 5.8% of the whole endotoxin and hexoses identified by paper chromatography included D-glucose and D-galactose.

LPS prepared by phenol water extraction was separated into two fractions by chromatography on Sepharose 4B. The higher molecular weight fraction was identified as LPS. It exhibited a single precipitin arc close to the antigen well in electrophoresis, and was highly toxic for mice. The second fraction showed a single fast migrating cathodal precipitin arc which was precipitated by antibodies to PWE. Its composition indicated that it corresponded to the O-specific side chain. It was not toxic for mice.

Further studies on the O-specific side chains were carried out on two strains: *S. marcescens* O:8 and *S. marcescens* strain Bizio (Tarcsay *et al.*, 1973; Wang and Alaupovic, 1973). Hydrolysis of endotoxin from both strains in acetic acid yielded partially degraded polysaccharide moieties, from which the O-specific side chain and an oligosaccharide core could be separated by gel filtration.

The O-specific chain of strain O:8 consisted of repeat units of a branched tetrasaccharide: D-glucose, D-galactose and N-acetylglucosamine in a molar ratio of 1 to 1 to 2. In contrast, the side chain of strain Bizio was a linear polysaccharide which contained as a repeat unit a D-glucose-L-rhamnose disaccharide (Tarcsay *et al.*, 1973).

Isolated oligosaccharide cores of *S. marcescens* strains O:8 and Bizio were analysed by Wang *et al.* (1974). The two strains differed quantitatively in their content of D-glucose and D-glucosamine. Each intact core contained two residues of KDO, one of these was bound through an acid-labile and the other through an acid-stable glycosidic linkage.

The *S. marcescens* cores studied so far consist of D-glucose, D-galactose, D-glucosamine, heptose and KDO, and therefore are similar to the cores of *Salmonella*, *Shigella* and *Escherichia coli*. However, *S. marcescens* cores appear to contain more D-glucose and heptose but less KDO than *Salmonella* and *E. coli* (Wang *et al.*, 1974).

B. O-antigen schemes

The O-antigen type scheme in universal use today is that of Ewing *et al.* (1962), which was developed at the Communicable Disease Center (CDC) Atlanta, Georgia, USA, and now called the Center for Disease Control. Fifteen O-antigen groups were recognized. Sedlák *et al.* (1965) used the first nine serotype strains of Ewing *et al.* (1959) and added two further types of their own. It is not known how the latter correspond to the six O-antigens not included in the study of Sedlák *et al.* The standard strains and code numbers of the CDC O-serotype set are listed in Table I.

Le Minor and Pigache (1978) described five new O-antigens from France. These had not reacted with the antisera of the CDC scheme. These antigens proved to be distinct by cross-absorption tests. Details of these strains are listed in Table II. It would seem logical to add the French type strains to the American O-antigen groups and workers should use all 20 *Serratia* O-antigens.

C. Preparation of antisera

Different authors have used various vaccines to stimulate O-antibodies in rabbits. As a rule higher titres of agglutinating antibody have been obtained with vaccines prepared from organisms grown in broth as opposed to agar.

There was general agreement that cultures should form smooth colonies on agar and give stable suspensions in saline before they were used for vaccines. Trypticase Soy Broth was universally used (Ewing *et al.*, 1962; Traub and Kleber, 1978; Pitt *et al.*, 1980), but the time of incubation varied with individual preference. Edwards and Ewing (1972), Traub and Kleber (1978) and Le Minor and Pigache (1978) recommended the use of young broth

TABLE I

O- and H-serotypes of strains of *Serratia marcescens* and code
numbers of the reference strains of the O-serotype scheme of
Ewing *et al.* (1962)[a]

O-antigen group	H-antigen group	CDC strain No.
1	5	866–57
2	1	868–57
3	1	863–57
4	4	864–57
5	1	867–57
6	3	862–57
7	10	843–57
8	3	1604–55
9	11	4534–60
10	8	1289–59
11	NM[b]	1914–63
12	9	6320–58
13	4	3607–60
14	12	4444–60
15	8	4523–60

[a] Available from Dr J. J. Farmer, Center for Disease Control,
Atlanta, Georgia, USA.
[b] Not motile.

TABLE II

O- and H-serotypes of strains of *Serratia marcescens* and code
numbers of the additional O-antigen group reference strains of
Le Minor and Pigache (1978)[a]

O-antigen group	H-antigen group	Strain No.
16	NM[b]	687
17	4	374
18	16	333
19	14	451
20	12	785

[a] Available from Dr S. Le Minor, Service des Enterobacteries,
Pasteur Institute, Paris, France.
[b] Not motile.

cultures (6–7 h at 37°C) inoculated with many single colonies of the type strain; Pitt *et al.* (1980) found that an 18-h culture at 32°C was satisfactory. Some variation in the time of heating at 100°C to destroy heat-labile components has been reported.

As a rule earlier studies favoured the boiling of cultures for 2.5 h, but both Traub and Fukushima (1979b) and Pitt *et al.* (1980) preferred 1-h treatment for the preparation of vaccine.

The immunization schedule was short, seldom exceeding three weeks in which time five injections were given at equal intervals. The CDC method recommended intravenous injection, and most workers agreed with this choice of route. However, Le Minor and Pigache (1978) advocated the use of an initial subcutaneous injection followed by intravenous injections.

At the Central Public Health Laboratory (CPHL), London, we used a method very similar to the CDC method except that we found the final two doses of 4 ml caused much local inflammation in the rabbits' ears, sometimes accompanied by general malaise and even death. We modified the procedure after experimentation (Pitt *et al.*, 1980) and used smaller final doses. Recently in our laboratory (unpublished data) we compared various ways of preparing broth vaccines in relation to time and temperature of incubation, doses injected and length of the immunization period. Our results showed that the two most important factors to consider were the number of cells per millilitre of the vaccine and the type of medium used to grow the strain. Nutrient Broth (Oxoid No. 2) was superior to culture in Trypticase Soy Broth although the strains grew less luxuriantly. The broth culture was concentrated by centrifugation. The reason for the excellence of the nutritional broth antigen may be due to production of less O-masking substances such as capsules. This feature was borne out by the colony appearance of type strains on nutrient agar or Trypticase Soy Agar (TSA). On the latter the colonies grew to give a glossy appearance, and capsules were easily seen under the light microscope, whereas on nutrient agar colonies appeared relatively dull and capsules, although evident in some strains, were seen on fewer cells.

Traub and Kleber (1978) compared a variety of O-antigen preparations for the immunization of rabbits. These were (1) TSA grown, 30-min boiled cells; (2) Trypticase Soy Broth grown, 2.5-h boiled cells; (3) ethanol-dried cells; (4) acetone-dried cells; (5) ethanol-acetone dried cells (Roschka, 1950) and (6) lipopolysaccharides extracted with trichloracetic acid (TCA-LPS). They found that marginally higher titres were obtained with anti-LPS sera, but in relation to the published data for other enterobacteria titres were generally poor. The agglutinin response of rabbits varied considerably depending on the antigen used for immunization, but no single preparation consistently gave higher titres for all the type strains than any other. Indeed, some type strains were poorly immunogenic in that they stimulated the production of only low titres of O-agglutinins regardless of the method used to prepare the vaccine.

The main features of the published methods for the preparation of O-antisera are summarized here.

1. CDC method (Edwards and Ewing, 1972)

Cultures are plated on infusion agar with a blood agar base and incubated at 37°C overnight. Smooth colonies are picked and put in Trypticase Soy Broth and incubated for 5 h at 37°C. Then they are boiled for 2.5 h. Rabbits are injected in the marginal ear vein at three- or four-day intervals with the following volumes: 0.25, 0.5, 1.0, 4.0 and 4.0 ml. The rabbits are bled one week after the last injection.

2. CPHL method

Twenty millilitres of Nutrient Broth is inoculated with five colonies from a nutrient agar plate and incubated at 32°C for 18 h. The culture is boiled for 1 h and centrifuged after which the cells are resuspended in 10 ml of saline. Rabbits are injected intravenously twice weekly with 0.5, 1.0, 1.5, 2.0 and 2.0 ml. The rabbits are test bled and if the titre is satisfactory (i.e. 160 or greater) the animal is bled out the same day. If necessary, another injection of 2 ml is given and after three days the serum is tested for elevated antibody titre and the animal is exsanguinated.

3. Method of Traub and Fukushima (1979b)

The type strain is plated on chocolate blood agar and incubated at 35°C for 18 h. Ten smooth colonies are inoculated into 30 ml of Trypticase Soy Broth and incubated at 35°C for 5–7 h. The culture is boiled for 1 h and the cells are centrifuged and resuspended in 6 ml of broth to give a five times concentration of the antigen. Rabbits are given five intravenous injections of 0.1, 0.2, 0.4, 0.8 and 0.8 ml at four- to five-day intervals. The animals are bled out seven to eight days after the last injection.

4. Method of Le Minor and Pigache (1978)

Rabbits are immunized with five injections of a broth culture (2×10^9 cells per millilitre) which has been grown in shaker culture for 5 h at 37°C and then boiled for 2.5 h. A primary injection of 0.5 ml of vaccine is given sub-cutaneously followed by four intravenous injections of 0.5, 1.0, 2.0 and 4.0 ml at four-day intervals. A test bleed is made seven days after the last injection and if the titre is satisfactory, the rabbit is bled out three days later. Otherwise a booster injection of 2.0 ml is given intraveneously and the serum is collected ten days later.

D. Agglutination test (O-antigen)

The method employed by a particular worker for the standard O-agglutination test will usually depend on the materials and equipment available in the laboratory. As a general rule it is good practice to prepare suspensions of the standard strains on a regular basis and to run periodical checks on the titres of the stock sera. Furthermore, the method used to determine the group of an unknown strain should mirror closely that used to characterize the stock sera. Stocks of sera should be kept frozen (preferably at $-20°C$), but an in-use aliquot of 2–5 ml may be kept at 4°C if preservative is added.

Edwards and Ewing (1972) stated that tube agglutination was more accurate than slide agglutination in *Serratia*. By their procedure, single tube agglutination tests were carried out on field strains. O-antiserum pools were used for the primary test in single tubes with a boiled log-phase culture as antigen; the final dilution of each antiserum component in the pool was standardized. Subsequently antigen suspensions were tested in individual sera contained in the pool in which the reaction occurred. Few cross-reactions were found and if present, confirmation of the O-antigen group could be obtained by simple titration in unabsorbed serum for the indicated groups.

Traub and Fukushima (1979b) carried out an extensive comparison of methods for the determination of O-antigens. As expected, their results confirmed the choice of a tube method utilizing a heated broth culture for routine O-antigen typing. They raised sera with five times concentrated Trypticase Soy Broth (heated 1 h at 100°C) and found that tube O-agglutination compared favourably with an indirect haemagglutination (HA) test with LPS-coated sheep red blood cells. Sera to seven of the 15 strains gave higher titres by the HA method. Tube agglutinations were superior to test in micro-titre trays with a cell suspension stained with saffranin, and fared equally well as triphenyltetrazolium chloride (TTC) stained cells.

An interesting innovation was utilized by Le Minor and Pigache (1978). They used a centrifugation method in which serum dilutions and bacterial suspensions were mixed in a ratio of 1 to 9 without prior incubation. The mixture was centrifuged and the cells were resuspended by giving the tube a sudden flick with the finger nail. If the cell deposit was not disturbed the cells were considered to be agglutinated. They found that this method was more sensitive than the conventional tube test.

Following the centrifugation tests with unabsorbed sera, the titres were determined with the appropriate individual serum by the same method. If the titre was within a single doubling dilution of the homologous titre of the serum and no cross-reactions were observed, this was taken as the serogroup. When cross-reactions were found in sera of known relatedness, the tests were repeated in absorbed sera; should this procedure still fail to reveal a specific type, serum was raised with the test strain and cross-absorption was performed with the serotype strain to determine the type identity of the O-antigen of the test strain.

Most workers found slide agglutination to be less reliable than tube tests (Edwards and Ewing, 1972; Sedlák et al., 1965) but Traub and Kleber (1978) obtained comparable results with both methods.

At CPHL we used to use reusable perspex agglutination trays (WHO trays) for the routine determination of O-antigens (Pitt et al., 1980). More recently the trend towards the use of disposable materials made it necessary for us to evaluate a micro-titre system. In practice this technique worked well for *Pseudomonas aeruginosa*, but some interesting results emerged from our initial experiments to determine the most suitable preparation for the typing of *Serratia*.

We compared three broth cultures, Brain Heart Infusion Broth (BHIB), Trypticase Soy Broth (TSB) and Nutrient Broth (NB). The O-antigen strains were grown overnight at 32°C and the whole culture boiled for 1 h. The suspensions were divided; one aliquot was retained untreated and the other was centrifuged and the cells were resuspended in saline to half of the original volume. Tests were performed in micro-titre trays with conventional U-shaped wells and all suspensions were incubated with serial serum dilutions for 3 h at 37°C and then at 4°C overnight.

Table III shows that washed NB cultures consistently agglutinated to higher titres. Satisfactory agglutination was produced in micro-titre plates and tests were easy to read with the aid of a plate viewer. However four sera gave low titres with all types of antigens, although in earlier plastic tray tests they exhibited titres in excess of 160 (Pitt et al., 1980). In general, the growth in NB was less dense than in the other two media but the reference strain for O:11 grew poorly and appeared granular in each type of broth. Some strains grew sparsely in BHIB (O:1 and O:2 in particular), but this was not true for all strains. Washing the cells produced a marked increase in titre regardless of the broth used. This suggests that masking substances were removed by washing.

The methods for performing agglutination are summarized here.

1. *Edwards and Ewing (1972)*

A TSB culture of the strain to be tested is grown at 37°C for 6 h and heated at 100°C for 1 h. One millilitre of diluted serum (1:100) is added to an equal volume of antigen in a 75 × 12 mm test tube and incubated at 48–50°C in a water bath overnight.

TABLE III

Comparison of broth cultures of *Serratia marcescens* O-antigen reference strains to provide a suitable antigen for micro-titre tray tests: homologous titres of agglutination of serum with unwashed and washed cells of grouping strains

Serum	Strain	Broth[a]	Titre Before wash	Titre After wash
O:1	O:1	BHIB	IG[b]	IG[b]
		TSB	<20	20
		NB	80	160
O:2	O:2	BHIB	IG[b]	IG[b]
		TSB	20	40
		NB	160	320
O:3	O:3	BHIB	320	1280
		TSB	160	1280
		NB	640	1280
O:4	O:4	BHIB	20	40
		TSB	<20	40
		NB	40	80
O:5	O:5	BHIB	20	20
		TSB	<20	20
		NB	40	160
O:6	O:6	BHIB	20	40
		TSB	20	40
		NB	40	160
O:7	O:7	BHIB	40	80
		TSB	20	40
		NB	80	320
O:8	O:8	BHIB	1280	1280
		TSB	1280	1280
		NB	1280	1280
O:9	O:9	BHIB	80	80
		TSB	40	80
		NB	80	1280
O:10	O:10	BHIB	1280	1280
		TSB	1280	1280
		NB	1280	1280
O:11	O:11[c]	BHIB	40	40
		TSB	80	80
		NB	1280	1280
O:12	O:12	BHIB	<20	20
		TSB	<20	20
		NB	20	160
O:13	O:13	BHIB	40	80
		TSB	20	40
		NB	80	320
O:14	O:14	BHIB	40	160
		TSB	20	40
		NB	40	160
O:15	O:15	BHIB	40	40
		TSB	20	40
		NB	40	160

[a] BHIB, Brain Heart Infusion Broth (Difco); TSB, Trypticase Soy Broth (Difco); NB, Nutrient Broth (Oxoid).
[b] IG, insufficient growth for antigen preparation, test not done.
[c] Granular sparse growth.

2. CPHL method

A single colony of the strain is inoculated into 3 ml of NB and incubated at 32°C overnight. The broth is boiled for 1 h, the cells are centrifuged and then resuspended in 1 ml of 0.85% saline; 0.025 ml of this antigen is mixed with 0.025 ml of serum dilution in a micro-titre tray. The tray is vigorously shaken and incubated at 37°C for 2 h and then at 4°C overnight. The tray wells are read according to the presence (not agglutinated) or absence (agglutinated) of a button of cells.

3. Le Minor and Pigache (1978)

A culture is grown in TSB with shaking at 37°C for 5 h and then boiled for 2 h. The opacity is adjusted by dilution with saline to approximately 7×10^8 cells per millilitre, 0.1 ml of serum diluted 1 : 200 is mixed with 0.9 ml of cell suspension in a test tube and centrifuged for 8–9 min at a speed sufficient to sediment the bacteria. The cells are resuspended by giving the tube a sudden flick with the finger nail. If the cells do not resuspend they are considered to be agglutinated.

4. Indirect haemagglutination test (Traub and Fukushima, 1979b)

The strain is grown on two TSA slopes at 35°C overnight and the cells are harvested in 0.85% saline and heated for 2 h to release LPS from the bacterial cells. The heated suspension is centrifuged and the supernatant is diluted 1:10 in phosphate-buffered saline, pH 7.5 (PBS). To 5 ml of 2.5% sheep red blood cells is added 5 ml of the 1:10 dilution of LPS. The mixtures are held at 35°C for 1 h after which the sensitized cells are washed three times in PBS. The O-antiserum is absorbed by adding 0.1 ml of unsensitized cells to 0.9 ml of serum and incubated at room temperature for 2 h before centrifugation. To 0.05 ml of absorbed serum dilutions in V-well micro-titre plates is added 0.05 ml of LPS-sensitized red blood cells. The trays are incubated at 35°C for 1 h followed by overnight incubation at 4°C, and the plates are read macroscopically for haemagglutination.

5. Slide agglutination (Traub and Fukushima, 1979b)

Slide agglutinations are carried out on glass slides which contain ten wells. To 0.05 ml of serum dilution or control saline is added 0.05 ml of diluted cell suspension (prepared from TSA incubated overnight at 37°C, washed once and boiled for 30 min). The slides are rotated mechanically (150 r/min) at room temperature for 5 min and then examined for agglutination.

6. *CPHL absorption of O-sera method*

Five colonies of the strain to be used for absorption of serum are inoculated into 5 ml of NB and incubated at 32°C for 5 h. Ten 9-cm diameter nutrient agar plates are seeded confluently with broth culture applied with a cotton wool swab. The plates are incubated at 32°C for 24 h and then washed off in 2–3 ml of saline per plate with the aid of a bent glass rod. The suspension is pooled, heated at 100°C for 2 h and centrifuged at 5000 g for 20 min. The cell deposit is resuspended in 4.5 ml of saline and mixed thoroughly with 0.5 ml of undiluted serum. After incubation at 37°C in a water bath for 4 h, the mixture is held at 4°C overnight and centrifuged. The supernatant is filtered through a sterilizing grade (0.45-μm) membrane filter and stored at 4°C.

7. *Pooled O-sera*

The recommended construction of pooled O-sera for agglutination (B. R. Davis, personal communication) is shown in Table IV. In preparing polyvalent sera, minor cross-reactions should be recognized as the combined effect of different cross-reacting antibodies may lead to confusing results in tests on field strains. Therefore single serum components should not necessarily be added to the pooled serum in volumes calculated on the basis of their individual working dilutions (Lányi and Bergan, 1978). Strains of *S. marcescens* which have reacted with pooled O-serum have in some cases not reacted with any of the specific constituent sera on subsequent testing.

TABLE IV

Composition of *Serratia marcescens* pooled O-sera

Pool number	Serogroups included
I	O:1
	O:2
	O:3
	O:4
	O:5
II	O:6
	O:7
	O:12
	O:13
	O:14
III	O:8
	O:9
	O:10
	O:11
	O:15

8. *Antibody class of agglutinin in O-sera*

Traub and Fukushima (1979b) attempted to identify the class of immunoglobulin responsible for the O-agglutination in rabbit sera. Both 2-mercaptoethanol (2-ME) or dithiothreitol (DTT), two reagents which mediate cleavage of disulphide bonds in immunoglobulin IgM and thus inactivate IgM agglutinogens, failed to reduce the O-agglutinin titres of rabbit sera. In contrast, the O-titre of a human control serum was reduced by treatment with either reagent. Furthermore, multiple absorptions with heat-killed *Staphylococcus aureus* strain Cowan I, which is rich in protein A and therefore binds immunoglobulin IgG, did not decrease the homologous O-antigen titre of the rabbit serum.

E. Cross-reactions of O-antisera with heterologous type strains

To determine the extent of cross-reacting antibody in O-sera, the sera are titrated by agglutination with the homologous and all heterologous type strains. Although some workers have reported extensive cross-reactions amongst the O-antigens of *S. marcescens* (Traub and Fukushima, 1979b), others have found comparatively few (Pitt *et al.*, 1980). Obviously some differentiation should be made between significant and insignificant reactions, and it is a good policy to note only those sera which exhibit reactions at the in-use dilutions of the serum. In this way the number of heterologous reactions is minimized and the number of sera that need to be absorbed is reduced.

Table V lists the cross-reactions reported by various authors and those shown in parentheses occurred within two doubling dilutions of the homologous titre or necessitated absorption of the type serum. Table V confirms the lack of agreement amongst laboratories as to the extent of cross-reactions given by particular sera. The reason for the major discrepancies between laboratories may be the different methods of preparation of antigen for immunization of rabbits and the various test procedures employed.

1. *Cross-reactions between O:6- and O:14-sera*

The antigens of the type strains O:6 (O:6;H:3,CDC 862-57) and O:14 (O:14;H:12,CDC 4444-60) are related but not identical. Traub and Kleber (1978) found that antisera to both groups could be absorbed to render specific sera of low titre, but these absorbed sera failed to agglutinate clinical strains originally designated to either group.

Subsequently, they examined strains representative of serogroups O:6 and O:14 to determine which could be used to prepare monospecific sera of higher titre. They include other serogroup reference strains in the Edwards and

TABLE V

Cross-reactions of O-antisera within heterologous type strains as reported by various authors

O-serum	Cross-reactions of stated sera with type strains as reported by			
	Edwards and Ewing (1972)	Traub and Fukushima (1979b)	Le Minor and Pigache (1978)	CPHL[a]
O:1	None	(O:4)	None	None
O:2	O:3	O:3	(O:3)	O:3
O:3	O:2	O:2	(O:3)	None
O:4	None	(O:1), O:6, O:7	None	O:3
O:5	None	O:4, O:13, O:15	None	None
O:6	O:7	O:4, (O:7), O:14	(O:14)	None
O:7	O:6	O:4, O:6	(O:6)	O:14
O:8	None	None	None	None
O:9	O:2, (O:3)	(O:4), O:15	None	None
O:10	O:8, O:9, O:15	None	None	None
O:11	(O:10)	None	None	O:10
O:12	O:14	O:4, O:14	(O:14)	O:4
O:13	O:8	(O:5)	(O:19)	O:8
O:14	O:6, (O:12)	(O:6)	(O:6)	(O:6)
O:15	None	None	None	O:3
O:16	NS	NS	None	NS
O:17	NS	NS	(O:19)	NS
O:18	NS	NS	(O:10)	NS
O:19	NS	NS	(O:1), (O:13)	NS
O:20	NS	NS		NS

The parentheses indicate significant reactions exhibiting titres within two doubling dilutions of the homologous titre or those which necessitated absorption of the serum.
NS, not studied.
[a] By modified micro-titre method adopted since Pitt *et al.* (1980).

Ewing (1972) set, e.g. O:6;H:8, O:6;H:10 and O:14;H:9 (H-type strains CDC 877-57, 2420-57 and 1783-57 respectively). None of these strains achieved the objective of providing monospecific O:6- and O:14-sera which would agglutinate all the original isolates of these groups. Some success was obtained with a clinical isolate Sli, as antiserum to this strain absorbed with the original serogroup reference strain O:14;H:12 agglutinated strain O:6 but not O:14, but it still failed to react with one of two representative field strains.

Part of Traub and Kleber's problem was their inability to prepare a high titre O:6-serum. This was attributed to two possible factors (a) the O-antigen of this group might have been immunosuppressive as has been observed in mice (Field *et al.*, 1970) and (b) the O-agglutinability of the group could have been impaired by prolonged boiling. Treatment of suspensions at 100°C for 2.5 h significantly reduced the homologous titre in comparison to 100°C for 30 min.

Le Minor and Pigache (1978) supported the conclusion that the two groups were related but as a result of the superior titres of their sera (eightfold higher) were able to clearly define the antigenic structure of the serogroup reference strains. Based on cross-absorption they proposed the following antigenic formulae for the serogroup reference strains:

Strain O:6;H:3   (CDC 862-57) $= O:6_1 6_2$
Strain O:6;H:8   (CDC 877-57) $= O:6_2 6_3$
Strain O:14;H:12   (CDC 4444-60) $= O:6_2$

It appears therefore that a single minor O-antigen factor common to both groups ($O:6_2$) is responsible for the cross-reaction. In practice, a laboratory intending to prepare typing sera for routine use should obtain both O- and H-antigen reference strains in order to differentiate between strains of O:6 and O:14.

By the system of Le Minor and Pigache (1978), a strain was considered to be O:14 if it was agglutinated by both unabsorbed O:6-sera and O:14-sera and not agglutinated by O:6;H:3-serum absorbed with strain O:14;H:12. They concluded that the definition of O:14 depended on the strain chosen for the preparation of vaccine.

In summary, the O-sera of Le Minor and Pigache (1978) for the subdivision of O:6/O:14 are as follows.

1. Factor 6_1-serum is prepared with strain O:6;H:3 (CDC 862-57) and absorbed with strain O:6;H:8 (CDC 877-57).
2. Factor 6_2-serum is prepared with strain O:14;H:12 (CDC 4444-60) and absorbed with strain 572 (Pasteur Institute, Le Minor and Pigache, 1978).
3. Factor 6_3-serum is prepared with strain O:6;H:8 and absorbed with strain O:14;H:12.

2. Subdivision of O:9-antigen group

Le Minor and Pigache (1978) described two factors within this O-antigen group. By cross-absorption with sera prepared against the standard reference strain O:9;H:11 (CDC 4534-60) and a neotype strain O:9;H:8 (CDC 2870-67) they found that the strains, although related, contained distinct minor factors. Strain O:9;H:11 had another surface antigen called k which was not present on O:9;H:8. This antigen was stable at 100°C but was destroyed at 120°C. The proposed formula of the two strains was:

O:9;H:11 $= O:9_1 9_2 k;H:11$
O:9;H:8 $= O:9_1;H:8$

3. *Heat-stable common antigen*

In the same study Le Minor and Pigache (1978) found an antigen which was common to O serogroup strains O:12 and O:14 and a strain with the serotype O:13;H:17. Agar cultures of types O:12 and O:14 dissociated into two colonial types: iridescent and non-iridescent. The latter when heated at $100°C$ was agglutinated by serum to the common antigen, whereas heat-treated iridescent colonies were not. Antiserum to a particular strain (no. 572) failed to agglutinate $100°C$ heated suspensions of types O:12 or O:14 but agglutinated cells of these strains when they had been heated at $120°C$. This was in direct contrast to the k-antigen described in strain O:9;H:11.

Traub (1978) observed phenotypic variation amongst a small number of clinical isolates originally designated to the serogroup O:6/O:14, which had undergone O:6- to O:14-antigen variation. He also found that a strain of the serotype O:14;H:12 (CDC 874-57) dissociated into colonial variants, which were agglutinated by either O:6- or O:12-antisera. These cross-reactions, which are not normally found with the serotype strains O:12 and O:14, may well be due to the presence of a common antigen. Pitt *et al.* (1980) found a small number of strains that were designated O:12/O:14 because the specific type could not be determined simply by titration in the appropriate antisera or by cross-absorption.

This phenomenon has been recognized for some years in *P. aeruginosa* where isolates from clinical material that react with two or more antisera prepared against antigenically unrelated groups have been called polyagglutinable (PA). Pitt and Erdman (1977) demonstrated that the specific type of PA strains of *P. aeruginosa* could be determined using antisera absorbed with a strain rich in the common antigen. Interestingly, these workers found that the class of immunoglobulin directed against the common antigen was of only the IgM class, whereas specific antibody to the O-antigen was present in classes IgG and IgM.

4. *Cross-reactions of O-antigens with other genera*

Kauffmann (1954) described an O-antigenic relationship between the antigens of *E. coli* serogroup O:19a and a strain of *Serratia*. This cross-reaction was investigated by Ewing *et al.* (1959) who also identified other cross-reacting antigens between the two genera. These included *S. marcescens* O:10 and O:6 cross-reacting with *E. coli* O:19a and *S. marcescens* O:8 with *E. coli* O:18a-, 18c- and 38-antigens.

190   T. L. PITT AND Y. J. ERDMAN

Furthermore Ewing and co-workers prepared broth cultures of a number of *Aerobacter* (now *Enterobacter*), *Hafnia* and aberrant coliforms and tested them with *Serratia* O-sera. Their results are summarized in Table VI. They did not pursue these studies further in view of the lack of information at that time regarding the antigenic structure of members of the *Aerobacter* group in particular. However, considering the recent emergence of *Serratia* and *Enterobacter* in hospital-acquired infections, there is a need for definitive studies in this area.

TABLE VI

Cross-reactions of *Serratia marcescens* O-sera with strains of *Enterobacter, Hafnia, S. liquefaciens* and other coliforms

Species[a]	Cross-reaction detected with *Serratia* O-serum
E. aerogenes	2, 5, 9
E. cloacae	9
Enterobacter sp.	1, 2, 3, 5, 6, 7, 8, 9
Hafnia	1, 9
Serratia liquefaciens	2, 3, 6, 7, 9
Coliforms	None

[a] *Enterobacter* described by Ewing *et al.* (1959) as *Aerobacter*; *Serratia liquefaciens* was termed *A. liquefaciens*.

F. Distribution of O-antigen groups in clinical specimens

An accurate assessment of the distribution of O-antigen serogroups in clinical isolates is difficult as few large series of cultures from national centres have been studied. Many publications report the distribution of types amongst cultures isolated locally which invariably contain a high proportion of cultures from a particular outbreak; the frequency of particular groups such as O:14 is therefore artificially increased.

A brief summary of the occurrence of the more common O-antigen groups is shown in Table VII. This illustrates the uneven distribution of the serogroups, and the high rates of typability experienced at the centres cited.

Serogroup O:14 was most frequently found and on average accounted for over 35% of all strains typed. The distribution of the other groups varied considerably in different countries but serogroups O:2, O:3, O:4, O:6 and O:11 were relatively common. Le Minor and Pigache (1978) found that over 75% of their strains were agglutinated by only three sera, O:14, O:13 and O:3. This lack of discrimination of O-serogrouping had also been experienced by other workers. Wilfert *et al.* (1970) found that their three most common groups

TABLE VII

Percentage distribution of the *Serratia marcescens* O-antigen groups of Ewing *et al.* (1962) amongst clinical specimens in different countries[a]

Country (Reference)	Number of strains	O-antigen serogroups															Typable (%)
		1	2	3	4	5	6	7	8	9	10	11	12	13	14	15	
USA (Wilfert *et al.*, 1970)	95	26			20	9	3					19			6		88
Rumania (Negut *et al.*, 1975)	116		5	6		5									51		96
USA (Rubin *et al.*, 1976)	86		10		4	5					8	3			56		93
UK (Anderhub *et al.*, 1977)	178			14		3	26		13						35		93
France (Le Minor and Pigache, 1978)	583			9						5				25	43		>95[b]
Europe (Pitt *et al.*, 1980)	273		13		9								4		69		99
USA (Young *et al.*, 1980a)	220	14	4		20										51	3	98

[a] Most common types only reported.

[b] Includes the five additional O-antigen groups of these authors (Table II).

accounted for over two-thirds of all isolates and Anderhub *et al.* (1977) reported that 75% of their isolates fell into three groups.

At CPHL we have attempted to subdivide the serogroup O:14 by subtyping with minor O-antigens distinguished by cross-absorption of sera prepared against four strains of this group isolated in different situations. This approach has not been successful because of the poor residual titres (20–40) of sera after absorption and has led us to conclude that O:14 is a homogeneous group, although the work of Le Minor and Pigache (1978) suggested that subdivision is possible given high-titred sera.

V. Heat-labile antigens

S. marcescens is a peritrichous flagellate organism which is actively motile in broth and soft agar cultures. Like other enterobacteria, the flagella are composed of a helical array of identical protein subunits or flagellin. The molecular weight of flagellin from a strain of *S. marcescens* has been reported to be approximately 40 000 by McDonough and Smith (1976).

Enterobacterial flagella are highly antigenic and stimulate their own specific antibody. Flagella are heat- and acid-labile being inactivated by temperatures in excess of 60°C and by 0.1 M hydrochloric acid. Their agglutinogenic properties are also destroyed by alcohol and protein-solubilizing agents such as urea and sodium dodecylsulphate.

Fimbriae are also present in *Serratia* (Fig. 1) and elicit a high level of antibody response in rabbit H-antisera prepared against whole cells. No attempt has been made to type *Serratia* on the basis of types of fimbriae, but the antibody to fimbriae in H-antisera might interfere with the determination of H-types (Pitt *et al.*, 1980).

Ewing *et al.* (1959) described 13 H-type antigens in *S. marcescens* and seven additional types were described by Traub and Kleber (1977) and Le Minor and Pigache (1978). The type numbers and strain designation of the 20 H-types are given in Tables VIII and IX. To avoid confusion between the two sets of additional H-types, the type strains for seven new H-antigens were agreed by the authors and comparison of their strains is shown in Table IX.

A. Preparation of H-antisera

A necessary prerequisite for the preparation of high-titred sera for *S. marcescens* is the enhancement of motility and development of H-antigen by passage through semi-solid agar. Ewing *et al.* (1959) passed cultures intended for serum production first through 0.2% soft nutrient agar and subsequently through 0.4% agar.

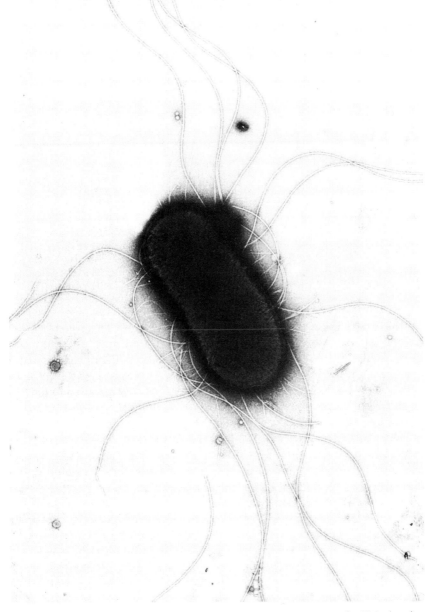

Fig. 1. Electron micrograph of *Serratia marcescens* serotype strain H:1 showing petrichous flagella and many fimbriae (× 23 000) (reproduced by courtesy of M. A. Gaston, CPHL).

TABLE VIII
H-serotype numbers and strain designation of the H-type strains
of *Serratia marcescens* studied by Ewing *et al.* (1959)[a]

H-type	O-type	CDC strain No.
1	3	863–57[b]
2	5	836–57
3	6	862–57[b]
4	4	864–57[b]
5	1	866–57[b]
6	5	680–57
7	4	841–57
8	6	877–57
9	14	1783–57
10	6	2420–57
11	5	827–57
12	14	874–57
13	5	2436–57

[a] Supplied by J. J. Farmer, Center for Disease Control, Atlanta, Georgia, USA.
[b] Same strain as in O-antigen group in Table I.

Most workers accept the general recommendation of CDC, which is based largely on the work of Ewing and collaborators, and utilize a log or stationary-phase broth culture of the vaccine strain to which formalin or phenol is added before it is injected intravenously into the rabbit. However, Traub and Kleber (1977) observed that phenolized saline (0.25% final concentration of phenol) failed to kill the majority of their strains, whereas a 0.3% concentration of formalin in saline was an efficient fixative.

The published techniques for raising H-sera vary only in the type of broth used and the amount of antigen injected; the resulting titres are usually high and, although these differ in magnitude between laboratories, it would appear that these factors do not significantly influence the quality of the H-sera obtained with the standard type strains.

However, H-sera raised against whole cell vaccines exhibit significant cross-reactions with heterologous strains and therefore need to be absorbed. At CPHL we have observed numerous cross-reactions with such sera which have proved difficult to remove by absorption. An alternative procedure in which flagella are detached from the bacterial cell by mechanical means and partially purified by ultracentrifugation has been employed to prepare H-specific sera. The details of the two major methods of preparing sera are as follows.

TABLE IX

Serotype H numbers and strain designation of the new H-antigens of Traub and Kleber (1977) and Le Minor and Pigache (1978)[a]

Previous designation		Present designation (agreed between authors)[a]
Traub and Kleber	Le Minor and Pigache	
H:14 (S326=O:13; H:14)	H:16 (451=O:19; H:16)	H:14 (S326=O:19; H:14)[b]
	H:15 (389=O:5; H:15)	H:15 (IP389=O:5; H:15)
H:17 (Schweiz=O:12; H:17)	H:14 (333=O:18; H:14)	H:16 (Schweiz=O:12; H:16)
H:16 (LFMuenchen=O:13; H:16)	H:17 (215=O:9; H:17)	H:17 (IP421=O:13; H:17)
	H:18 (250=O:12; H:18)	H:18 (IP250=O:12; H:18)
	H:19 (97=O:5; H:19)	H:19 (IP416=O:5; H:19)
H:15 (Sli=O:6; H:15)		H:20 (Sli=O:6; H:20)

[a] From data cited by Le Minor and Pigache (1978).
[b] Supplied by the Pasteur Institute, Paris.

1. *Whole cell vaccine from Ewing* et al. *(1959)*

The serotype strains are passaged through semi-solid nutrient agar (0.4% agar w/v) to enhance motility on six consecutive days. TSA is seeded with culture from the last passage and incubated at 37°C for 18 h. Phenol or formalin is added to a final concentration of 0.25% or 0.3% respectively, and rabbits are injected in the same manner as described for the preparation of O-antisera.

2. *Purified flagellar antigen*

An alternative procedure is to prepare antisera with purified flagella of the type strains as follows: six polypropylene trays of dimensions $30 \times 25 \times 5$ cm are used for each of the serotype strains. The trays are washed in a dilute detergent solution, rinsed thoroughly with distilled water and dried. They are covered with sheets of aluminium foil (35×30 cm) which are secured with autoclave tape and autoclaved at 121°C for 15 min. Each tray is filled with 750 ml of nutrient agar to a depth of about 3 cm. The agar surface is dried at 37°C for 1 h and 10 ml of a 4-h broth culture of the motile bacteria – after six subcultures in soft agar – is spread over the surface. The excess fluid is removed and the trays are incubated at 25°C for 72 h. The bacterial growth is harvested in 150 ml of 0.5% (w/v) phenol saline and left to stand for 1 h in an ice bath. The suspension is homogenized in a pre-cooled stainless steel chamber for 45 s at low speed and the cells are removed by centrifugation twice at 16 000 g for 45 min. Flagella and other small bodies are deposited at 40 000 g for 1 h at 4°C. The clear gelatinous pellet is covered with 5 ml of Tris-HCl buffer pH 8.0, left to disperse overnight at 4°C, and centrifuged at 40 000 g for 1 h. The final deposit is dispersed in 5 ml of the same buffer and examined to assess purity in the electron microscope by a negative stain with 2% (w/v) sodium phosphotungstate solution. The flagellar suspension is freeze-dried and reconstituted in saline to a concentration of 0.5% (w/v). The protein content is measured by the method of Lowry *et al.* (1951) with bovine serum albumin as a standard and rabbits are given four injections at three- to four-day intervals of 50, 100, 200 and 400 μg of protein suspended in 1 ml of saline. Rabbits are bled after 21–25 days.

B. Agglutination and immobilization tests

Two procedures have been used to demonstrate H-antibody in sera raised against either whole cells or purified flagella: agglutination of formalin- or phenol-treated cells or immobilization of the bacteria in soft agar.

1. Agglutination

(a) *Slide agglutination from Traub and Kleber (1977)*. The test is performed with serum dilutions in PBS pH 7.5 and an overnight TSB culture diluted with an equal volume of 0.6% (w/v) formol saline. One drop (0.05 ml) of serum dilution is mixed with 1 drop of broth culture on a glass microscope slide which is gently rotated for 3 min. Agglutination is observed with the naked eye.

(b) *Tube H-agglutination from Edwards and Ewing (1972)*. The strain is cultivated in TSB for 5 h having been diluted with an equal volume of 0.5% (w/v) phenol saline. Equal volumes (0.3 ml) of the bacterial suspension and serum dilutions are mixed in Dreyer's tubes and incubated for 2 h at 50°C in a water bath. The water level in the bath should not exceed one-third of the tube depth as convection currents generated during incubation facilitate mixing of the bacteria in the serum. Agglutinated cells settle to give a cotton wool like precipitate and titres are expressed as the highest serum dilution that gives easily visible agglutination.

(c) *Micro-titre tray agglutination from CPHL*. Tray tests for H-agglutination are performed by a method similar to that for O-agglutination described previously. For H-antigen tests, however, 3 ml of an uncentrifuged overnight nutrient broth culture is mixed with 0.04 ml of formalin and this suspension (approximately 1×10^8 cells per millilitre) is added to serum dilutions in trays.

2. Immobilization

(a) *Procedure from Le Minor and Pigache (1977)*. Tests are carried out in a soft agar medium containing mannitol. Nitrate is included to inhibit the production of gas during fermentation of the sugar and a pH indicator is incorporated to monitor the formation of acid as the bacteria migrate throughout the medium. The constituents of the medium are:

Tryptic peptone	20 g
Agar	4.5 g
Mannitol	2.0 g
Potassium nitrate	1.5 g
Phenol red (1%, w/v, aqueous)	4.0 ml
Distilled water	1 litre

The pH is adjusted to 7.4 and the medium is dispensed in 2 ml volumes in sterile capped tubes (92 × 13 mm). The dispensed medium is melted in boiling water and cooled to 50°C before a standard volume of diluted serum is added. When set, the medium is inoculated centrally using a platinum wire. A tube not containing serum is also seeded with culture to serve as a control. The tests are incubated at 37°C for 24 h and immobilization of the culture by H-antibody is evident by the restriction of growth to the inoculum line and lack of change of pH. Motile bacteria (control tube) grow confluently throughout the agar and the medium changes from red to yellow.

(b) *Procedure and modification of Pitt* et al. *(1980)*. One-millilitre volumes of 0.4% nutrient agar is pipetted into sterile tubes and allowed to set at 4°C. Equal volumes of soft agar containing dilutions of H-serum are carefully layered over the control butt of agar. After gelation, cultures are seeded directly through both layers of agar. After incubation at 35°C overnight, immobilization of the culture in the upper serum agar layer is directly compared with the motile culture in the lower layer.

(c) *Procedure and modification of Traub and Kleber (1977)*. Three millilitres of TSA (0.4%, w/v, agar) is dispensed into a metal-capped tube (100 × 13 mm) and held at 56°C in a water bath for 1 h. Heat-inactivated unabsorbed H-sera are diluted in PBS, pH 7.5 and 0.05 ml of the appropriate dilution is added to the soft agar aliquots. The tubes are shaken gently and transferred to 4°C for 20 min. Cultures are inoculated with a straight wire to a depth of about 2 mm and the tubes are incubated at 35°C overnight. Immobilization of the culture by a particular serum is defined as restriction of migration of the culture to less than half the total height of the medium in comparison with a control tube not containing serum.

C. Agglutination titres and cross-reactions of H-sera

In general, workers have found that the agglutination titres of H-sera raised against whole cells of *S. marcescens* were high and ranged from about 3000 to greater than 30 000. The method used by individuals probably influenced the magnitude of titres and the highest titres reported were those given by slide agglutination (Traub and Kleber, 1977).

By tube agglutination, Ewing *et al.* (1959) at CDC originally found significant cross-reactions between H:8-serum and strains H:9, H:10 and H:3; and H:13 and H:6. In Table X showing cross-reactions issued by the CDC in 1973, additional reactions were reported between serum H:9 and H:6; and H:10 and H:8. The cross-reactions cited by other authors are compared in Table X, which shows relatively good agreement. Pitt *et al.* (1980) reported extensive

TABLE X

Cross-reactions of H-antisera of *Serratia marcescens* within heterologous type strains as reported by various authors

	Cross-reactions of stated sera with type strains as reported by		
Serum	Ewing *et al.* (1959	Traub and Kleber (1977)[a]	Le Minor and Pigache (1977)
H:1	H:11	None	H:12, H:18
H:2	H:13	H:4, H:11	H:11, H:19
H:3	H:10	(H:10)	H:10
H:4	H:13	(H:12)	(H:7), (H:12)
H:5	None	None	None
H:6	H:13	H:13	H:11, (H:13)
H:7	H:11	None	H:16
H:8	H:6, (H:9), H:10	H:5, (H:9), H:10, H:13	H:5, (H:9), (H:10)
H:9	H:8	(H:8), H:10	H:2, H:6, (H:8), H:14
H:10	H:3, H:8	(H:3), H:6, H:8	H:3, H:8, H:9
H:11	H:2	H:2, H:9	H:2
H:12	None	H:3, H:4	H:4, H:9
H:13	(H:6), H:7	(H:6), H:9, H:10, H:11	H:6
H:14	NS	NS	None
H:15	NS	NS	H:11, H:17, H:19
H:16	NS	NS	H:7
H:17	NS	NS	H:15
H:18	NS	NS	H:17
H:19	NS	NS	H:2, H:7, H:16

The reactions in parentheses indicate significant reactions exhibiting titres within four doubling dilutions of the homologous titre or those which were above the lower range of titres quoted.

[a] By immobilization test.

NS, not studied.

cross-reactions in nine of their 13 H-sera. Three cross-reacted with five strains. After absorption with suspensions of heterologous strains, the homologous titres of some sera were drastically reduced. Surprisingly, most sera were rendered type-specific by a single absorption with strain H:1. These results appeared to indicate that the sera contained antibody in high titre to other antigens in addition to flagella. Antibody to fimbriae was present in titres of similar magnitude to antibodies to flagella.

Recent studies at CPHL using nutrient broth suspensions in micro-titre trays did not confirm the earlier findings of Pitt *et al.* (1980) who used dense suspensions (5×10^9 cells per millilitre) in perspex (WHO) agglutination trays (i.e. different cell numbers per millilitre for micro-titre and perspex trays). By the micro-titre tray method using the same batches of H-sera as those used in

TABLE XI

Comparison of the cross-reactions of H-sera to the 13 type strains of *Serratia marcescens* (Ewing *et al.*, 1959) obtained by agglutination in (WHO) plastic trays and micro-titre trays

Titre of agglutination (micro-titre trays in parenthesis)

Serum	H:1	H:2	H:3	H:4	H:5	H:6	H:7	H:8	H:9	H:10	H:11	H:12	H:13
H:1	3200 (12 800)	1600	—[a]	1600	1600	—	—	1600	800	—	— (800)	—	—
H:2	1600 (400)	3200 (25 000)	—	—	—	—	—	1600	1600	—	— (400)	—	—
H:3	800	1600	3200 (12 800)	—	—	—	—	800	—	800 (1600)	—	—	—
H-4	—	—	—	3200 (1600)	1600 (800)	—	—	—	—	—	—	—	—
H:5	1600	—	—	3200	6400 (25 000)	—	—	—	—	—	—	—	—
H:6	1600	—	—	—	800	3200 (25 000)	—	—	—	—	—	—	—
H:7	800	—	—	1600	400	—	1600 (6400)	—	—	—	—	—	—
H:8	1600	1600	800	—	—	—	—	6400 (6400)	1600 (3200)	—	—	800	—
H:9	—	1600 (400)	—	—	—	—	—	3200 (3200)	6400 (12 800)	—	—	—	—
H:10	—	—	800 (800)	—	—	—	—	800	—	1600 (6400)	—	—	—
H:11	—	—	—	—	—	—	—	—	—	—	3200 (3200)	—	—
H:12	—	—	—	—	—	—	—	—	—	—	—	3200 (6400)	—
H:13	—	—	—	—	—	—	—	—	—	—	—	—	3200 (6400)

[a] —, less than 200.

the earlier study, we found significantly less cross-reactions. Titres with homologous suspensions were usually higher than those obtained in perspex trays and heterologous titres were likewise lower (Table XI). Strain H:1 was agglutinated by six heterologous sera in plastic trays but by only one serum in micro-titre trays. These results suggested that the use of more concentrated suspensions of the type strains (used in the perspex tray method) led to increased cross-reaction possibly because of the presence of other antigens in detectable amounts.

The additional H-type reference strains H:14 to H:20 were compared with the 13 CDC type strains by both Le Minor and Pigache (1977) and Traub and Fukushima (1979a). No significant cross-reactions were found.

A considerable improvement in the specificity of H-agglutination can be achieved by using H-sera prepared against purified flagella of the type strains. We have evaluated a selection of sera prepared in this way and found titres comparable with those obtained with sera raised against whole cells. Antibody titre to non-motile variants of the vaccine strains was low and unlikely to interfere in tests at the in-use dilution of the serum. Of five sera prepared against flagella (of strains H:1, H:2, H:3, H:5 and H:6) we found significant cross-reactions only with H:3-serum and strain H:10 which differ from cross-reactions of sera produced with whole cell antigens (Table XI).

D. Specificity of immobilization tests

The immobilization test originally devised by Le Minor and Pigache (1977) provides the most satisfactory method for the specific detection of H-antibody in sera containing antibodies to other cellular constituents as well. It was claimed that the test could be performed more quickly than agglutination but had the same specificity as the latter. The same conclusion was made by Traub and Kleber (1977) who have adopted it for routine use.

Pitt *et al.* (1980) reported that H:1, H:6, H:9 and H:12 titration end-points in soft agar were clear, but they found that four type strains grew poorly in the presence of H:1- to H:11- and H:13-sera. With hindsight, this may have been due to enhanced sensitivity of certain strains to preservatives such as thimerosal, and as a consequence preservative should not be added to H-sera intended for use in immobilization tests. We found the test to be more specific than agglutination, although this may have been due to the type of cell suspension used previously.

The obvious advantage of the immobilization test is that unabsorbed H-sera can be used on a routine basis making the accurate identification of H-antigens of clinical isolates a relatively simple task.

Traub and Fukushima (1979a) carried out tests similar to their earlier studies on O-antisera. They concluded that H-immobilizing antibody, although resistant to treatment with 2-ME and DTT reagents, was in the IgM class as dual absorptions of rabbit immune H-sera with *S. aureus* Cowan I did not significantly reduce the H-antibody titre of the sera.

E. Phase variation of H-antigens

Recently, Young *et al.* (1980b) demonstrated a phenomenon which appeared to resemble flagellar phase variation among some strains of *S. marcescens*. These workers determined the H-antigen type of field strains by tube agglutination using H-antiserum commercially prepared by Lee Laboratories (Grayson, Georgia, USA). This company has since discontinued the marketing of O- and H-sera (personal communication). After the H-type of the strain was determined, a Gard plate (Gard, 1937) that contained antiserum against the homologous H-type in a final dilution of 1:250 was prepared. The plate was inoculated with a TSB culture of the strain and incubated for 24 h at room temperature. If the culture was immobilized the H-type obtained by agglutination was confirmed.

If motility was not inhibited and growth away from the point of inoculation was evident, a broth was inoculated with bacteria from the outermost edge of the growth, incubated at 37°C for 5 h and then treated with phenol. Agglutinations were then performed to determine the antigenic type of the swarming outgrowth. Sera to the first H-type of the strain and the H-type of the swarm were both incorporated into a Gard plate and this was point-inoculated with the original culture. If motility was inhibited, two H-antigens were considered to be present. On rare occasions, when motility persisted even in the presence of two antisera, agglutination was repeated with the suspension of the motile bacteria and a third antiserum was incorporated with the other two sera into the Gard plate. They demonstrated that 21 of the 241 strains (8.7%) had more than one H-antigen. Of these, 19 were diphasic and two were triphasic. Approximately half of the strains with more than one phase antigen reacted with H:1-antiserum and one or more other type sera. The phases of strains appeared to be readily reversible as in *Salmonella* (Andrewes, 1922). Furthermore when single colonies were streaked onto plates and resulting colonies were tested separately, both phases could be demonstrated.

After storage in semi-solid agar for about six months, repeated tests showed that reversion to the original H-antigen type could only be demonstrated with difficulty. Young and co-workers concluded that this late reversion resembled the induction of phase described by Fife *et al.* (1960) rather than a simple alteration of phase.

Previously flagellar phase variation in *S. marcescens* had been discounted by Traub and Fukushima (1979a) who found that four clinical isolates originally reported by Young's laboratory to be phase variants consisted of cell populations of two distinct O-serotypes and bacteriocin types. They stated that during the examination of over 500 isolates of *S. marcescens* by tube immobilization, they never observed changes in H-antigen type indicative of diphasic variation.

At CPHL we have observed anomalous results in the course of H-antigen typing of clinical isolates which could possibly be explained by phase variation. Moreover, many of our H-antigen type strains were strongly agglutinated by H:1-serum, in particular when dense suspensions were used in perspex tray agglutination tests. It may be speculated that the increased cell concentration of our H-antigen suspensions facilitated the detection of phase variant antigens which may have been expressed less than the "true" H-antigen type. Clearly, this phenomenon requires further study in order to confirm the findings of Young *et al.* (1980b) and to document, if relevant, the phase antigens found in each serogroup in a manner similar to the situation for *Salmonella*.

F. H-typing from clinical material

1. *Frequency of types*

Like O-antigen serogrouping, H-antigens of clinical strains of *S. marcescens* has only been determined at relatively few specialist centres. In the studies reported, it is generally agreed that H-antigen typability is high if sera to the 13 CDC H-antigen types are used (Ewing *et al.*, 1959; Negut *et al.*, 1975; Pitt *et al.*, 1980). The original scheme of Ewing appears to be comprehensive and little would be gained by the addition of further types. In France, however, it has been necessary to introduce new H-antigen types since a considerable proportion of their strains do not belong in the 13 original types (Le Minor and Pigache, 1977; Minck *et al.*, 1980).

Ewing *et al.* (1959) were able to determine the H-antigen of all except two of the 115 motile strains from clinical specimens using their 13 antisera. Serogroups H:1 and H:2 were most frequent and accounted respectively for 27% and 21% of their cultures. The other H-antigen types were equally divided amongst the remainder of the strains, but types H:4, H:5, H:7 and H:11 were rare. A similar level of typability (99%) was reported by Negut *et al.* (1975) from their study of Rumanian strains, but type H:1 occurred in only about 10% of isolates, and types H:8 and H:12 were represented by more than half of the strains. Le Minor and Pigache (1977) reported type H:12 in excess of 40% in 427 motile strains. H:1 and H:4 were each equally represented (8%), but with the exception of type H:7 relatively few other strains were grouped by sera to the 13 CDC types; approximately two-thirds of their strains overall were typable with these sera. One of the new H-antigen factors (H:14 to H:19 in the 1977 study), H:17 accounted for 28% of all strains typed; the other new types were uncommon.

Eleven different H-antigen factors were detected by Pitt *et al.* (1980) amongst isolates from British and European hospitals (excluding France) and H:4 was the most frequent type found. Although the percentage typability of H-antigen typing was significantly lower than that obtained for O-antigen typing (84% versus 96%), the former distinguished more often between strains.

2. *Distribution of OH-serotypes*

Unlike O-antigen serogrouping, H-antigen typing of *S. marcescens* is unlikely to be carried out alone as most workers agree that O-antigen grouping forms the primary antigenic division. It is usually unnecessary to determine the flagellar antigen type of a strain of some uncommon O-antigen groups, as these tend not to be distinguished by H-antigen typing, amongst cultures from a single outbreak of infection. In this context therefore H-antigen typing is most important for further subdivision of strains within the more frequent O-antigen groups.

In most of the investigations reported one or two serotypes have been predominant. Type O:14;H:12 is ubiquitous and in five of the references cited in Table XII, it represented 16–50% of all the strains. In contrast, many serotypes are rare. It should be stated that the figures in Table XII are weighted in certain cases because large numbers of strains from a particular outbreak of infection with *S. marcescens* were included by some authors. This was especially the case with 1499 strains isolated from hospitals within the city of Strasbourg alone (Minck *et al.* (1980). It is not entirely surprising to find that most of these fell into two types: O:14;H:12 and O:13;H:17. Le Minor and Pigache (1978) also included a large number of strains from Strasbourg and most of these were also O:13;H:17.

There is insufficient data to indicate that any serotype may be especially associated with a certain type of disease or clinical syndrome. However, it was suggested by Minck *et al.* (1980) that serotype O:14;H:12 was mostly recovered from sputa, whereas serotype O:13;H:17 was mostly recovered from urine. Rubin (1980) commented that the high incidence of O:14 strains in sporadic infections, as well as in outbreaks may indicate an enhanced virulence among these strains. An alternative view is that the high isolation rate of serogroup O:14 merely reflects the overall predominance of this serogroup in the genus as a whole.

TABLE XII

Percentage distribution of *Serratia marcescens* O- and H-antigen types amongst clinical specimens[a]

Serotype	USA[b]	Rumania[c]	USA[d]	France[e]	UK[f]	USA[g]	France[h]
				Country			
O:1; H:4						6	
O:1; H:7						5	
O:2; H:1		5	7			3	
O:2; H:4	22						
O:2; H:5			3				
O:3; H:12		6		7			5
O:4; H:1	18		5	3		3	
O:4; H:4						10	
O:4; H:7						3.5	
O:4; H:12					8	3	
O:5; H:1	3	5	3				
O:5; H:3	4						
O:9; H:17				3.5			
O:10; H:11			3.5				
O:11; H:1			3.5				
O:11; H:4	12						
O:11; H:13	5						
O:12; H:9					3		
O:13; H:17				21			33
O:14; H:1			8				
O:14; H:2			4		21		
O:14; H:3						3	
O:14; H:4				7	36		4
O:14; H:5			10				
O:14; H:8		25					
O:14; H:10			3			4	
O:14; H:12	3	25	16	26		38	50
O:14; H:20				7			4
H NT[i]	6	<1	13	15	16	6	4
Number of strains	95	116	86	583	273	220	1499

[a] Only the most common types reported.
[b] Wilfert *et al.* (1970).
[c] Negut *et al.* (1975)
[d] Rubin *et al.* (1976).
[e] Le Minor and Pigache (1978).
[f] Pitt *et al.* (1980).
[g] Young *et al.* (1980a).
[h] Minck *et al.* (1980).
[i] NT, non typable by H-antisera.

G. Fimbriae

Fimbriae or pili (see preface to Volume 14 of this series) are filamentous appendages of bacteria, which are usually thinner and more numerous than flagella. Duguid (1968) described several different types of fimbriae based on their dimensions, distribution on the bacterial cell, haemagglutination properties and phage receptor characteristics. In *S. marcescens* fimbriae are distributed peritrichously and can be demonstrated by direct haemagglutination.

1. Haemagglutination test (Duguid et al., 1966)

A colony of the culture under test is grown in 5 ml of a nutrient broth at 37°C for 24 h. The cells are deposited by centrifugation and the supernatant is discarded. Phenol saline (0.1 ml) is added and the cells are resuspended. Fowl erythrocytes are washed three times in saline and adjusted by dilution to a concentration of 2% (v/v) in saline. A drop (0.05 ml) or erythrocytes is mixed with an equal volume of bacterial suspension on a white ceramic tile. The tile is rocked gently for 2 min and clumping of the erythrocytes indicates haemagglutination and is compared to a negative control with a known non-fimbriated strain. A negative haemagglutination test, however, does not confirm the absence of fimbriae. Further tests with tannic acid treated cells are necessary (Duguid et al., 1966).

To test for mannose-resistant haemagglutinin (Duguid et al., 1966) a drop of 2% (v/v) D-mannose in saline is mixed with the erythrocyte suspension before the bacteria are added. Failure to observe haemagglutination following the addition of mannose indicates mannose-sensitive fimbriae (type 1 fimbriae), whereas the haemagglutination of type 4 fimbriae is mannose-resistant.

2. Antibody to fimbriae in H-sera

Pitt et al. (1980) examined the H-antigen type strains of Ewing et al. (1959) for the presence of fimbriae by haemagglutination. Fimbriae were detected on 11 of the 13 type strains (Table XIII). Four bore the mannose-sensitive type 1 fimbriae and seven possessed the mannose-resistant type 4 fimbriae. Two strains (H:4 and H:10) did not agglutinate fowl erythrocytes, but the possibility of non-haemagglutinating fimbriae, e.g. type 2 of Duguid, was not investigated further. Repeated subculture on excessively dried agar plates rendered seven of the 11 strains non-fimbriated as judged by the loss of haemagglutination, but some strains continued to develop fimbriae after numerous subcultures.

TABLE XIII

Agglutination of fowl red blood cells by broth suspensions of the H-antigen type strains of *Serratia marcescens* before and after the addition of mannose

Haemagglutination of red blood cells by suspensions of type strains

Method of test	H:1	H:2	H:3	H:4	H:5	H:6	H:7	H:8	H:9	H:10	H:11	H:12	H:13
Without mannose	+[a]	+	+	−	+	+	+	+	+	−	+	+	+
With mannose	±	−	+	−	−	+	±	+	±	−	−	+	−
Fimbria type[b]	4	1	4	−	1	4	4	4	4	−	1	4	1

[a] +, strong agglutination; ±, weak agglutination; −, no agglutination (explained in text).
[b] Duguid *et al.* (1966).

To demonstrate that antibody to fimbriae was stimulated by the injection of whole cells into rabbits, as in the preparation of H-sera, the cross-reactivity of *fim*+ and *fim*− suspensions of selected type strains were compared by agglutination with the H-sera that had exhibited cross-reactions. The results (Table XIV) confirmed that antibody to fimbriae was responsible for some of the heterologous type reactions. For example, H:2-serum failed to agglutinate the *fim*− variant of strain H:1 and the agglutination titre of this variant with H:3-serum was significantly lower than the titre with the fimbriated variety of strain H:1. However, the *fim*− variant of strain H:5 exhibited the same agglutination reactions as the fimbriated form suggesting that neither flagella

TABLE XIV

Cross-agglutination of fimbriated and non-fimbriated variants of selected H-antigen type strains of *Serratia marcescens* by homologous and heterologous H-antisera (Pitt *et al.*, 1980)

| | Agglutination by stated antiserum of type strains | | | | | |
| | H:1 | | H:5 | | H:8 | |
Serum	*fim*+[a]	*fim*−[a]	*fim*+	*fim*−	*fim*+	*fim*−
H:1	+[b]	+	+	+	+	−
H:2	±	−	ND[c]	ND	+	−
H:3	+	±	ND	ND	+	−
H:5	±	−	+	+	ND	ND
H:6	±	−	+	+	ND	ND
H:8	+	−	ND	ND	+	+

[a] *fim*+, fimbriated; *fim*−, non-fimbriated.
[b] +, strong agglutination at titre; ±, weak agglutination at titre; −, no agglutination.
[c] ND, not done.

(immobilization data in Table XI) or fimbriae were responsible for the cross-reaction of this strain.

That antibody to fimbria could contribute to the cross-reactions in agglutination with H-sera was expected. Nowotarska and Mulczyk (1977) have demonstrated that high titres of antibody to fimbriae were obtained in sera raised against unheated whole cells of *S. marcescens*. They found no antigenic relationship between the fimbriae of *Serratia* and other members of the Enterobacteriaceae.

VI. Summary and proposal

It is relatively simple to prepare specific O-agglutinating sera of adequate titre and most strains of *S. marcescens* are typable by conventional tests. The results are reproducible when pairs of strains from the same patient are tested, but the relative predominance of a few serogroups (in particular O:14) suggests that O-antigen typing alone is insufficient to discriminate between isolates in epidemiological studies.

H-antisera prepared against whole bacterial cells may exhibit many heterologous agglutination reactions amongst the type strains, but not all of the reactions are due to flagella-specific antibody. Immobilization in agar tests by H-serum provides a convenient and reliable way for the accurate determination of the H-antigen type of clinical strains.

Bacteriophage and bacteriocin typing have also been used to subdivide the species. However, Pitt *et al.* (1980) showed that although the typability of clinical isolates was high, no single method was sufficiently discriminating to be used alone. The reasons for this were that some types could not be further divided or because less than complete reproducibility made it necessary to introduce "reaction differences rules" to allow for the method variability (Anderhub *et al.*, 1977). Therefore we propose that a combination of methods be used to type strains of *S. marcescens* in a hierarchical fashion as advocated by Meitert and Meitert (1966) who used O-antigen serogrouping for the primary classification of *P. aeruginosa* strains and subdivided the serological groups by phage typing.

Our results obtained from the typing of many routine isolates throughout the UK suggest that either H-antigen serotyping by immobilization or phage typing are the best alternatives as secondary methods to divide further strains of the same O-antigen group. Bacteriocin typing will suffice for this purpose, but it defines broad groups and it has not been as discriminating as the foregoing methods in this laboratory.

References

Anderhub, B., Pitt, T. L., Erdman, Y. J. and Willcox, W. R. (1977). *J. Hyg.* **79**, 89 102.
Andrewes, F. W. (1922). *J. Pathol. Bacteriol.* **25**, 505 521.
Ball, A. P., McGhie, D. and Geddes, A. M. (1977). *Quarterly J. Med.* **46**, 63 71.
Brooks, H. J. L., Chambers, T. J. and Tabaqchali, S. (1979). *J. Hyg.* **82**, 31 40.
Bruynoghe, R. (1936). *C. R. Seances Soc. Biol.* **121**, 891 893.
Davis, B. R. and Woodward, J. M. (1957). *Can. J. Microbiol.* **3**, 591 597.
Dodson, W. H. (1968). *Arch Intern. Med.* **121**, 145 150.
Duguid, J. P. (1968). *Arch. Immunol. Ther. Exp.* **16**, 173–188.
Duguid, J. P., Anderson, E. S. and Campbell, I. (1966). *J. Pathol. Bacteriol.* **92**, 107–138.
Edwards, P. R. and Ewing, W. H. (1972). "Identification of Enterobacteriaceae", 3rd edn. Burgess, Minneapolis, Minnesota.
Ewing, W. H., Davis, B. R. and Reavis, R. W. (1959). "CDC Laboratory Manual". Center for Disease Control, Atlanta, Georgia.
Ewing, W. H., Johnson, J. G. and Davis, B. R. (1962). "The Occurrence of *Serratia marcescens* in Nosocomial Infections". CDC Publication, Center for Disease Control, Atlanta, Georgia.
Farmer, J. J., Davis, B. R., Hickman, F. W., Presley, D. B., Bodey, G. P., Negut, M. and Bobo, R. A. (1976). *Lancet* **2**, 455 459.
Field, C., Allen, J. L. and Friedman, H. (1970). *J. Immunol.* **105**, 193 203.
Fife, M. A., Edwards, P. R., Sakazaki, R., Nozawa, M. and Murata, M. (1960). *Jpn. J. Med. Sci. Biol.* **13**, 173 178.
Freer, J. H. and Salton, M. R. J. (1971). *In* "Microbial Toxins" (G. Weinbaum, S. Kadis and S. J. Ajl, Eds), Vol. IV, pp. 67 126. Academic Press, New York and London.
Gard, S. (1937). *Z. Hyg. Infektionskr.* **120**, 59 65.
Grimont, P. A. D. and Grimont, F. (1978). *Annu. Rev. Microbiol.* **32**, 221 248.
Hefferan, M. (1906). *Zentralbl. Bakteriol. Parasitenkd. Infektionskr. Abt. Orig.* **41**, 553 562.
Jan, A. (1939). *Bull. Soc. Sci. Bretagne* **16**, 27 34.
Kaufmann, F. (1954). "Enterobacteriaceae", 2nd edn. Munksgaard, Copenhagen.
Lányi, B. and Bergan, T. (1978). *In* "Methods in Microbiology" (T. Bergan and J. R. Norris, Eds), Vol. 10, pp. 93 168. Academic Press, London and New York.
Le Minor, S. and Pigache, F. (1977). *Ann. Microbiol.* **128B**, 207 214.
Le Minor, S. and Pigache, F. (1978). *Ann. Microbiol.* **129B**, 407 423.
Lowry, O. H., Rosebrough, N. J., Farr, A. L. and Randall, R. J. (1951). *J. Biol. Chem.* **193**, 265 275.
McDonough, M. W. and Smith, S. E. (1976). *Microbios* **16**, 49 53.
Maki, D. G., Hennekens, C. G., Phillips, C. W., Shaw, W. V. and Bennett, J. V. (1973). *J. Infect. Dis.* **128**, 579 587.
Meers, P. D., Foster, C. S. and Churcher, G. M. (1978). *Br. Med. J.* **1**, 238 239.
Meitert, T. and Meitert, E. (1966). *Arch. Roum. Pathol. Exp. Microbiol.* **25**, 427 434.
Minck, R., Le Minor, S., Pigache, F., Dammron, A., Reeb, E., Kohen, M. and Le Faou, A. (1980). *Pathol. Biol.* **28**, 501 507.
Negut, M., Davis, B. R. and Washington, J. A. (1975). *Archs. Roum. Pathol. Exp. Microbiol.* **34**, 33 39.
Nikaido, H. (1973). *In* "Bacterial Membranes and Walls" (L. Leive, Ed.), pp. 131 208. Dekker, New York.

Nowotarska, M. and Mulczyk, M. (1977). *Arch. Immunol. Ther. Exp.* **25**, 7–16.
Pitt, T. L. and Erdman, Y. J. (1977). *J. Med. Microbiol.* **11**, 15–23.
Pitt, T. L., Erdman, Y. J. and Bucher, C. (1980). *J. Hyg.* **84**, 269–283.
Ringrose, R. E., McKown, B., Felton, F. G., Barclay, B. O., Muchmore, H. G. and Rhoades, E. R. (1968). *Ann. Intern. Med.* **69**, 719–729.
Roschka, R. (1950). *Klin. Med.* **5**, 88–93.
Rose, H. D. and Schreier, J. (1968). *Am. J. Med. Sci.* **255**, 228–236.
Rubin, S. J. (1980). *In* "The Genus *Serratia*" (A. von Graevenitz and S. J. Rubin, Eds), pp. 101–118. CRC Press, Boca Raton, Florida.
Rubin, S. J., Brock, S., Chamberland, M. and Lyons, R. W. (1976). *J. Clin. Microbiol.* **3**, 582–585.
Schaberg, D. R., Alford, R. H., Anderson, R., Farmer, J. J., Melly, M. A. and Schaffner, W. (1976). *J. Infect. Dis.* **134**, 181–188.
Sedlák, J., Dlabac, V. and Motlikova, M. (1965). *J. Hyg. Epidemiol. Microbiol. Immunol.* **9**, 45–53.
Tarcsay, L., Wang, C. S., Li, S. C. and Alaupovic, P. (1973). *Biochemistry* **12**, 1948–1955.
Traub, W. H. (1978). *Zentralbl. Bakteriol. Parasitenkd. Infektionskr. Abt. Orig.* **240**, 57–75.
Traub, W. H. and Fukushima, P. I. (1979a). *J. Clin. Microbiol.* **10**, 56–63.
Traub, W. H. and Fukushima, P. I. (1979b). *Zentralbl. Bakteriol. Parasitenkd. Infektionskr. Abt. Orig.* **244**, 474–493.
Traub, W. H. and Kleber, I. (1977). *J. Clin. Microbiol.* **5**, 115–121.
Traub, W. H. and Kleber, I. (1978). *Zentralbl. Bakteriol. Parasitenkd. Infektionskr. Abt. Orig.* **240**, 30–56.
Tsang, J. C., Wang, C. S. and Alaupovic, P. (1974). *J. Bacteriol.* **117**, 786–795.
Wang, C. S. and Alaupovic, P. (1973). *Biochemistry* **12**, 309–315.
Wang, C. S., Burns, R. K. and Alaupovic, P. (1974). *J. Bacteriol.* **120**, 990–993.
Wilfert, J. N., Barrett, F. F., Ewing, W. H., Finland, M. and Kass, E. H. (1970). *Appl. Microbiol.* **19**, 345–352.
Young, V. M., Moody, M. R. and Morris, M. J. (1980a). *J. Med. Microbiol.* **13**, 333–339.
Young, V. M., Moody, M. R. and Morris, M. J. (1980b). *J. Med. Microbiol.* **13**, 341–343.

Addendum

Since this chapter was written Le Minor and Sauvageot-Pigache (1981) have described five new H-antigen factors in *S. marcescens* H:21 to H:25.

Factor H:21 was associated with strains of serogroup O:18 and nine isolates from France and Tunisia were described. The H-serum exhibited some cross-reaction below the homologous titre with the reference strain H:17. Factor H:22 was found in a single pigmented strain of serogroup O:10 and factor H:23 occurred in nine strains of serogroups O:5, O:7, O:15 and O:6/O:14. Eight strains with factor H:24 were described and it was associated with serogroup O:5. A cross-reaction was found between this factor H and H:16. Factor H:25 was found in two strains isolated in Brazil, and the serum was specific for this factor alone.

These workers also reported the subdivision of serogroups O:5 and O:10 each into two subgroups respectively, namely 5a, 5b; 5a, 5c; 10a and 10a, 10b. Serogroup O:16 was further divided into three factors, 16a, 16b; 16a, 16c, 16d; and 16a, 16c. A new serogroup O:21 (IP-1521) was recognized. The reference strain was agglutinated by O:2- and O:3-sera but a serum to the serotype strain O:21 did not agglutinate suspensions of strains O:2 or O:3. The new O group was found in 44 strains isolated in France, Tunisia and Portugal, 17 of which were non-motile and the others possessed flagellar factor H:21. Moreover an antigen common to serogroups O:2 and O:3 was recognized. The authors recommended that O:2-serum should be absorbed with strain O:3 for routine typing and O:3-serum used unabsorbed.

Roig *et al.* (1979) demonstrated that *S. marcescens* produces two extracellular antigens which can be detected by immunoprecipitation. One antigen corresponds to the lipopolysaccharide specific for each O serotype and the other was identified as an antigen carried by all strains of the species, but not by other Gram-negative bacteria. Recently, Roig *et al.* (R. Diaz, personal communication) have described a rapid method for identification of *S. marcescens* by coagglutination of clinical strains with a serum prepared against the common antigen.

Cultures of *S. marcescens* serotype O:5 (CDC-40-1573) were grown for 18 h at 35°C in shaker flasks containing Trypticase Soy Broth (TSB) (BBL). Cells were removed by centrifugation and the antigen was precipitated from the supernatant with 60% saturated ammonium sulphate (final concentration). The precipitate was extensively dialysed against distilled water and lyophilized. Rabbits were injected subcutaneously with 2 mg of antigen per millilitre in Freunds incomplete adjuvant in doses of 1 ml twice a week for four weeks and the animals were bled one week after the last injection. Staphylococci, rich in protein A, were sensitized by the method of Kronvall (1973) with the antiserum (1 ml of 10% v/v staphylococci and 0.2 ml of undiluted serum).

Each of the 20 O serotypes of *S. marcescens* were agglutinated by this reagent and the test proved to be specific for *S. marcescens* in a collection of 500 clinical isolates of Gram-negative bacteria.

References

Kronvall, G. (1973). *J. Med. Microbiol.* **6**, 187–190.
Le Minor, S. and Sauvegeot-Pigache, F. (1981). *Ann. Microbiol. (Paris)* **132A**, 239–252.
Roig, J. M., Dorronsoro, I. and Diaz, R. (1979). *Laboratorio* **68**, 507–527.

4
Serological typing of *Edwardsiella tarda*

R. SAKAZAKI

Enterobacteriology Laboratories, National Institute of Health, Tokyo, Japan

I. Introduction

The organisms known as *Edwardsiella* were independently reported by several investigators early in the 1960s. Sakazaki and Murata (1962) isolated the organisms from snakes while performing a survey of the distribution of *Salmonella*, and suggested a vernacular name, Asakusa group, for this group of organisms in Enterobacteriaceae. Hoshina (1962) recognized the bacteria as a causative agent of diseases in eels and called it *Paracolobactrum anguillimortiferum*. King and Adler (1964) described the isolation of a culture of the organism and proposed the name the Bartholomew group. On the other hand, Ewing *et al.* (1965) suggested *Edwardsiella tarda* for a group of bacteria that were very similar to the Asakusa and Bartholomew groups and has been known as biotype 1483–59 in their laboratory since 1959.

According to the International Code of Bacterial Nomenclature (1953), *Paracolobactrum anguillimortiferum* by Hoshina is a legitimate name for this group. However, because the specific epithet *anguillimortiferum* has priority over the epithet *tarda*, Sakazaki and Tamura (1975) proposed the name *Edwardsiella anguillimortifera* instead of *Edwardsiella tarda*. Against this, Farmer *et al.* (1976) proposed keeping *Edwardsiella tarda* since the latter name has been more widely used than the former name. Thus, two scientific names, *E. tarda* and *E. anguillimortifera*, are given for this organism in the Approved Lists of Bacterial Names (Skerman *et al.*, 1980).

METHODS IN MICROBIOLOGY
VOLUME 15 ISBN 0-12-521515-0

More recently, Grimont *et al.* (1981) recognized the second species of *Edwardsiella* among the organisms formerly described as *E. tarda*, and suggested the name *Edwardsiella hoshinae* for this species. In addition, Hawke *et al.* (1981) reported a bacterium causing enteric septicaemia in catfish and designated the name *Edwardsiella ictalura*.

II. Biochemical characteristics

Edwardsiella organisms are members of the family Enterobacteriaceae. Most of the organisms require nicotic acid for growth. Although they grow on ordinary media, growth is more profuse on media containing yeast extract than on ordinary media. They grow on, for example, *Salmonella–Shigella* agar, XLD agar and Hektoen enteric agar, but not on brilliant green agar and bismuth sulphite agar.

The organisms are motile rods, conforming to the definition of Enterobacteriaceae. *E. tarda* produces hydrogen sulphide in Kligler iron agar and triple sugar iron agar. The organism produces indole, lysine and ornithine decarboxylases, but does not produce arginine dihydrolase, phenylalanine deaminase and urease. It does not utilize citrate (Simmons' citrate agar). Most strains of *E. tarda* do not ferment lactose, arabinose or xylose. *E. tarda* does not ferment mannitol and sucrose, whereas *E. hoshinae* attacks these sugars. Malonate is utilized by *E. hoshinae* but not by *E. tarda*.

The biochemical reactions given by 780 strains of *E. tarda* are summarized in Table I and compared to *E. hoshinae* and *E. ictalura*.

III. Serology

A. General principles

Serology of *Edwarsiella* was first reported by Sakazaki (1967) and McWhorter *et al.* (1967) independently. Sakazaki established an antigenic scheme including 17 O-antigen groups, 11 H-antigens and 18 serovars based on 256 cultures. Later, the antigenic scheme was extended by K. Tamura and R. Sakazaki (1977, unpublished data) and 31 O-antigen groups, 25 H-antigens and 63 serovars were reported among 780 cultures. On the other hand, McWhorter *et al.* (1967) presented a provisional antigenic scheme of 49 O-antigens, 37 H-antigens and 148 serovars, although their paper presented at the annual meeting of the American Society for Microbiology was not published. Unfortunately, the two antigenic schemes have not been compared. Although only one serotyping system developed by Sakazaki (1967)

TABLE I
Biochemical characteristics of *Edwardsiella tarda* and related species

Characteristics	Sign	E. tarda %+	E. tarda %(+)	E. hoshinae	E. ictalura
Indole	+[a]	99.0	0.6	+	−
Voges–Proskauer	−	0		−	−
Citrate (Simmons)	−	0		−	−
H$_2$S (TSI)	+	98.6	1.4	+	−
Urease (Christensen)	−	0		−	−
Gelatinase	−	0		−	−
Phenylalanine deaminase	−	0		−	−
Lysine decarboxylase	+	100.0		+	+
Arginine dihydrolase	−	0		−	−
Ornithine decarboxylase	+	99.2	0·8	+	d
Malonate	−	0		+	−
D-Tartrate	−	0		−	−
Mucate	−	0		−	−
Esculin	−	0		−	−
ONPG[b]	−	0		−	(+)
Fermentation					
Arabinose	−	2.0	0.4	−	−
Lactose	−	0		−	−
Maltose	−	99.8	0·2	+	+
Raffinose	−	0		−	−
Rhamnose	−	0		−	−
Sucrose	−	1.2	0·6	+	−
Trehalose	−	0	0.4	+	−
Xylose	−	0		−	−
Adonitol	−	0		−	−
Dulcitol	−	0		−	−
Mannitol	−	0	1.6	+	−
Sorbitol	−	0		−	−
Inositol	−	0		−	−
Salicin	−	0	1·8	d	−

[a] Symbols: +, 90% or more positve; −, 90% or more negative; d, different reactions (11–89% positive); (+), delayed positive reactions.
[b] ONPG, Ortho-nitrophenol-β-D-galactopyranoside (β-galactosidase test).

and by K. Tamura and R. Sakazaki (1977, unpublished data) will be described in this chapter, it would be useful if an antigenic scheme combining the above schemes could be established for international use. A study with this purpose in mind is now in progress.

Antigens of *Edwardsiella* for serovar determination consist of two fractions, somatic (O) and flagellar (H) antigens. The O-antigens are lipopolysaccharide in nature and heat-stable, whereas the H-antigens are protein in nature and destroyed by heating at 100°C for 2 h. The presence of capsular or masking (K) antigens in *Edwardsiella* cultures has not been demonstrated. Sakazaki (1967) reported that living cultures as well as boiled cultures were agglutinated in a homologous titre of the corresponding O-antiserum in all strains employed.

B. Preparation of diagnostic antiserum

1. O-antiserum

After plating on ordinary agar medium, selected smooth colonies are inoculated into Brain Heart Infusion Broth (BHIB) at pH 7.6–7.8, incubated at 35°C for 18–24 h. Growth of most *E. tarda* strains is slow on those media, because they require nicotic acid for growth. The growth may be improved by adding 0.3% yeast extract to those media. The BHIB cultures are then heated for 2.5 h in boiling water to inactivate H-antigens. After cooling, the broth culture may be preserved with 0.3% formalin.

Rabbits receive four intravenous injections of the boiled culture at intervals of four days. The amounts usually given are 0.5 ml, 1.0 ml, 2.0 ml and 3.0 ml. Four injections usually result in antisera of satisfactory titres of 1:800, or more, against the homologous antigen. If the titre is low on test bleeding, an additional one or two injections may be given. If a satisfactory titre has then not been attained, the rabbit is discarded, since further injections are generally ineffective. The rabbits are bled on the sixth day after the last injection.

2. H-antiserum

Since H-antigenicity is poor in many cultures, serial passages through a semi-solid medium are necessary to promote motility and H-antigen development. For the passage through semi-solid medium, it is advisable to use Craigie's method. Approximately 3 ml of motility test medium containing 0.3% yeast extract in a tube (3 × 100 mm) is recommended. The passage should be repeated until the organisms inoculated into the small inner tube reach the surface of the medium of the outer tube, within 15 to 18 h when incubated at 35°C. Usually one or two, occasionally more, passages are necessary before satisfactory H-antigens are obtained.

After enhancement of motility, one loopful of the cultures from the surface of the semi-solid medium of outer tube is inoculated into BHIB containing 0.3% yeast extract (pH 6.6–6.8), which is then incubated overnight at 30°C and subsequently preserved by adding 0.3% formalin. It is desirable to use the broth adjusted at pH 6.6–6.8 to produce the best H-antigens.

The immunization schedule is as mentioned above for O-antiserum production. A titre of 1:10 000, or more, against the homologous H-antigen strain is satisfactory.

3. *Preservation of antiserum*

O- and H-antisera are preserved by adding an equal volume of glycerol and stored at 4°C.

4. *Absorption of antiserum*

Reciprocal or unilateral reactions disturbing the results are occasionally observed between heterologous O- or H-antisera. Then absorption should be carried out to remove the minor agglutinin.

For O-agglutinin absorption, a 24-h culture of five agar plates is washed off by saline, boiled for 1 h and centrifuged. Five millilitres of the O-antiserum diluted 1:5 are added to the packed cells and mixed thoroughly. For absorbing H-agglutinins, growth on approximately ten plates of soft agar is washed off in saline containing 1% formalin and centrifuged. Then 10 ml of the H-antiserum diluted 1:10 are added to the packed cells. In all instances, the mixture is incubated for 2 h in a water bath at 50°C and allowed to stand overnight at 4°C. After centrifugation, the supernatant serum is preserved with thimerosal.

C. Determination of antigens

1. *O-antigen*

For routine determination of O-antigens of *E. tarda*, slide agglutination is the method of choice. Slide agglutination is carried out using overnight agar cultures and the working dilutions of O-antisera. Although living cultures, as well as boiled cultures, may be agglutinated in a homologous O-antiserum in almost all instances (Sakazaki, 1967), boiled cultures are more appropriate for reliable results. Antisera are customarily diluted 1:10 for slide O-agglutination, which is better performed with sera diluted to give strong agglutination within a few seconds.

The results of reciprocal agglutination with boiled cell suspensions of each O-antigen group tested in each of 41 O-antisera are given in Table II. Reciprocal or unilateral O-antigen relationships are seen between some O-antigen groups, although cultures of most O-antigen groups react specifically. In such antisera, absorption may be required for improved determination of O-antigens from cultures.

TABLE II

Edwardsiella tarda O-antigen relationships

O-antiserum	Homologous titre	Titre with other O-antigens[a]	
1	2000	2:100	*29:100*
3	1000	1:200	
5	2000	1:200	
8	2000	9:100	
12	1000	1:100	4:100
14	1000	23:1000	
29	2000	1:200	

[a] Reciprocal relationships are in italic.

2. H-antigen

The formal-treated broth cultures prepared in a manner described for H-antiserum preparation is not satisfactory for H-agglutination. The pH of the broth and preservative may influence H-agglutinability. A pH above 7.0 of the medium into which organisms are inoculated is unsatisfactory for antigens possessing good H-agglutinability, and the formalin-treated antigen also may reduce its agglutinability (Sakazaki, 1967). To obtain satisfactory H-antigen cultures, the actively motile organisms passing through semi-solid medium are inoculated into BHIB containing 0.3% yeast extract at pH 6.6–6.8, incubated overnight at 35°C and then added to an equal volume of saline containing 0.2% thimerosal.

Phase variation occurring in H-antigens of *Salmonella* cultures is not observed in those of *E. tarda*. So far all cultures of *E. tarda* have possessed monophasic H-antigens.

Determination of H-antigens is best carried out by tube agglutination. Although the dilution employed may be different with particular H-antisera, the unabsorbed antisera are usually used in a dilution of 1:100. 0.1 ml of a 1:100 dilution of antiserum is transferred into a small tube and 1.0 ml of the broth culture is added. The test is incubated in a water bath at 50°C for 1 h and read.

H-agglutination can also be performed by the slide technique using antisera diluted 1:100 from which O-agglutinins have been removed by absorption.

Reciprocal and unilateral reactions exhibited between the H-antisera produced by the author are shown in Table III. Antisera with extensive cross-reactions should be absorbed to remove minor agglutinins. In other instances in which the relationships are minor, unabsorbed antisera may be used at a dilution of 1:1000.

TABLE III
Edwardsiella tarda H-antigen relationships

H-antiserum	Homologous titre	Titre with other H-antigens[a]		
1	40 000	*10:4 000*	*11:4 000*	
2	20 000	*5:500*		
4	10 000	*10:1 000*	*17:1 000*	
7	10 000	*5:2 000*		
8	5 000	*20:1 000*		
10	20 000	*1:4 000*	3:500	*4:500*
11	10 000	*1:500*		
12	10 000	20:500		
13	40 000	8:2 000	20:500	
14	40 000	3:20 000		
17	10 000	*4:10 000*		
19	10 000	*20:1 000*		
20	40 000	8:2 000	*19:2 000*	21:10 000
25	10 000	21:2 000		

[a] Reciprocal relationships are in italic.

D. Antigenic Scheme

Seventeen O-antigen groups and 11 H-antigens of *E. tarda* were established by Sakazaki (1967). K. Tamura and R. Sakazaki (1977, unpublished data) expanded the number to 31 O-antigens and to 25 H-antigens. It has been possible to delineate 63 serovars among 780 cultures of *E. tarda* from a variety of sources. Table IV shows serovars of *E. tarda* characterized in our laboratory. Table V shows test strains.

McWhorter *et al.* (1967) also presented a provisional antigenic scheme. A collaborative study is now in progress between the Center for Disease Control, USA and the National Institute of Health, Japan in order to establish a single antigenic scheme which could be internationally acceptable.

1. *Inter-generic and extra-generic relationships of the antigens*

Antigen relationships of *E. tarda* to other two species of *Edwardsiella*, *E. hoshinae* and *E. ictalura*, have not been investigated. It is probable, however, that some antigen relationships may be expected between *E. tarda* and *E. hoshinae* because the latter species were involved in studies of K. Tamura and R. Sakazaki (1977, unpublished data) as a biovar of *E. tarda*.

Sakazaki (1967) reported that no significant reactions were observed when each O- and H-antigen of *E. tarda* in his antigenic scheme was tested with 45

TABLE IV
Antigenic scheme for *Edwardsiella tarda*

O-group	H-antigen			
1	1	10	13	
2	2	8	10	
3	2	3		
4	4	5	10	11
5	4	5	10	
6	3	8	17	19
7	6	15		
8	3			
9	1	8	19	
10	3	19	20	
11	7	12		
12	1			
13	1	2		
14	9	21		
15	1	8	19	
16	1	10		
17	11	19		
18	3			
19	9	14		
20	16			
21	8	10		
22	8			
23	18			
24	11	13		
25	10			
26	11	24		
27	22			
28	8	10		
29	1	2	10	
30	25			
31	23			

O- and 56 H-antisera of *Salmonella*, 146 O- and 41 H-antisera of *Escherichia coli*, 32 O- and 71 H-antisera of *Citrobacter freundii*, 49 O- and 17 H-antisera of *Proteus vulgaris* and *P. mirabilis*, 53 O- and 23 H-antisera of *Enterobacter cloacae*, and 29 O- and 23 H-antisera of *Hafnia alvei*.

IV. Epidemiology of *Edwardsiella tarda* and occurrence of its serovars

According to present knowledge, *E. tarda* seems to be distributed in all over the world. Although a number of papers have reported the isolation of *E.*

TABLE V
Edwardsiella tarda O- and H-antigenic test strains[a]

Antigenic formula		
O	H	Strain
1	*1*	223–61
2	*2*	1032–60
3	*3*	1058–60
4	*4*	1068–60
5	*5*	387–60
6	*17*	1070–62
7	*6*	407–58
8	*3*	1079–60
9	*19*	1031–62
10	*20*	86–61
11	*7*	471–62
11	*12*	482–62
12	1	219–63
13	1	1042–63
14	*9*	1071–63
14	*21*	490–62
15	1	1027–60
16	*10*	1004–55
17	*11*	160–63
18	*3*	1031–63
19	*14*	1053–60
20	*16*	1001–68
21	10	87–63
22	*8*	237–73
23	*18*	489–72
24	*13*	334–62
25	*15*	1001–61
26	*24*	1049–70
27	*22*	29–73
28	10	49–71
29	10	1059–70
30	*25*	31–73
31	*23*	395–68

[a] Arranged by O-antigen numbers.
Numbers in italic indicate test antigens.

tarda from human materials, it would appear that this organism rarely causes human disease. Bhat *et al.* (1967) isolated only four *E. tarda* strains from 513 children with diarrhoea. Makulu *et al.* (1973) reported that only 51 strains of this species were isolated from human materials during a seven-year survey in which a yearly average of 3000 stool cultures were performed. Koshi and

Lalitha (1976) isolated 12 strains of the organism from different clinical materials screened for *E. tarda* during a three-year period. Kourany *et al.* (1977) reported *E. tarda* from ten patients with diarrhoea during a seven-year period. In healthy persons Onogawa *et al.* (1976) isolated *E. tarda* from 0.001% of 97 704 adults and 0.01% from 255 896 children.

Although the relative importance of *E. tarda* in diarrhoea cannot be assessed at present, a number of recent papers from tropical countries such as Madagascar (Fourquest *et al.*, 1975), Tahiti (Fourquest *et al.*, 1975), Panama (Kourany *et al.*, 1977), India (Sakazaki *et al.*, 1971; Bhat *et al.*, 1967), Thailand (Bockemuehl *et al.*, 1971), Vietnam (Nguyen-Van-Ai *et al.*, 1975), Malaysia (Gilman *et al.*, 1971) and Zaire (Makulu *et al.*, 1973) have emphasized the aetiological relationship of *E. tarda* with diarrhoeal disease. Some of them considered *E. tarda* as a major enteropathogen. There was a tendency for *E. tarda* to be isolated from stool specimens together with well established enteropathogens such as *Salmonella, Shigella, Vibrio cholerae, Vibrio parahaemolyticus, Entamoeba histolytica* and intestinal parasites (Bockemuehl *et al.*, 1971; Sakazaki *et al.*, 1971; Kourany *et al.*, 1977; Makulu *et al.*, 1973). Intestinal infection with *E. tarda* was also reported by King and Adler (1964) and Chatty and Gavan (1968) in the USA.

Extra-intestinal infections, such as septicaemia, meningitis and wound infection, have also been reported (Sonnenwirth and Kallus, 1968; Okubadejo and Alausa, 1968; Pankey and Seshul, 1969; Gonzales and Ruffolo, 1966; Field *et al.*, 1967; Jordan and Hadley, 1969; Bockemuehl *et al.*, 1971; Koshi and Lalitha, 1976). In most extra-intestinal infections, *E. tarda* has been an opportunist in underlying diseases or in conditions with predisposing factors. Bockemuehl *et al.* (1971) pointed out that infants and the aged tend to be most susceptible to infection with *E. tarda*, but Makulu *et al.* (1973) reported no apparent predilection for a given age group.

E. tarda was infrequently isolated from warm-blooded animals other than man, such as dog, pig, opossum, monkey, cattle, panther, sea lion, seal, and birds (D'Empaire, 1969; Wallace *et al.*, 1966; Sakazaki, 1967; Ewing *et al.*, 1965; Chamoiseau, 1967; Tocal and Mezez, 1968; Owen *et al.*, 1974; Kourany *et al.*, 1977; Arambulo *et al.*, 1967). The isolation of this organism was associated with disease in some of those animals, but pathogenicity of *E. tarda* to warm-blooded animals is still unknown.

Two ecological groups of hosts, reptiles and fishes, could be considered as possible reservoirs of *E. tarda*. Since Sakazaki and co-workers (Sakazaki 1965, 1967; Sakazaki and Murata, 1962) first considered *E. tarda* as a normal intestinal inhabitant of snakes, it has been demonstrated that a wide range of reptiles and amphibians, such as snakes, crocodiles, toads, frogs, turtles, skinks and lizards, are possible natural reservoirs for *E. tarda* (D'Empaire, 1969; Iveson, 1971; Kourany *et al.*, 1977; Jackson *et al.*, 1969; Richard *et al.*,

1978; Wallace *et al.*, 1966; Sharma *et al.*, 1974; Van Der Waaiji *et al.*, 1974; Meyer and Bullock, 1973; Roggendorf and Müller, 1976; Makulu *et al.*, 1973; White *et al.*, 1969). On the other hand, *E. tarda* has been known as an important causative agent of diseases of eels similar to red disease, since Hoshina (1962) found this organism in eels with the disease. Although the organism was named *Paracolobactrum anguillimortiferum*, Sakazaki and Tamura (1975) and Wakabayashi and Egusa (1973) concluded that Hoshina's bacterium was a senior synonym of *E. tarda*. Hoshina (1962) demonstrated the pathogenicity of the organism to eels and he described that the organism was an aetiological agent of red disease, but Wakabayashi and Egusa (1973) confirmed that the disease of eels caused by *E. tarda*, which is common in cultured eels at many ponds during the summer season, can be distinguished from those causing red disease. Also, *E. tarda* was considered as a pathogen for the channel catfish by Meyer and Bullock (1973). Kusada *et al.* (1977) reported it as causing an outbreak in cultured crimson sea bream. Van Damme and Vandepitte (1980) reported the isolation of *E. tarda* from various kinds of river fish in Zaire. They concluded that freshwater fish seem to constitute the natural habitat of *E. tarda* and to be the most probable source of human infection in the tropics, at least.

There has been no known study on the association between serovars and human disease. In 756 strains studied by the author, no relationship has been found between serovars and their source (R. Sakazaki and K. Tamura, unpublished data). Serovars found in human diseases, especially intestinal infections, have also been isolated from snakes, fishes and animal sources. Many serovars causing disease in eels have also been found in reptiles and warm-blooded animals. It is clear that the determination of serovars could be a useful epidemiological tool for human infections and for outbreaks of disease in fish.

References

Arambulo, P. V., Westerlund, N. C., Sarmiento, R. V. and Abaga, A. S. (1967). *Far East Med. J.* **3**, 385–386.

Bhat, P., Meyers, R. and Carpenter, K. P. (1967). *J. Hyg.* **65**, 293–298.

Bockemuehl, J., Pan-Urai, R. and Burkhardt, F. (1971). *Pathol. Microbiol.* **37**, 393–401.

Chamoiseau, G. (1967). *Rev. Elev. Med. Vet. Pays. Trop.* **20**, 493–495.

Chatty, H. and Gavan, T. (1968). *Cleveland Clin. Q.* **35**, 223–228.

D'Empaire, M. (1969). *Ann. Inst. Pasteur* **116**, 63–68.

Edwards, P. R. and Ewing, W. H. (1972). "Identification of Enterobacteriaceae", 3rd edn. Burgess, Minneapolis, Minnesota.

Ewing, W. H., McWhorter, A. C., Escobar, M. R. and Lubin, A. H. (1965). *Int. Bull. Bacteriol. Nomencl. Taxon.* **15**, 33–38.

Farmer, J. J., III, Brenner, D. J. and Clark, W. C. (1976). *Int. J. Syst. Bacteriol.* **26**, 293–294.

Field, B., Uwaydah, M., Kunz, L. and Swartz, M. (1967). *Am. J. Med.* **42**, 89–106.

Fourquest, R., Coulanges, P., Boehrer, J. L. and Delavaud, J. (1975). *Arch. Inst. Pasteur Indochine* **44**, 31–48.

Gilman, R. H., Madasamy, M., Gan, E., Mariappan, M. Davis, C. E. and Kyser, K. A. (1971). *Southeast Asian J. Trop. Med. Public Health* **2**, 186–189.

Gonzales, A. B. and Ruffolo, E. H. (1966). *South. Med. J.* **59**, 340–342.

Grimont, P. A. D., Grimont, F., Richard, C. and Sakazaki, R. (1981). *Current Microbiol.* **4**, 347–351.

Hawke, J. P., McWhorter, A. C., Steigerwalt, A. G. and Brenner, D. J. (1981). *Int. J. Syst. Bacteriol.* (in press).

Hoshina, R. (1962). *Bull. Jpn. Soc. Sci. Fish.* **28**, 162–164.

Iveson, J. B. (1971). *J. Hyg.* **69**, 323–330.

Jordan, G. W. and Hadley, W. K. (1969). *Ann. Intern. Med.* **70**, 283–288.

Jackson, M. M., Jackson, C. C. and Fulton, M. (1969). *Bull. Wildlife Dis. Assoc.* **5**, 328–329.

King, B. M. and Adler, D. L. (1964). *Am. J. Clin. Pathol.* **41**, 230–232.

Koshi, G. and Lalitha, M. K. (1976). *Ind. J. Med. Res.* **64**, 1753–1759.

Kourany, M., Vasquez, M. A. and Saenz, R. (1977). *Am. J. Trop. Med. Hyg.* **26**, 1183–1190.

Kusada, R., Itami, T., Munekiyo, M. and Nakajima, H. (1977). *Bull. Jpn. Soc. Sci. Fish.* **43**, 129–134.

McWhorter, A. C., Ewing, W. H. and Sakazaki, R. (1967). *Bacteriol. Proc.* **89**.

Makulu, A., Gatti, F. and Vandepitte, J. (1973). *Ann. Soc. Belge Med. Trop.* **53**, 165–172.

Meyer, E. P. and Bullock, G. L. (1973). *Appl. Microbiol.* **43**, 155–156.

Nguyen-Van-Ai, Nguen-Duc-Hank, Le-Tien-Van, Nguen-Van-Le and Nguen-Thi-Lan-Huong (1975). *Bull. Soc. Pathol. Exot.* **68**, 355–359.

Okubadejo, O. A. and Alausa, K. O. (1968). *Br. Med. J.* **3**, 357–358.

Onogawa, T., Terayama, T., Zen-Yoji, K., Amano, Y. and Suzuki, K. (1976). *Kansensho-gaku Zasshi* **50**, 10–17 (in Japanese).

Owen, D. R., Nelson, S. L. and Addison, J. B. (1974). *Appl. Microbiol.* **27**, 703–705.

Pankey, G. A. and Seshul, M. B. (1969). *J. la St. Med. Soc.* **121**, 41.

Richard, C., Lhuillier, M. and Laurent, B. (1978). *Bull. Inst. Pasteur Paris* **76**, 187–200.

Roggendorf, M. and Müller, H. E. (1976). *Zentralbl. Bakteriol. Parasitenkd. Infektionskr. Orig. A* **236**, 22–35.

Sakazaki, R. (1965). *Int. Bull. Bacteriol. Nomencl. Taxon.* **15**, 45–47.

Sakazaki, R. (1967). *Jpn. J. Med. Sci. Biol.* **20**, 205–212.

Sakazaki, R. (1980). "Enterobacteriaceae", III. Kindai Shuppan Publisher, Tokyo (in Japanese).

Sakazaki, R. and Murata, Y. (1962). *Jpn. J. Bacteriol.* **17**, 616–617 (in Japanese).

Sakazaki, R. and Tamura, K. (1975). *Int. J. Syst. Bacteriol.* **25**, 219–220.

Sakazaki, R., Tamura, K., Prescott, L. M., Bencic, Z., Sanyal, S. C. and Sinha, R. (1971). *Indian J. Med. Res.* **59**, 1025–1034.

Sharma, V. K., Kaura, Y. K. and Singh, I. P. (1974). *Antonie van Leeuwenhoek J. Microbiol. Serol.* **40**, 171–175.

Skerman, V. B. D., McGowan, V. and Sneath, P. H. A. (Eds) (1980). *Int. J. Syst. Bacteriol.* **30**, 225–420.

Sonnenwirth, A. C. and Kallus, B. A. (1968). *J. Clin. Pathol.* **49**, 92–95.

Tocal, J. V., Jr and Mezez, C. F. (1968). *Philipp. J. Vet. Med.* **7**, 143–145.

Van Damme, L. R. and Vandepitte, J. (1980). *Appl. Environ. Microbiol.* **39**, 475–479.

Van Der Waaiji, D., Cohen, B. J. and Nace, G. W. (1974). *Lab. Anim. Sci.* **24**, 307–317.

Wakabayashi, H. and Egusa, S. (1973). *Bull. Jpn. Soc. Sci. Fish.* **39**, 931–936.

Wallace, L. J., White, F. H. and Gore, H. L. (1966). *J. Am. Vet. Med. Assoc.* **149**, 881–883.

White, F. H., Neal, F. C., Simpson, C. F. and Wash, A. F. (1969). *J. Am. Vet. Med. Assoc.* **155**, 1057–1058.

5

Biochemical and Serological Characterization of *Erwinia*

M. B. SLADE* and A. I. TIFFIN

Department of Immunology, Royal Postgraduate Medical School, Hammersmith Hospital, London, England and Department of Microbiology, University of Reading, Reading, England

* *Present address:* AMD Ltd, Artaman, NSW, Australia.

METHODS IN MICROBIOLOGY
VOLUME 15 ISBN 0–12–521515–0

I. Introduction

The genus *Erwinia* contains a heterogeneous group of bacteria including plant pathogens and epiphytic bacteria, occurring in a wide range of habitats. Some of the plant pathogens are of considerable economic importance while many of the epiphytic bacteria are opportunistic pathogens for man and animals. For historical reasons, the genus *Erwinia* has not been as intensively studied as other members of the Enterobacteriaceae and knowledge of some aspects of the group, e.g. serology, is only fragmentary. An understanding of the historical background of the genus and the disparate taxonomic relationships between *Erwinia* spp. is essential for any serious study of the *Erwinia*.

A. History

The genus *Erwinia* was established in 1917 (Winslow *et al.*, 1917) specifically to contain plant pathogenic bacteria and so contained strains which were biochemically dissimilar from the type species *Erwinia amylovora*. The formation of the genus was condemned by the phytopathologist Erwin Frank Smith, after whom the genus was named, and has been the subject of much debate and confusion ever since (Starr, 1959; Starr and Chatterjee, 1972). Over 60 species of *Erwinia* have been named, many on the basis of the "new host, new species" fallacy (Starr and Chatterjee, 1972). Many of these species were poorly characterized and completely lacked host specificity studies. It is now recognized that some of the species named were not plant pathogens, but belonged to the epiphytic flora that had invaded areas of damaged tissue. This resulted in the *Erwinia* containing not only the plant pathogens included in the original definition, but also a group of similar epiphytic organisms.

By definition, *Erwinia* contained organisms of particular interest to plant pathologists so it remained ignored by the medical bacteriologists who were more concerned with the Enterobacteriaceae. Thus it was not until over 40 years after its creation that strains of *Erwinia* were directly compared with the Enterobacteriaceae (Starr and Chatterjee, 1972).

B. Taxonomy

The majority of taxonomists appear to favour the division of the genus into three taxonomic groups: the *amylovora* group or true erwinias; the *carotovora* or soft rot group and the *herbicola* group (Dye, 1968, 1969a,b). A fourth group containing atypical erwinias (Dye, 1969c) has now been excluded from the genus (Lelliott, 1974). This classification, described in "Bergey's Manual of Determinative Bacteriology" (Lelliott, 1974), is used here except that *E. rhapontici* is included in the *herbicola* group in accordance with a recent numerical taxonomic study (Slade and Tiffin, 1980). Numerical taxonomic studies show that the *herbicola* and *carotovora* groups are distantly related, but neither is related to the *amylovora* group (Goodfellow *et al.*, 1976; Grimont *et al.*, 1977; Slade, 1978).

Current views on the taxonomy of the *Erwinia* have been profoundly influenced by DNA homology studies (Brenner *et al.*, 1972, 1973, 1974; Garder and Kado, 1972; Murata and Starr, 1974). They cause the genus to be viewed as a collection of species with relationships as close to each other as individual species are to other members of the Enterobacteriaceae (Murata and Starr, 1974; Billing, 1976a; Fonnesbech, 1974). Genetic clusters were found within the three *Erwinia* groups. In the *amylovora* group, *E. salicis*, *E. rubrifaciens* and *E. nigrifluens* were distantly related but did not share significant DNA homology with either *E. amylovora* or *E. quercina* (Murata and Starr, 1974). In the *carotovora* group, *E. carotovora* and *E. chrysanthemi* were shown to be related, but Brenner *et al.* (1973) found that *E. cypridedii* had a greater homology with *E. herbicola* than with any member of the carotovora group. *E. rhapontici* was assumed to form an independent genetic cluster within the *carotovora* group, although it had not been compared with *E. chrysanthemi* or any member of the *herbicola* group. The *herbicola* group formed an extremely loose cluster with only 17–30% DNA homology between some strains of *E. herbicola* compared with a core homology of 10–15% found throughout the *Erwinia* and other genera (Murata and Starr, 1974).

1. Proposals for changes in nomenclature

The two proposed changes in nomenclature concern the *carotovora* group and the *herbicola* group.

Waldee (1945) recognized that the soft rot *Erwinia* were distinct from the *Erwinia* spp. now contained in the *amylovora* group and proposed that the soft rot pathogens should be placed in a new genus *Pectobacterium*, distinguished by the production of pectinases. The species Waldee called *Pectobacterium* are now combined into one species, *E. carotovora*, but *E. chrysanthemi* also fits the description of *Pectobacterium*. The genus *Pectobacterium* was initially accepted (Dowson, 1957; Starr, 1959; Graham, 1964; Starr and Chatterjee, 1972), but in the last decade evidence has suggested that pectinase production is not of overriding taxonomic importance. It is now recognized that pectinases are also produced by some members of the *amylovora* group (Graham, 1971; Lelliott, 1974) and other members of the Enterobacteriaceae, e.g. *Enterobacter aerogenes* (Dias, 1967), *Yersinia* spp. and *Klebsiella pneumoniae* (Starr *et al.*, 1977; Von Riesen, 1976). Whilst the definition of the genus *Pectobacterium* has been the subject of criticism, the name would have priority for the soft rot pathogens *E. carotovora* and *E. chrysanthemi* in the event of any reclassification of *Erwinia* spp.

Ewing and Fife (1972) proposed that the species *E. herbicola* should be reclassified as *Enterobacter agglomerans*. They claimed that the specific epithet *agglomerans* had priority over *herbicola*. Numerical taxonomic studies have variously described the *herbicola* and *carotovora* groups as: closely related to *Enterobacter cloacae* (Grimont *et al.*, 1977); not closely related to *Enterobacter*, but within the tribe *Klebsielleae* (Bascomb *et al.*, 1971; Gavini *et al.*, 1976; Thurner and Busse, 1978); not closely related to *Enterobacter* nor any other members of the Enterobacteriaceae (Goodfellow *et al.*, 1976; Sakazaki *et al.*, 1976). Other taxonomic studies have produced little evidence in favour of classifying some *Erwinia* spp. within *Enterobacter*.

The use of *Enterobacter agglomerans* was not accepted by plant pathologists or by some taxonomists in the medical field (Cowan and Steel, 1974).

It has been proposed that the genus *Erwinia* should be dismembered and the species distributed amongst the other genera of the Enterobacteriaceae (Starr and Chatterjee, 1972; Murata and Starr, 1974), but in spite of general dissatisfaction with the taxonomy of the group, the genus appears to have remained in its present form through lack of agreement about a change. Dye (1981) has performed numerical taxonomy on *Erwinia*.

2. Biochemical identification

The biochemical reactions reported for some *Erwinia* spp. vary considerably even when apparently similar methods have been used (Tables I and II). This is because the reactions may be very dependent upon the method used, the time and temperature of incubation and the criteria adopted by different observers to distinguish between negative and weak positive results. It is important that unknown strains are tested in parallel with control strains of known identity. Most medical laboratories test isolates at 37°C, whereas laboratories interested in non-clinical isolates of *Erwinia* use incubation temperatures of 25, 27 and 30°C, since some strains do not grow at 37°C. When comparing different genera it should be noted that the biochemical reactions of other members of the Enterobacteriaceae (e.g. *Serratia*, *Hafnia*) may alter if tested at 25°C (Cowan and Steel, 1974).

The *amylovora* group, and to a lesser extent the *herbicola* group, give much weaker acid production from carbohydrates than do other members of the Enterobacteriaceae. The erwinias tend generally to give very weak reactions in common biochemical tests. 1% peptone water is frequently used as the basal medium for the detection of breakdown of carbohydrates by members of the Enterobacteriaceae. Some *Erwinia* spp. produce only low levels of acid, and so it is appropriate to use a modified test for all *Erwinia* spp. A weakly buffered mineral salts medium, supplemented with 0.1% yeast extract has been found to give clear cut results with few problems of alkaline reversion (Billing *et al.*, 1960, 1961; Dye, 1968; Webb, 1974). Members of the *herbicola* group give very

TABLE I

Comparison of published descriptions of *Erwinia herbicola*[a]

	No. strains studied	Yellow pigment	Indole	Methyl red	Acetoin	Gas from glucose	Urease	Nitrate reduction	Gluconate, reducing compounds from	Sucrose, reducing compounds from	Casein hydrolysis	Growth at 37 C	H_2S production	Tributyrin hydrolysis	Lactose	Dulcitol	Salicin	Raffinose	Inositol	Sorbitol	Rhamnose	Cellobiose	Glycerol	Melibiose	Melezitose	Starch	Inulin	Dextrin	Tartaric acid	Galacturonic acid
Erwinia herbicola subsp. *herbicola*																														
Komagata *et al.* (1968)	29	+	41[b]	97	86	17	11	90	31	19	—	92	77	11	19	—	+	7	+	+	+		+						±	—
Graham and Hodgkiss (1967)	26	+	4	15	85	—	—	89	19	±			+		±	—	±	±	—	+	+	—	—	—	—	±	+	+		
Dye (1968b)	45	+	—	±	+	±	—	+					—		—	+	+	—	—	—	+	±	±			—				
Bascomb *et al.* (1971)	10	+			±	—					±																			
Goodfellow *et al.* (1976)	27	+	—		±	5		—																						
Grimont *et al.* (1977)	5	+	—	—	60	—	—	+		+																				
Slade (1978)	20	70	—	—	65		—	+	70	5	—	55	+	+	60	35	+	80	60	+	+	80	70	60	+	90	55	95		+
Thurner and Busse (1978)	54		—	31	78								—	—	22	—		56	+	20	+	69	—	62	11				85	
Erwinia herbicola subsp. *ananas*																														
Graham and Hodgkiss (1967)	4	+	+	75	50	—	—	—			—	+	+	+	+	—	+	+	+	+	50	+	+	+						+
Dye (1969b)	9	+	+	—	+	—	—	—			—	+	±		+		+	+	+	+	±		+		—	+	±		+	±
Goodfellow *et al.* (1976)	3	+		—	—			±		+	—	+	+		50	—	+	+	+	+	+	+	+	+	+	+	—	—	+	±
Slade 1978	4	+	75	75	75		—	25	75	75	+	+	+	25	75	25	+	+	+	+	+	+	75	75	+	+	+	+	—	+

[a] Table I shows only those tests for which conflicting results were obtained using broadly comparable methods. The original publications should be consulted for details on the experimental methods.

[b] Figures indicate the percentage of strains giving a positive result.

TABLE II

Comparison of published descriptions of *Erwinia rhapontici*[a]

	Levan	Pink pigment	Motility	Growth at 36°C	Phosphatase	Gluconate	Reducing compounds from sucrose	Phenylalanine deaminase	Acetoin	Methyl red	Galacturonic acid	Tartaric acid	Melezitose	Dulcitol	Dextrin	Xylose	Lactose	Inositol	No. strains studied
											Acid production from carbohydrates →								
Dye (1969a)	±	±	+	—	+	+	+	±	+	—	—	—	+	+	—	—	+	+	3
Graham (1971)		+	+	+	±	+	+	—	±	±	+	+	±	±		+	+	+	3
Roberts (1974)		+		50			±	—		±							+	+	11
Sellwood and Lelliott (1978)		+		83	33	33	—		83		67	—	83	67	+	50		83	6
Slade (1978)	—		83	83	33	83	+	50	+	17	+	—	+	+	+	+	83	83	6

[a] Table II shows only those tests for which conflicting results have been obtained using comparable methods. The figures indicate the percentage of strains giving a positive result. The methods are described by Dye (1968) or Roberts (1974).

TABLE III

Serological differences between laboratory strains of *Erwinia amylovora*[a]

Strain	TV	GA5[b]	GA1	LPS1	LPS2	Capsule	Virulence[c]	Phage sensitivity			
								L3H	23	4S	4L
T(typical)	+[d]	+	+	+	−	+	+	+	±[e]	+	+
AT(atypical)	+[d]	+	+	+	−	+[d]	+	+	±[e]	+[d]	+[d]
E8	−	+	+	+	−	−	−	+	+	−	−
AT1, AT4	+[d]	+	−	−	+	+[d]	+	−	±[d]	+[d]	+[d]
AT6	−	±[c]	−	−	−	−	−	−	−	−	−
AT7	±	+	−	−	+	−	−	−	−	−	+
AT9	−	+	−	−	+	−	−	−	−	−	−
AT19	+	−	+	+	−	+	+	+	−	+	+
AT23	+	−	+	+	−	+	+	+	±	+	+
P66	+	−	+	+	−	+	−	+	±	+	+

[a] Data mostly from Slade (1978). These cultures and bacteriophages are deposited with the National Collection of Plant Pathogenic Bacteria, Hatching Green, Harpenden, Hertfordshire, England.

[b] Virulence was detected in immature pears (R. A. Bennett, personal communication).

[c] GA5 is a two-component antigen—only one component was produced by strain AT6.

[d] These results are negative under conditions in which atypical strains do not produce capsules.

[e] Atypical strains are sensitive to phage 23 when grown on media on which they do not produce capsules.

weak acetoin production with some positive reactions visible only after 1 h. The temperature of incubation may influence acetoin production (Cowan and Steel, 1974). Phenylalanine oxidase tests may be positive after incubation for 24 h and give negative reactions after 48 h at 25°C (Graham and Hodgkiss, 1967). In our hands, some *E. herbicola* and *E. rhapontici* strains have failed to give reproducible results in the alkaline phosphatase test despite the use of several methods (Lelliott, 1974; Cowan and Steel, 1974) and prolonged incubation times. The influence of pH on the detection of pectolytic enzymes is discussed in Section III,C,2. The detection of urease production by *E. herbicola* strains requires a sensitive test, such as that employing Christensen's medium. *Erwinia* strains may produce indole only in shaken cultures. The use of triphenyltetrazolium chloride in the KCN test inhibits many *Erwinia* spp.

Some commercial rapid identification kits do not appear to be suitable for distinguishing between *Erwinia* spp. which are lumped together as *E. agglomerans* with a few strains assigned to other genera.

An additional problem in the identification of *Erwinia* spp. is that many published descriptions are based on the reactions of only a few strains and so may not reflect the full range of biochemical reactions which a species possesses. For example, the most comprehensive biochemical study of *Erwinia* spp. (Dye, 1968, 1969a) included four strains of *E. rhapontici*, and only three strains each of *E. salicis* and *E. cypripedii*. *Erwinia* spp. isolated from different environments may possess distinct biochemical properties, e.g. clinical isolates of *E. herbicola* frequently possess urease and produce gas from glucose. These are properties which occur infrequently among strains isolated from plants.

A group of biochemical tests, useful for the identification of *Erwinia* spp., is the utilization of organic acids in Dye's OY medium (Dye, 1968) using bromothymol blue as indicator. More clear cut reactions with proteolytic strains are given by a modification of this method where the initial pH is reduced from 6.8 to 6.4 and the yeast extract concentration reduced from 0.08% to 0.05% (Slade, 1978). The modified method gives results identical with the original method except that it does not detect acetate utilization by *E. salicis*.

C. Serology

1. *General structure*

The serological structure of *Erwinia* spp. is comparable with the most complex examples from other genera in the Enterobacteriaceae. A notable feature of *Erwinia* spp. is the frequent production of rough lipopolysaccharide (R-LPS) or a mixture of both rough and smooth (S-LPS). It is usually not possible to select clones producing only S-LPS because most strains produce capsules or micro-capsules. It may be difficult to identify the LPS antigens by the usual agglutination techniques because of heat-stable capsular antigens in some members of the *herbicola* group and, possibly also, in the *carotovora* group.

The heat-labile antigens have been little studied, but most species are motile and the flagellae of *E. carotovora* have been reported to exhibit phase changes (Section III.D.2) Fimbriae have been demonstrated in *E. carotovora* and *E. rhapontici* (Slade, 1978; Christofi *et al.*, 1979).

More detailed discussion of serology will be found under individual species.

Serological studies of LPS are best carried out by indirect haemag-glutination or immunofluorescence. If the S-LPS is of interest, antibodies to R-LPS may be removed by suitable absorption.

Many antisera raised against *Erwinia* spp. react with an antigen common to most *Erwinia*. This antigen is released by heating sonicated cell suspensions at 100°C for 30 min and can be detected using immunodiffusion and passive haemagglutination techniques (Slade, 1978). This *Erwinia* antigen may interfere in other serological tests, but antibodies to it can be removed by absorption with cells of another *Erwinia* species. It has not been established whether this antigen is the same as the enterobacterial common antigen.

2. *Production of antisera*

Most *Erwinia* spp., apart from *E. carotovora* subsp. *atroseptica*, have been reported to be highly toxic for rabbits undergoing immunization (Graham, 1963; Lazar, 1971a). The reason for this toxicity is unknown since endotoxin from *E. herbicola* and washed, nutrient broth grown cells of a number of *Erwinia* spp. have toxicity for mice similar to that of preparations from other genera of Gram-negative rods (Dutkiewicz, 1976; Slade, 1978). However, Slade (1978) found that the extracellular slime from an *E. herbicola* strain grown on solid medium was considerably more toxic for mice than the bacterial cells. It appears preferable to immunize with washed cells grown in nutrient broth.

Lazar (1971a) developed immunization procedures which overcame the toxicity of *Erwinia* spp. for rabbits and these methods have been simplified (Slade, 1978). The following general method has been found satisfactory. Cells are grown in nutrient broth at 25°C for 18 h, washed twice with saline and resuspended in 0.3% (v/v) formalin in saline. For intravenous injection, the opacity of the suspension is adjusted to Brown's tube 2 and to Brown's tube 10 if it is to be used with Freund's adjuvant. One millilitre of vaccine, emulsified in 1 ml Freund's complete adjuvant, is injected subcutaneously into eight sites along the back of a rabbit and the injections are repeated 21 days later with the vaccine emulsified in Freund's incomplete adjuvant. After a further 21 days, the rabbit is given intravenous injections of 0.2, 0.5, 1.0 and 2.0 ml vaccine at three- to four-day intervals, followed by two 2 ml injections of live cells (suspended to Brown's tube 2). The rabbit is bled six to eight days after the last injection.

3. Production of antisera to antigens of special interest

(a) *Antigen AP(H) (Section II.A.6.d)*. This antigen is produced as an extracellular slime by most strains of *E. herbicola* and *E. rhapontici*. Purified preparations of the slime are non-immunogenic. Antibodies to AP(H) may be produced in rabbits immunized with washed cells of *E. herbicola* AL7 grown in nutrient broth (Slade, 1978; Slade and Tiffin, 1978).

(b) *Antigen TV (Section II.A.6.a)*. *E. amylovora* strains with the typical colonial morphology (Billing, 1960) contain antigen TV and suspensions of washed cells grown in nutrient broth may be used for production of antibody. Antisera raised against preparation of atypical strains usually do not contain antibody to TV. Antigen TV is non-immunogenic if a preparation of cells and slime from growth on a solid medium is injected (Slade, 1978).

(c) *Antigen GA5 (Section II.A.6.a)*. Antibodies to this extracellular antigen have been detected only in rabbits immunized with the cells and slime of strains grown on 2% glycerol nutrient agar (Slade, 1978).

II. The *amylovora* group

All members of this group are plant pathogens and there are no reports of their causing disease in mammals; indeed, four of the species are unable to grow at 37°C. These organisms produce wilts or dry necrotic diseases and usually have a restricted host range. The extent to which these organisms have been studied is directly related to their economic importance. Little is known about the diseases of low economic importance. The identification of members of the *amylovora* group is aided by their low degree of natural variation – some of the species appear to be nearly homogeneous with respect to their pathogenicity, biochemistry, serology and sensitivity to bacteriophage. As most species in this group are found on different hosts, the main problem is distinguishing them from saprophytic *Erwinia* spp., e.g. *E. herbicola*, rather than distinguishing between members of the group.

A. *Erwinia amylovora*

1. *Introduction*

E. amylovora causes fireblight, a necrotic disease of apple, pear and ornamental shrubs. This disease causes very serious economic losses and in some areas has completely destroyed pear crops. The name fireblight refers to the brown, shrivelled appearance of leaves and shoots on affected branches similar to the effect of a fire in close proximity to the tree. The symptoms of the disease and rate of progress may vary according to the host, but the first obvious signs are dead blossoms or dark brown leaves hanging from a truss (Eden-Green and Billing, 1974). In pears the disease spreads along the branches and on reaching the trunk it rapidly invades the other branches and kills the tree, but in some less susceptible hosts, e.g. *Pyracantha*, it is unusual for the disease to progress beyond killing the blossom trusses. Shoot infection is characteristic of fireblight in apples. Golden or whitish droplets of bacterial ooze resembling gum are frequently produced from infected tissues and in dry weather this may appear as a dry silvery film. The disease usually becomes dormant in autumn and spreads further along branches in spring (Eden-Green and Billing, 1974; Ministry of Agriculture, Fisheries and Food, 1969).

Fireblight appears to have been endemic among wild hosts in the eastern USA and was spread by the movement of highly susceptible cultivated species (Eden-Green and Billing, 1974). The pathogen was probably spread to California on nursery materials in the 1870s where it caused very severe epidemics in the early 1900s. *E. amylovora* was first isolated in Canada in 1904 and has been reported from Mexico, Guatemala and Chile (Van der Zwet, 1970). Outbreaks of fireblight occurred in New Zealand in 1919 and in England in 1957 (Crosse *et al.*, 1958). Fireblight has since become well established in Western Europe, being recognized in the Netherlands and Poland in 1966, Denmark in 1968, West Germany in 1971, Belgium in 1973, Northern France in 1972 and South West France in 1978 (Eden-Green and Billing, 1974; Baker, 1971; Paulin and Lachaud, 1978). There is considerable concern that the disease will spread to the warmer fruit growing centres of Southern Europe and the Mediterranean. In the USA fireblight is estimated to have caused annual losses of apple and pear crops of six million dollars in the period 1951–1960 and has largely confined the production of high quality pears to the arid climates of the Pacific coastal states (US Department of Agriculture, 1965).

In some areas of the USA fireblight has been controlled since the early 1950s by spraying orchards with streptomycin at three- to four-day intervals during the blossom period. Not surprisingly, *E. amylovora* strains resistant to streptomycin emerged in California during the early 1970s and have since spread to other States (Miller and Schroth, 1972; Schroth *et al.*, 1979). The resistance is not carried on a plasmid (Panopoulos, 1978; Schroth *et al.*, 1979). Streptomycin-resistant strains are still present in orchards seven years after they had been sprayed with streptomycin (Schroth *et al.*, 1978). Orchards have been sprayed with tetracyclines to control fireblight.

2. *Pathotypes and host range*

Two distinct pathotypes of *E. amylovora* have been described, both causing diseases of rosaceae, but with different host ranges. The pathotype of primary economic importance causes severe disease in members of the subfamily Pomoideae. These hosts include species of *Amelanchier, Chaenomeles, Cotoneaster, Crataegus, Cydonia, Eriobotrya, Malus, Mespilus, Pyracantha, Pyrus* and *Sorbus* (Billing *et al.*, 1959; Eden-Green and Billing, 1974). Natural infections of apricot and plums have been reported, but only when close to heavily infected pome fruits. Within this host range, *E. amylovora* does not show any host specificity.

The second pathotype, *E. amylovora* subsp. *rubi*, has been isolated from raspberries (*Rubus iclaeus*), blackberries (*Rubus allegheniensis*) and various hybrid thornless blackberries in the USA (Starr *et al.*, 1951; Ries and Otterbacher, 1977). The raspberry isolates do not cause disease when inoculated into apple, and the raspberry is not infected by apple and pear isolates. However, the raspberry isolates will cause rot and bacterial ooze on the immature pear fruit used for testing the pathogenicity of *E. amylovora* (Ries and Otterbacher, 1977; Section II.A.5.b).

3. *Epidemiology*

E. amylovora overwinters in dormant hold-over cankers at the junction of healthy and diseased tissues. In spring these cankers produce a bacterial ooze which serves as inoculum for dissemination by wind-blown rain and insects. The relative importance of these two vectors probably varies according to climate and from year to year (Schroth *et al.*, 1974; Eden-Green and Billing, 1974). Bees can rapidly spread the infection among blossom. Fireblight can be spread between countries on infected propagation materials and it has been speculated that migrating birds may have spread the disease in Europe (Schroth *et al.*, 1974).

The spread of *E. amylovora* is favoured in warm, humid conditions particularly when the trees are in blossom. Warm weather increases the growth rate of *E. amylovora* and reduces the inoculum required to infect blossom (Lelliott, 1978). In addition to the transmission of the disease by rain splashes, high humidity increases the production of bacterial ooze, dilutes inhibitory substances in nectar and increases the likelihood of infection through stomata, lenticels and mechanical wounds (Eden-Green and Billing, 1974; Schroth *et al.*, 1974). Conditions suitable for a fireblight outbreak are particularly likely to occur if the trees blossom late and if late flowers (rattails) overlap with the flowering of wild hosts, e.g. hawthorn and pyracantha. Methods for assessing potential fireblight activity using meteorological data have been published (Billing, 1976b, 1978, 1979, 1980a,b; Møller *et al.*, 1978).

4. *Isolation*

E. amylovora is usually easily isolated from diseased blossom, shoots, bacterial ooze and active cankers. The leading edge of the infection is most likely to yield pure cultures. Small pieces of diseased tissue (preferably from the cortical layer) may be macerated in sterile water and streaked onto 5% sucrose nutrient agar (Billing *et al.*, 1960). For rapid diagnosis of fireblight, the immature pear fruit used to test pathogenicity (Section II.A.5.b) can be inoculated with the diseased tissue. The population of *E. amylovora* on healthy blossom can be assessed by washing several hundred flowers (0.5 ml sterile water per flower) and plating dilutions on selective media. Selective media are unnecessary for isolating *E. amylovora* from fireblight outbreaks, but can be used for isolation from healthy blossom and tissues or hold-over cankers in winter, all of which are likely to yield saprophytic bacteria. *E. amylovora* has characteristic colonies on two semi-selective media. The medium of Miller and Schroth (1972) is more selective than that of Ritchie and Klos (1978), but requires both a longer incubation time and microscopic examination of the colonies. Since any selective medium is likely to be toxic for senescent cells, it is desirable to check the counts using 5% sucrose nutrient agar. *E. amylovora* has also a characteristic colonial morphology on 0.3% yeast extract (Oxoid)–0.5% peptone (Oxoid) agar (Billing *et al.*, 1960; Billing, 1960; Crosse and Goodman, 1973). This medium is easy to prepare and gives results quickly.

After two days incubation at 30°C on 5% sucrose nutrient agar, *E. amylovora* forms large, dome-shaped, mucoid colonies due to the production of levan. The colonies are 3–4 mm in diameter, uniform in structure and cream coloured (Billing *et al.*, 1960). A few non-pigmented *E. herbicola* strains also produce levan on this medium, but most produce a watery mucoid growth (Slade, 1978). Colonies of *Pseudomonas syringae*, which can also cause a blossom blight of pears resembling the early stages of fireblight, are less mucoid and have a definite mottled surface (Billing *et al.*, 1960). *P. syringae* can be distinguished by the production of a green fluorescent pigment on King's medium B (King *et al.*, 1954; Lelliott, 1967).

5. *Bacteriological diagnosis of fireblight*

Colonies which are positive in a slide agglutination test are inoculated into immature pear fruit to test their pathogenicity and also streaked onto sucrose nutrient agar to obtain pure cultures for phage testing and biochemical tests (Lelliott, 1967; Miller, 1979).

(a) *Slide agglutination.* A positive slide agglutination test using putative *E. amylovora* colonies from 5% sucrose nutrient agar isolation plates is reliable presumptive evidence for fireblight (Lelliott, 1967). The antiserum should be raised against an unwashed suspension grown on 2% glycerol nutrient agar, in which the capsular antigen TV [which cross-reacts with the exopolysaccharide AP(H) produced by *E. herbicola* (Section IV:B.5.a)] is non-immunogenic for rabbits (Slade, 1978). The antiserum is diluted in saline until further dilution fails to give marked agglutination of the homologous strains within 30 s. The isolates, suspended in saline, are mixed with drops of both diluted *E. amylovora* antiserum and normal rabbit serum as a negative control (Lelliott, 1967). The results can be confirmed using pure cultures, which should give titres within 50% of the homologous titre, in tube agglutination tests.

The success of the slide agglutination test is due to the fact that *E. amylovora* strains are almost serologically homogeneous and there are no cross-reactions with *E. herbicola* when live cells, in which the lipopolysaccharide antigens are masked with capsular material, are used with antiserum that does not react with the *E. herbicola* exopolysaccharide AP(H) (Slade, 1978). Cross-reactions between *E. amylovora* and *P. syringae* (Israilskij *et al.*, 1966) have been reported not to interfere with the slide agglutination test.

(b) *Pathogenicity tests.* The pathogenicity of *E. amylovora* is easily demonstrated by inoculation of immature pear fruit (Billing *et al.*, 1960). The pear fruit should be collected before the end of July and can be stored at 4°C in plastic bags until the following season. Mature fruit is unsuitable. The pears or pear slices are inoculated by stabbing with a needle and incubated at 27°C in a moist atmosphere in plastic boxes, or on damp filter paper in Petri dishes, if very small fruit or slices are used. *E. amylovora* produces ooze within two to four days and in young fruit and slices this appears over the surface as milky beads which later turn amber. Young fruit are completely necrotized in five to six days. *P. syringae* produces dry black lesions without ooze.

Pathogenicity can also be tested for by inoculating succulent, actively growing shoots. The cut shoots are stood in water under a bell jar and should show lesions and ooze within two to six days (Billing *et al.*, 1960).

(c) *Phage testing.* A considerable number of bacteriophages have been isolated which lyse *E. amylovora* (Crosse *et al.*, 1958; Billing, 1960; Baldwin and Goodman, 1963; Erskine, 1973; Ritchie, 1978; Ritchie and Klos, 1977,

1979), but most also infect some *E. herbicola* strains (Ritchie and Klos, 1979; Billing *et al.*, 1960; Billing and Baker, 1963). Some of these phage preferentially lyse capsulated or non-capsulated forms of *E. amylovora* (Billing, 1960; Ritchie and Klos, 1979; Section II.A.6.a). Phage L3H will lyse capsulated and non-capsulated forms (E. Billing, personal communication, 1978) and has not been observed to infect other *Erwinia* spp. (Slade, 1978). However, fresh preparations of this phage can produce areas with reduced turbidity on *E. herbicola* due to the presence of an enzyme which degrades the exopolysaccharide AP(H).

Stocks of phage L3H can conveniently be produced by inoculating 5 ml of broth containing 0.3% yeast extract (Oxoid) and 0.5% peptone (Oxoid) with 0.5 ml of a 24-h culture of *E. amylovora* and a drop of phages suspension. After 24 h at 25°C the culture is centrifuged to remove cell debris and the supernatant fluid stored at 4°C in the presence of excess chloroform. Higher titre stocks can be produced from confluent lysis on glycerol nutrient agar, using the methods of Adams (1959). Phage L3H was found to be unstable if stored in a broth containing 0.5% sodium chloride (Slade, 1978). The phage test is most conveniently carried out by spreading 0.1 ml of a 24-h broth culture of the test organism on a well-dried 2% glycerol nutrient agar plate and spotting with a phage suspension containing about 2×10^2 plaque forming units (PFU) per drop (i.e. enough to give confluent lysis of a sensitive strain) and incubating at 25°C for two days.

(d) *Biochemical tests.* *E. amylovora* can most readily be distinguished from saprophytic bacteria by its inability to reduce nitrate and its ability to produce acid from sucrose but not from salicin, in a weakly buffered mineral salts medium containing nicotinic acid and using bromocresol purple as indicator (Billing *et al.*, 1960; Lelliott, 1967). *E. herbicola* produces acid from salicin and most strains reduce nitrate. Most *Erwinia* spp., apart from *E. amylovora*, hydrolyse aesculin (Slade, 1978). Biochemical differences between *E. amylovora* strains appear more quantitative than qualitative (Billing *et al.*, 1961).

6. *Serology*

(a) *Serological structure.* *E. amylovora* is capsulated with a rough lipopolysaccharide (R-LPS). The LPS can be partially extracted by the phenol–chloroform–petroleum ether method of Galanos *et al.* (1969), which does not extract smooth lipopolysaccharide (S-LPS), and also precipitates in the presence of magnesium ions (Slade, 1978). Non-capsulated strains and cells which have been heated at 100°C for 30 min frequently autoagglutinate. The R-LPS contains heptose, glucose, galactose, glucosamine and rhamnose (Chatterjee *et al.*, 1977), the same composition as the R-LPS of *Escherichia coli* K12 (Rapin and Mayer, 1966; Lüderitz *et al.*, 1971) when grown at the supra-optimal temperature of 30°C. The cell wall of *E. amylovora* is unusually permeable to antibacterial agents and periplasmic enzymes in a manner similar to the R-LPS mutants from other genera (Chatterjee *et al.*, 1977). The LPS antigen could not be detected in sonicated cell suspensions of *E. amylovora* using immunodiffusion techniques (Slade, 1978).

Another neutral, heat-stable antigen, GAl (standing for group amylovora), could readily be detected in both capsulated and non-capsulated strains and in cells that had been heated at 100°C for 30 min. Antigen GAl was present in the supernatant fluid after centrifuging extracted R-LPS at 100 000 **g** and so is thought to be a cell wall polysaccharide distinct from the R-LPS. However, GAl could not be detected in any mutants with a modified LPS structure and was regained when mutants reverted to the normal LPS, so the syntheses of GAl and LPS appear to be linked (Slade, 1978).

Virulent *E. amylovora* strains produce a high molecular weight, acidic polysaccharide TV (standing for typical virulent) which forms a large capsule and extracellular slime. Cells grown in nutrient broth and heated at 100°C for 30 min do not absorb antibodies to the capsular antigen (Slade, 1978). Antigen TV is the principal constituent of the ooze or gum exuded from plant tissue infected with *E. amylovora* (Slade, 1978). Antigen TV loses antigenic determinants (probably O-acyl groups) when treated with 0.25 M sodium hydroxide at 56°C for 2 h and is then called AP(A) (standing for acidic polysaccharide from amylovora). AP(A) is antigenically indistinguishable from AP(H) produced by *E. herbicola* NCPPB 1264 (Slade and Tiffin, 1978). Antigen AP(A) has approximately the same molecular weight and charge as antigen TV as determined chromatographically. Antigen AP(H) contains glucose, galactose and glucuronic acid (E. Percival, personal communication) with a 2 to 1 ratio between neutral and acidic sugars (Slade, 1978). The same sugars are found in antigen TV together with a variable proportion of mannose (0.8–5.1%) (Eden-Green and Knee, 1974; Eden-Green, 1972; Bennett and Billing, 1980). The presence of mannose may be due to the presence of contaminants since the samples analysed had been purified only by ethanol precipitation.

Using immunodiffusion techniques, an antigenic difference can be detected between purified antigen TV and the antigen TV in the extracellular slime of *E. amylovora* cultures. The antigenic properties of the slime become identical with those of the purified antigen if the slime is heated at 100°C, ethanol precipitated or even dialysed against distilled water, but not if dialysed against phosphate-buffered saline (PBS) (Slade, 1978). These treatments are unlikely to alter the polysaccharide, and so it is probable that the antigenic changes are due to the denaturation of basic proteins complexed with the acidic polysaccharide. The formation of labile complexes could have important implications in the debate on whether the slime is a phytotoxin (Section II.A.6.c). *E. amylovora* is motile, but there are no reports on the nature of the flagellar antigen.

Another antigen, called GA5 by Slade (1978), might be related to virulence and is found in the extracellular slime of all virulent strains of *E. amylovora*. It is a complex antigen with at least two distinct components which can be detected by immunodiffusion techniques. One capsulated but avirulent strain lacking antigen GA3 is described in Table III, p. 233.

The function of *E. amylovora* antigens has been studied using mutants with altered serological structures (Table I). Large numbers of mutants which differ in antigenic composition can be obtained by selection for resistance to bacteriophages (Slade, 1978).

Phages 4S and 4L (NCPPB 1307 and 1508) infect only capsulated cells and so can be used to select non-capsulated mutants. Phage 23 preferentially infects non-capsulated cells resulting in the isolation of capsulated mutants and mutants with a modified LPS structure (and lacking antigen GA1) which are completely resistant to phage 23. The majority of mutants with altered LPS antigen possess LPS 2 which cross-reacts strongly with the wild type antigen LPS 1 in passive haemagglutination tests. Mutants with LPS 2 are frequently virulent, but LPS 2 has not been observed in wild type strains. Mutants AT6 and AT7 are thought to contain LPS antigens different from LPS 1 and LPS 2 (Slade, 1978).

Ayers *et al.* (1979) selected avirulent non-capsulated variants using different bacteriophages. The serological structure of these strains has not been studied, but some of their bacteriophages appear to have receptors different from those of phages 4S, 4L, 23, and L3H.

(b) *Serological classification.* Naturally occurring strains of *E. amylovora* exhibit little serological variation apart from loss of the capsule. No serological classification has been described suitable for epidemiological studies.

Samson (1972) detected slight serological differences between *E. amylovora* strains from different countries. The strains could be subdivided into five groups according to the presence of three antigenic factors on cells which had been heated to 121°C for 2 h. The agglutinating antibody was not absorbed by live cells of the homologous organism. There are no reports of the use of this classification by other workers.

Slade (1978) compared the heat labile antigens of English and American isolates using immunoelectrophoresis and could not detect any differences.

Billing (1960) recognized typical and atypical forms of *E. amylovora* depending on their colonial morphology on 0.3% yeast extract, 0.5% peptone agar. Only the typical form produces a capsule on this medium. Typical and atypical forms differ in the regulation of the production of antigen TV (Section II.A.6.a), since both forms will produce a capsule if the medium is supplemented with a suitable sugar, e.g. sorbitol or fructose (sorbitol is the principal translocation sugar in Rosaceae) (Bennett and Billing, 1978). The difference in the regulation of antigen TV may be due to a difference in metabolism as many atypical strains require thiamin (R. A. Bennett, personal communication) and atypical strains produce more acid in the methyl red test (Slade, 1978).

The production of antigen TV by typical, but not atypical, variants of the same strain grown in nutrient broth (without yeast extract) provides a convenient method of producing antigens with and without antigen TV for the production of antisera. Atypical variants of typical strains can readily be obtained by selecting for non-capsulated cells with phages 4S and 4L under conditions where atypical strains do not produce capsules (Slade, 1978). The opposite change can be selected with phage 23. Cells of atypical strains, grown in nutrient broth, usually absorb antibodies to antigen TV.

(c) *Serology and virulence.* The pathogenetic mechanism causing fireblight is of particular interest because, unlike many other plant pathogens, *E. amylovora* does not produce any enzymes to degrade plant cell walls (Seemuller and Beer, 1976) and does not produce any protein toxins. Thus it offers the potential of studying the subtleties of pathogen–host interaction without the presence of the obvious pathogenetic mechanisms. As toxic proteins or enzymes are not involved, the possibility remains that the disease is mediated by the carbohydrates of the bacterium. Carbohydrate–lectin interactions have been implicated in the host specificity of legumes infected by *Rhizobium* and the binding of *Agrobacterium tumifaciens* to host cell walls (Dazzo and Hubbell, 1975; Dazzo *et al.*, 1976; Wolpert and Albersheim, 1976; Lippincott and Lippincott, 1969; Roberts and Kerr, 1974).

Two observations are of importance in the study of pathogen–host interaction. First, there is the observation that the production of a capsule is necessary for virulence (Bennett and Billing, 1978). The role of the capsule is unclear but it may protect the cell from host defence mechanisms or prevent the induction of host defence mechanisms by other bacterial components. The former possibility is supported by the observation that a non-capsulated strain had a markedly lower growth rate in plant tissue than a non-virulent capsulated strain which had the same growth rate *in vitro* (Bennett, 1980). Non-capsulated strains have been observed to induce the host's hypersensitive reaction (Huang *et al.*, 1975).

Eden-Green (1972) observed that apple shoots wilt if placed in a solution of the extracellular slime (antigen TV, Section II.A.6.a) produced by *E. amylovora*. This is probably due to the vascular tissues becoming blocked by the polysaccharide (Eden-Green, 1972; Sjulin and Beer, 1978), but Goodman *et al.* (1974) claimed that the polysaccharide was a specific toxin, produced only *in vivo*, and named it amylovorin. The polysaccharides produced in the slimes produced *in vivo* and *in vitro* are antigenically identical, have similar chemical composition and infra-red spectra and can not be distinguished by chromatography on Sepharose 4B or DEAE cellulose (Slade, 1978; Slade and Tiffin, 1978; Bennett and Billing, 1980). Although host-specific properties were initially attributed to the polysaccharide (Goodman *et al.*, 1974), further studies have demonstrated that there is no correlation between the sensitivity of hosts to *E. amylovora* and the wilt-inducing properties of amylovorin (Ayers *et al.*, 1977; Beer and Aldwinckle, 1976; Goodman *et al.*, 1978). The question of whether the extracellular slime is a phytotoxin is quite distinct from the correlation between capsulation and virulence although the slime and capsules always occur together and both consist of antigen TV.

The role of the polysaccharide in the aetiology of the disease is uncertain since *E. amylovora* is not usually present in the phloem (Eden-Green and Billing, 1974) and shoots inoculated with *E. amylovora* do not usually wilt until the later stages of the disease (Eden-Green, 1972; Wheeler, 1975).

Three factors have complicated studies on the toxicity of antigen TV. Shoot wilt is a very subjective test which can not be easily quantified. The polysaccharide preparations used by different workers are likely to have differed in their purity. The third factor is that solutions of highly acidic polysaccharides, like antigen TV and solutions of high molecular weight dextrans used as controls, are likely to differ widely in molecular conformation, viscosity and ability to form gels. The problem could be resolved by comparing the very similar polysaccharides produced by *E. herbicola* and *E. amylovora* (Slade and Tiffin, 1978) and using depolymerases produced by bacteriophage to alter the molecular weight of antigen TV. The possibility that antigen TV may act as a carrier for basic proteins has not been adequately studied.

Avirulent strains of *E. amylovora* producing both capsules and slime, indicate the existence of other pathogenetic mechanisms (Bennett, 1980). A potential candidate for a virulence-associated antigen is the extracellular antigen GA5 (Section II.A.6.a) which is produced by virulent strains but not by one avirulent capsulated *E. amylovora* strain (Slade, 1978). The loss of antigen GA1 (Section II.A.6.d) and modification of the lipopolysaccharide did not make strains avirulent, so cell wall antigens (which are masked by the capsule) may not be important for virulence.

(d) *Preparation of antigens TV and AP(H)*.   The relationships between antigen TV, AP(A) and AP(H) are discussed in Section II.A.6.a. These antigens are important in studies on the virulence of *E. amylovora* (see above) so it is appropriate to consider factors of importance in their preparation. Antigen TV may be prepared from any virulent strain of *E. amylovora*. *E. herbicola* NCPPB 1264 has been used to prepare antigen AP(H); other strains of *E. herbicola* probably differ in the amount of AP(H) produced, its molecular weight and in its content of O-acyl groups (Slade, 1978). The main problems encountered in the purification of these antigens are associated with their negative charge and high molecular weight, the latter being over five million, as measured by their exclusion from Sepharose 2B and 4B (Slade, 1978; Slade and Tiffin, 1978; Bennett and Billing, 1980). As solutions of these polysaccharides are very viscous, considerable dilution is essential before bacterial cells can be removed by centrifugation. Precipitation of the polysaccharide is best avoided since it is frequently difficult to re-dissolve it. The principal difficulty encountered during purification is the removal of basic proteins bound to the acidic carbohydrate by ionic forces.

Attention should be paid to the following steps in any purification procedure.

1. The bacteria should be grown on a defined solid medium (Bennett and Billing, 1978, 1980).
2. The mucoid bacterial growth should be diluted and the bacterial cells removed by centrifugation.
3. Pronase and EDTA should be added to aid the removal of protein and to reduce the viscosity.
4. Phenol may be used to extract protein and polypeptide, the phenol being subsequently removed from the aqueous phase by extraction with di-ethyl ether.
5. A final purification step of absorption and one step elution from DEAE cellulose using phosphate buffer (NaCl produces fumes during carbohydrate analysis).

Passive haemagglutination techniques are best used for a serological study of these polysaccharides. The polysaccharides need to be acylated with long chain fatty acids (Hammerling and Westphal, 1967) before they will adsorb to sheep erythrocytes (Slade, 1978). The polysaccharides may also be compared and estimated by two-dimensional immunoelectrophoresis (Laurell technique) (Slade and Tiffin, 1978; Ayers *et al.*, 1979), but relatively large amounts of antiserum are required. Low molecular weight fractions of the polysaccharide give precipitation lines in Ouchterlony immunodiffusion plates, but purified preparations of undegraded antigen may fail to diffuse through the gel.

B. *Erwinia salicis*

1. *Introduction*

E. salicis causes watermark disease of cricket bat willows, *Salix alba* subsp. *caerulea*, in which the wood becomes stained red-brown or black and considerably weakened (Preece, 1977; Preece *et al.*, 1979). This renders the wood useless for making e.g. cricket bats, which need lightweight, white wood with the ability to spring and recover after severe blows. Trees with watermark disease may show wilting of foliage, reddening and browning of leaves, and die-back of several branches near the top of the tree, but external symptoms may not develop for five to seven years after initial infection. The disease has been most intensively studied in East Anglia and in the Netherlands, but probably also occurs elsewhere in Europe.

Watermark disease can affect other species of willows, but the severity of the disease in cricket bat willows is probably due to both genetic and environmental factors. The cricket bat willow is a sterile hybrid which is reproduced vegetatively, and so genetically identical trees are grown repeatedly on the same sites in a continuous monoculture. The disease can unwittingly be transmitted during vegetative reproduction because of the long lag before symptoms of the disease are expressed. Artificial infection is usually only possible via deep wounds into the xylem or via exposed wood surfaces. Infected cutting tools are the most obvious method of cross-infection (Wong, 1974). *E. salicis* can survive for at least four years in stumps of felled trees and these stumps may root graft to healthy trees. Healthy cuttings may be infected if grown in soil heavily infected with *E. salicis* (Preece, 1977). In eastern counties of England, the disease is being successfully controlled by statutory inspection and destruction of infected trees (Wong *et al.*, 1974; Preece *et al.*, 1979).

2. *Isolation*

E. salicis is easiest to isolate when red leaf symptoms first become visible, since old die-back lesions additionally contain large numbers of saprophytic bacteria. Young branches and shoots are surface sterilized by flaming with alcohol and the bark removed. Sap from a fresh cut is then squeezed out using sterile pliers and plated on to 1% glycerol nutrient agar. *E. salicis* forms small (1–2 mm) translucent smooth colonies after five days incubation at 26°C. The colonies of most English isolates are golden brown in transmitted light and grey in oblique light, but isolates from the Netherlands lack yellow pigment (De Kam, 1976). *E. salicis* may also be isolated on 5% sucrose nutrient agar, on which it forms characteristic mucoid colonies. *E. salicis* gives very poor growth on nutrient agar and its growth in some media is inhibited by agar (Dye, 1968).

3. Serological tests

Slide agglutination is a good rapid diagnostic test for *E. salicis*. The test employs neat antiserum raised against *E. salicis* NCPPB 1466 and either live cells from isolation plates or sap expressed from infected shoots (Wong and Preece, 1973; De Kam, 1976; Preece *et al.*, 1979). The reaction, which is immediately visible, should be compared with a control using normal rabbit serum. Diagnosis by slide agglutination using sap should always be confirmed by culture of the organism, since sap from healthy willow shoots may occasionally give false-positive agglutination reactions (De Kam, 1976; Wong and Preece, 1973). There is no report of other organisms agglutinating in *E. salicis* antiserum.

E. salicis appears to exhibit little serological variation. A few strains give noticeably weak agglutination in antiserum to NCPPB 1466 (Preece *et al.*, 1979), but De Kam (1976) using cross-absorption techniques was unable to detect any serological differences between isolates from England and the Netherlands. Immunodiffusion has detected only quantitative strain differences (De Kam, 1976; Slade, 1978).

The lipopolysaccharide of *E. salicis* 1466 may be in the R form since it can be partially extracted by the petroleum ether–chloroform method of Galanos *et al.* (1969) (Slade, 1978). In passive haemagglutination tests, this LPS antigen gave very weak (at 1/40 or less serum dilution) reactions with antisera to other *Erwinia* spp. The other heat-stable antigens detected in sonicated cell suspensions of strain 1466 may be capsular polysaccharides since they were found in the supernatant fluid after centrifugation of phenol–water extracted LPS at $100\,000\,\text{g}$ (Slade, 1978).

It is not easy to prepare high titre antiserum to *E. salicis*, the homologous titre usually being in the range 1:160 to 1:640.

4. Pathogenicity tests

The best diagnostic test for a plant pathogen is the demonstration of its ability to reproduce the disease. However, testing the pathogenicity of *E. salicis* is difficult, due to the very long incubation time of the disease (Adegeye and Preece, 1978). Wong (1974) has overcome these problems using large inocula of *E. salicis* to produce watermark staining in willow cuttings. Cricket bat willow shoots (20 cm long and less than 2 cm in diameter) are wiped with ethanol and the end is cut off below the surface of a dish of sterile water. The cut end is covered with a small plastic tube containing water and transferred to a boiling tube containing 25 ml of a 48-h nutrient broth culture (approximately 10^9 colony forming units, CFU) of the isolate to be tested. The small tube is removed with sterile forceps. The boiling tube is sealed with plastic film and wrapped in light proof plastic. The cuttings are incubated in a controlled environment cabinet (75% relative humidity, 16-h day, 34 000 Lux) and the nutrient broth in the tubes replenished as required. After 10–14 days the discoloration of the wood due to the action of *E. salicis* can be observed by cutting the base of the shoot longitudinally and exposing it to the air for 5–10 h (Preece *et al.*, 1979).

5. *Variable characteristics*

A few characteristics suitable for typing *E. salicis* have been reported. *E. salicis* strains isolated in the Netherlands are unable to grow on either galactose or raffinose as sole carbon source and do not produce a yellow pigment on autoclaved potato tissue (De Kam, 1976). However, English isolates are able to grow on galactose, some can grow on raffinose and they usually produce a yellow pigment on isolation. Production of the yellow pigment may be an unreliable characteristic since it is lost on prolonged subculture *in vitro* (Preece *et al.*, 1979). English isolates are also variable in pectolytic activity and sensitivity to bacteriophage (Wong, 1974). An alkaline medium (pH 8.0) is required for the detection of the pectolytic enzyme of *E. salicis* and no strains liquefy pectate gels below pH 7.0 (Preece *et al.*, 1979; Section II.C.2).

C. *Erwinia nigrifluens* and *rubrifaciens*

Both these pathogens cause diseases of the Persian Walnut tree (*Juglans regia*) in California. *E. nigrifluens* produces necrotic cankers in the outer bark, whereas *E. rubrifaciens* produces long necrotic bands in the phloem. These diseases usually occur on mature trees, i.e. more than six to eight years old, but *E. rubrifaciens* can be transmitted to yearling trees by graft tissue (Wilson *et al.*, 1967; Teviotdale, 1979). *E. nigrifluens* does not usually kill infected trees, but when both pathogens occur in the same tree, extensive necrosis of the trunk and scaffold branches may result.

1. *Isolation*

The watery exudates from necrotic areas and pieces of infected bark may be streaked on to a modified EMB agar (Difco) containing 10 g per litre of glucose instead of lactose and sucrose, and incubated at 30°C (Wilson *et al.*, 1967). On this medium, colonies of *E. nigrifluens* are dark violet with a green metallic sheen in oblique light while *E. rubrifaciens* colonies are pink or orange-red and some strains have a metallic sheen. Neither organism grows well on nutrient agar.

2. Identification

E. rubrifaciens produces a diffusible red pigment when grown on yeast extract (1%)–glucose (2%)–chalk (2%) agar, produces levan on 5% sucrose nutrient agar, and produces small craters around colonies on pectate gels (Paton, 1959; Wilson et al., 1967). E. nigrifluens can be identified using a lytic bacteriophage (Zeitoun and Wilson, 1969), but high titres of this phage can cause clear areas in lawns of E. rubrifaciens by binding to the cells and inhibiting growth. Little is known of the serology of these organisms, although immunodiffusion techniques have been used to identify them (Zeitoun and Wilson, 1966; Wilson et al., 1967). There is no report of any serological variation within these species.

D. Erwinia quercina

This organism causes the drippy nut disease of oaks (Quercus agrifolia and Q. wislizenii) in California. Infection of the acorn is primarily through ovipositor wounds made by insects. The disease causes copious oozing of sap from the wound and later from the cup after the acorn has become detached (Hildebrand and Schroth, 1967).

E. quercina colonies are white, circular and raised with entire margins on potato–glucose–peptone–chalk agar and luxuriant growth is produced within 24 h at 30°C. The organism causes blight within seven days in shoots of oak seedlings incubated at 29°C and inoculated by puncturing the stem with a needle. E. quercina has the unusual ability of inducing slices of carrot, turnip and beetroot to produce profuse lateral root development within three to four days (Hildebrand and Schroth, 1967).

Erwinia spp. have also been implicated in die back of oaks in Eastern Europe, but the organisms concerned do not appear to have been well characterized (Petrescu, 1974).

E. Erwinia tracheiphila

This organism causes a vascular wilt of Cucurbita spp. It can be isolated on glucose–nutrient agar but grows very poorly on nutrient agar. There are very few references to recent work on this species. Evans and Stevenson (1980) have used immunofluorescent techniques to identify E. tracheiphila in artificially infected melon stems.

F. *Erwinia alfalfa*

This organism was first named *E. amylovora* subsp. *alfalfa* (Shinde and Lukezic, 1974b) following Dye's (1968) proposal that all members of the *amylovora* group were to be treated as subspecies of *E. amylovora*. However, the specific status of members of the *amylovora* group has now been restored (Lelliott, 1974) so, in the absence of any evidence for a close relationship between *E. amylovora* and *E. alfalfa*, the latter will be considered as a distinct species, comparable in rank with other members of the group.

E. alfalfa has been isolated from discoloured roots of alfalfa (*Medicago sativa*) in eastern Pennsylvania (Shinde and Lukezic, 1974b). *Fusarium* spp. and fluorescent pseudomonads are also associated with this condition (Shinde and Lukezic, 1974a). Inoculation of alfalfa plants with *E. alfalfa* caused slight to moderate necrosis of the roots and a yellowing of the leaves followed by death of the plants.

E. alfalfa colonies on King's B medium (King *et al.*, 1954) are medium in size, white, convex and smooth. A light pink, diffusible pigment is produced within seven days at 27°C. The pigment is easily distinguishable from the red pigment produced by *E. rubrifaciens*. Capsules are produced by both *E. alfalfa* and *E. amylovora*, but *E. alfalfa* can be distinguished by its ability to hydrolyse Tween 80, reduce nitrate and by its failure to produce levan (Shinde and Lukezic, 1974b).

Antiserum raised against the type strain, E3, agglutinated all isolates of *E. alfalfa*, but not strains of *E. amylovora* and *E. rubrifaciens*. Cross-absorption studies have not been carried out, but immunoprecipitation techniques demonstrated three serologically distinct types represented by strains E2, E3 and E4 (Shinde and Lukezic, 1974b).

G. *Erwinia mallotivora*

E. mallotivora is a recently described member of the *amylovora* group which causes leaf spot of *Mallotus japonicus* in Japan (Goto, 1976). Colonies on 5% sucrose nutrient agar are white, translucent, mucoid and domed. *E. mallotivora* is very similar to *E. amylovora*, but can be distinguished by the production of acid from xylose, glycerol and mannose, but not from sorbitol in 1% peptone water (Dye, 1968) and by its failure to liquefy gelatin or utilize formate. *E. mallotivora* is also able to produce acid from mannitol and cellobiose with which *E. amylovora* can occasionally give very weak acid production (Slade, 1978).

H. Other members of the *amylovora* group

The isolation of new members of the *amylovora* group will have important implications for the taxonomy of the group and studies of their pathogenetic mechanisms. Undescribed species are very likely to exist since the diseases of relatively few plants have been studied and non-pathogenic species in the poorly characterized epiphytic flora could easily have been overlooked. The presence of undescribed species in the epiphytic flora is illustrated by strain *AL2* which was isolated from apple by Billing and identified by numerical taxonomy as a member of the *amylovora* group (Slade, 1978).

Most members of the *amylovora* group may be recognized by their weak production of acid from sugars and by their failure to reduce nitrate using the methods of Dye (1968). The absence of nitrate reduction may depend on the method used, since some workers have detected nitrate reduction by *E. salicis* and *E. nigrifluens* (Preece *et al.*, 1979; Wilson *et al.*, 1967). However, *E. alfalfa* does reduce nitrate (Shinde and Lukezic, 1974b) along with the majority of other members of the Enterobacteriaceae (apart from a few strains in the *herbicola* group, Table IV). Some members of the *amylovora* group produce high domed colonies on 5% sucrose nutrient agar due to the production of levan, but this property is also possessed by a few strains of *E. herbicola*. No member of the *amylovora* group produces acid from lactose, a sugar found only in mammals.

III The *carotovora* group

A. Introduction

The *carotovora* group contains plant pathogens causing soft rot diseases. These pathogens are usually found associated with plant materials, but may occur transiently in associated habitats, e.g. irrigation water (Goto, 1979), and insects feeding on plants (Molina *et al.*, 1974; Harrison *et al.*, 1977; Kloepper *et al.*, 1979). There is little evidence that members of the *carotovora* group are pathogenic for animals and the presence of these organisms on vegetables is not usually regarded as a health hazard.

There are two *Erwinia* spp. causing soft rot diseases. *E. carotovora* causes soft rot in a wide range of plants, particularly attacking storage tissues, whereas *E. chrysanthemi* causes vascular wilts and parenchymatal necrosis in field crops in tropical and subtropical countries and occurs in glass houses in temperate climates.

TABLE IV

Tests differentiating between *Erwinia herbicola* subsp. *herbicola*, *E. herbicola* subsp. *ananas* and *E. rhapontici* [a]

Species	Cluster Method [b]	No. of strains studied	Amygdalin (A)	α-Methyl glucoside (A)	Glyceric acid (B)	Malonic acid (B)	Mucic acid (B)	Saccharic acid (B)	meso-Tartaric acid (B)	L-Tartaric acid (B)	Gelatin (C)	Casein (D)	Indole (E)	Nitrate reduction (F)	Phenylalanine deaminase (G)	Reducing compounds from sucrose (H)	Gas from glucose (A)	Potato rotting (I)	Growth at 38°C (J)	Growth in NB+10% NaCl (K)	High-domed colonies (Levan) (L)	Symplasmata (M)	Yellow pigment (N)	Pink pigment (O)
E. herbicola subsp. *herbicola*	H:1	10	1/10	—	+	+	+	+	+	9/10	+	—	—	+	+	—	—	—	+	2/10	—	+	+	—
	H:2	3	—	1/3	2/3	2/3	+	+	+	+	+	—	—	+	+	—	—	—	—	—	—	2/3	+	—
	H:3	4	—	—	2/3	1/3	+	+	+	—	2/4	—	—	+	+	—	—	—	—	—	—	+	—	—
Strains of low phenon cluster value ("unclustered")	2036	1	+	+	+	+	+	+	+	+	+	—	—	+	+	+	—	—	+	—	—	+	+	—
	1267	1	—	—	—	—	+	+	+	+	+	—	—	+	+	—	+	—	—	—	+	—	—	—
E. herbicola subsp. *ananas*		4	2/4	2/4	+	+	+	+	+	+	+	+	3/4	1/4	—	—	—	—	+	+	—	—	+	—
Erwinia sp.	E1	3	+	—	—	—	—	—	—	—	+	—	—	—	—	2/3	—	—	—	2/3	2/3	—	—	—
E. rhapontici		6	+	5/6	5/6	5/6	—	—	—	—	—	—	—	+	3/6	+	—	+	—	2/3	—	—	—	+

[a] The fractions indicate the proportion of strains which gave positive results. Data from Slade (1978).

[b] Methods: All cultures were incubated at 25°C unless otherwise stated. A: Acid production from 1% (w/v) sugar in 1% peptone water within 14 days. B: Utilization of organic acids in Slade's (1978) modification of Dye's (1968) OY medium within 21 days. C: Hydrolysis of nutrient gelatin within 30 days. D: Hydrolysis of casein detected on 2% (w/v) skimmed milk agar flooded with acidic mercuric chloride solution after four days incubation. F: Nitrate reduction tested by method 1 of Cowan and Steel (1974) after 24 h incubation. G: Phenylalanine deaminase tested by the method of Cowan and Steel (1974) after five days incubation. H: Production of reducing compounds from sucrose, tested by the methods of Dye (1968) in shaken cultures in 5% (w/v) sucrose nutrient broth after 48 h incubation. I: Potato rotting tested by the methods of Roberts (1974). J: Visible growth in nutrient broth incubated at 38°C for seven days. K: Visible growth in nutrient broth containing 10% (w/v) NaCl within seven days. L: Production of Levan on 5% (w/v) sucrose nutrient agar. M: Symplasmata (sausage shaped aggregates of cells) observed in condensate water of nutrient agar slopes after four days incubation. N: Yellow-pigmented colonies on nutrient agar. O: Production of a pink water soluble pigment on glycerol nutrient agar containing 0·1% (w/v) yeast extract (Difco).

B. Effect on plants

The soft rot diseases of plants caused by *E. carotovora* and *E. chrysanthemi* are completely different from the necrotic diseases caused by the *amylovora* group and are characterized by pectinolytic degradation of the middle lamella of host cell walls resulting in rapid tissue destruction (Codner, 1971). Tissue maceration, loss of electrolytes and death of plant cells can be reproduced by purified endopectate transeliminase (lyase) from *E. carotovora* (Dean and Wood, 1967; Mount *et al.*, 1970; Tseng and Mount, 1974) with the damage to the cell membrane resulting in the inability of the damaged cell wall to protect against turgor or osmotic pressures (Stephens and Wood, 1975; Basham and Bateman, 1975a,b). Enzymes which attack protein, phosphatides, cellulose, galactomannan and xylan do not macerate potato tuber tissue (Stephens and Wood, 1975; Tseng and Mount, 1974; McClendon, 1964). Mutants of *E. chrysanthemi*, lacking pectate transeliminase, but possessing the hydrolytic polygalacturonase, are unable to macerate plant tissues (Chatterjee and Starr, 1977). The nomenclature of pectolytic enzymes has been discussed by Codner (1971) and Bateman and Millar (1966).

C. Isolation and counting techniques

1. *Methods*

The procedures used for isolating soft rot erwinias depend on their probable population density. Specimens from soft rot lesions are streaked on to nutrient agar or King's medium B (King *et al.*, 1954) and a selective medium incorporating a pectate gel (Cuppels and Kelman's medium, 1974). The edge of the soft rot lesion is likely to contain the pathogen with the lowest proportion of saprophytic bacteria. For the isolation of other pectinolytic bacteria from soft rots (e.g. pseudomonads), a pectate gel overlay (Paton, 1959) on a non-selective medium such as nutrient agar (with the addition of 3 **g** per litre of $CaCl_2.6H_2O$) may be used.

Selective media are essential for isolating and enumerating soft rot *Erwinia* in soil since high numbers of saprophytic organisms can suppress the growth of *E. carotovora* on non-selective media (O'Neill and Logan, 1975) and delay the isolation of pure cultures. Where the concentration of *E. carotovora* in soil exceeds 10^3 CFU per gram of soil, counts may be done by plating serial dilutions of a 10% (w/v) suspension of soil in sterile distilled water on to a selective medium (Cuppels and Kelman, 1974; Meneley and Stanghellini, 1976). Counts of lower populations of *E. carotovora* (< 25 CFU per gram of soil) may be obtained using the quantal method (Miles and Misra, 1938) in which drops from dilutions are applied to selective media (Perombelon,

1971b). However, as the pits in the medium resulting from pectolysis have to be counted, the bacterial growth has to be washed off first and it then becomes more difficult to check the identity of the organisms. Small populations of soft rot *Erwinia* in soils can be enriched by incubating the soil anaerobically in a pectate–mineral salts solution (Meneley and Stanghellini, 1976). After 48 h at room temperature, the enrichment medium is plated on to selective media. The technique will reliably isolate *Erwinia* spp. from soils inoculated with > 10 CFU per gram (dry wt) of soil and most of the time from soils containing 2–7 CFU per gram (Meneley and Stanghellini, 1976). The method could be adapted to a most probable number method for estimating soil populations of soft rot coliforms.

Incubation of vegetables under conditions likely to promote soft rot is a method of enrichment used to assess the degree of contamination of a crop with soft rot bacteria and the losses likely to be due to soft rot during storage. Development of soft rot is favoured by a high relative humidity, when organisms in the surface film of water can enter lenticels and wounds, and under anaerobic conditions when the vegetables' resistance to *E. carotovora* is greatly reduced (Lund, 1979). The methods used for detecting the contamination of potato tubers with *E. carotovora* include sealing tubers wrapped in damp tissue paper in polythene bags (Perombelon, 1972) or polyvinylidene film (De Boer *et al.*, 1975), incubating tubers under water (Vruggink and Maas Geesteranus, 1975) and in a mist chamber (Lund and Kelman, 1977). Onset of rotting is accelerated by wounding the lenticels with sterile toothpicks (De Boer *et al.*, 1975). Perombelon (1979) found that the incubation of potato tubers at 15°C selected for *E. carotovora* subsp. *atroseptica* and incubation at 25°C favoured the isolation of *E. carotovora* subsp. *carotovora*. Incubation at 22°C was optimal for the isolation of both subspecies. The presence of *E. carotovora* in the induced soft rot is confirmed by plating on to selective media. These methods are easy to perform and amenable to large scale testing.

Selective media are incubated at 22°C for 48 h and any colonies of pectolytic bacteria subcultured on to the same medium. Colonies of soft rot *Erwinia* form deep cup-like depressions on Cuppels and Kelman's medium and the colonies are irridescent, translucent and criss-crossed with internal markings when examined under oblique light. An experienced worker can usually distinguish them from other pectolytic soil-borne organisms which usually form shallow wide depressions on Cuppels and Kelman's medium.

Isolated colonies are streaked onto nutrient agar or King's medium B and inoculated into Hugh and Leifson's glucose oxidation–fermentation medium and on to a raw potato slice to confirm the isolation of a soft rot *Erwinia*. Soft rot *Erwinia* are fermentative in Hugh and Leifson's medium, whereas all other Gram-negative pectolytic bacteria that are not members of the Enterobacteriaceae show oxidation. Soft rot *Erwinia* spp. would not be confused with other pectolytic members of the Enterobacteriaceae [e.g. *Erwinia salicis, E. rubrifaciens* (Lelliott, 1974), *Klebsiella pneumoniae* and *Yersinia* spp. (Starr *et al.*, 1977)] which produce only small amounts of pectolytic enzymes causing the colonies to sink slightly below the surface of the surrounding pectate gel. These organisms would not be found in soft rot lesions and will not rot potato slices.

Potato slices are prepared by surface sterilizing washed tubers with sodium hypochlorite solution or ethanol. One cm thick slices are aseptically placed in a Petri dish on sterile, moistened filter paper and a loop of bacterial suspension placed in a freshly made cut or nick in the upper surface. Rot production is recorded if most of the slice has rotted after incubation at 22 C for 48 h. Slight rot around the point of inoculation is ignored. At least one uninoculated slice from each tuber should be used as a negative control to check that the potato is not the source of the soft rot bacteria.

Tests differentiating *E. carotovora* from *E. chrysanthemi* are given in Table V.

2. Selective media

Most methods of isolating the soft rot pathogens rely on detecting their copious production of extracellular pectolytic enzymes. There are few reports of naturally occurring non-pectolytic strains. Pectate (polygalacturonic acid or its sodium salt) can be incorporated in either a single layer medium with a reduced (0.4% w/v) concentration of agar or it can be gelled as an overlay on a solidified basal medium containing increased calcium levels (to stabilize the pectate gel overlay; Paton, 1959). Single layer media, as described by Cuppels and Kelman (1974) and Beraha (1968), are easier to prepare, but have to be dispensed immediately after preparation as they can not be reheated.

O'Neill and Logan (1975) examined in detail the selective media for isolating *E. carotovora* and concluded that a modification of Cuppels and Kelman's (1974) medium with the addition of 0.8 g per litre of manganous sulphate and a silicone antifoam gave the highest *E. carotovora* counts and greatest inhibition of other soil bacteria. This medium is simple to prepare (unlike many others) and soft rot *Erwinia* growing on it have a characteristic colonial morphology. The medium contains crystal violet to suppress Gram-positive bacteria (approximately 93% of the bacterial population of the soil; Holding, 1960) and fungi, whereas the addition of manganous sulphate inhibits *Pseudomonas* spp. (Cuppels and Kelman, 1974). Cuppels and Kelman's medium containing manganous sulphate is reported not to be inhibitory to *E. chrysanthemi* (Cuppels and Kelman, 1974), although the plating efficiency of *E. chrysanthemi* on various selective media has not been comprehensively studied.

TABLE V

Differentiation of *Erwinia chrysanthemi* and *E. carotovora*[a]

		Method[b]	*E. carotovora*	*E. chrysanthemi*
Acid production from	Lactose	A	+	−
	Trehalose	A	+	−
Utilization of	Tartrate	B	−	+[c]
	Malonate	B	−	+
Production of	Indole	C	−	+
	Lecithinase	D	−	+[c]
	Phosphatase	E	−	+
	Blue pigment	F	−	+
Growth in 5% NaCl–peptone water		G	+	−
Sensitivity to erythromycin (50 μg disc)		H	R	S
Agglutination in *E. chrysanthemi* antiserum		I	−	+
Pectin methylesterase		J	−	+

[a] Symbols: +, >90% of strains positive; −, >90% of strains negative; R, organism shows no inhibition zone around antibiotic disc; S, clear inhibition zone visible around antibiotic disc;

[b] All cultures incubated at 27°C. A: acid production within seven days from 1% sugar in 1% peptone water (Dye, 1968). B: utilization of organic acids in Dye's OY medium (Dye, 1968; Slade, 1978). C: indole production in shaken cultures (Dye, 1968). D: lecithinase detected in egg yolk agar (Cowan and Steel, 1974) after seven days. E: phosphatase detected by the methods of Graham (1971). F: A blue insoluble pigment produced within five to ten days on yeast extract (1%)–glucose (2%)–chalk (2%) agar (Starr *et al.*, 1966). G: growth in peptone water containing 5% NaCl within seven days. H: sensitivity to erythromycin by the methods of Graham (1971). I: slide agglutination in antiserum raised against *E. chrysanthemi* NCPPB 1385 (Graham, 1971) and absorbed with *E. carotovora* (Section III.E.2). J: pectin methylesterase according to the methods of Bonnet and Vénard (1975).

[c] Isolates from some hosts give negative reactions in these tests (Table VI).

Beraha's (1968) medium gave the most rapid identification of isolates, with the pits due to pectolysis being visible after 24 h incubation at 26°C (O'Neill and Logan, 1975), but gave a lower isolation rate than Cuppels and Kelman's medium and allowed more soil organisms to grow.

The media of Stewart (1962) and Logan (1963) have been widely used and consist of McConkey agar and Simmons citrate agar respectively, with an overlay of pectate gel. However, *E. carotovora* has a markedly lower plating efficiency on these media in comparison with the media of Cuppels and Kelman (1974) and Beraha (1968). Logan's medium becomes toxic to *E. carotovora* during storage (O'Neill and Logan, 1975). Perombelon and Lowe (1971) demonstrated that *E. carotovora* (particularly *E. carotovora* subsp. *atroseptica*) was sensitive to the bile salts in Stewart's medium. Cells in the stationary growth phase (a condition similar to the physiological condition of cells in the soil) were particularly strongly inhibited by bile salts (Perombelon and Lowe, 1971). Perombelon's (1971a) selective medium consisted of Stewart's medium with the bile salts omitted and supplemented with gentian violet (to suppress Gram-positive bacteria) and Agral, a non-toxic surfactant (to enhance the absorption of inocula). Pure cultures of *E. carotovora* had an increased plating efficiency on Perombelon's medium, compared with Stewart's medium, but isolates from the soil may be masked due to faster growing bacteria with mucoid colonies (O'Neill and Logan, 1975; Perombelon, 1971b).

The media discussed above all rely on detecting the degradation of pectic substances, which are a heterogeneous group of polygalacturonic acids (with a linear α-1,4 linkage) which may vary in chain length, chemical composition and contaminating materials. In pectic acids the carboxyl groups are not esterified or are esterified only to a limited extent (< 5%) with methyl groups, while the term pectinic acids is applied when a higher proportion of the carboxyl groups is esterified (Codner, 1971). The polygalacturonase of *E. carotovora* is specific for non-methylated polygalacturonic acid (Nasuno and Starr, 1966), but the pectate transeliminase (lyase) is able to degrade both polygalacturonic acid and, to a lesser extent, pectinic acids (Morgan *et al.*, 1968). For this reason pectolysis is usually detected using polygalacturonic acid. The production of pectinmethyl esterases should be borne in mind in biochemical and genetical studies of pectin degradation and virulence.

The detection of pectolysis is influenced by pH since the optimum for pectate transeliminases from *Erwinia* spp. is pH 8.3–8.5 (Morgan *et al.*, 1968; Chatterjee and Starr, 1977), while the hydrolytic polygalacturonases have optima of pH 5.2–5.4 (Nasuno and Starr, 1966). Polygalacturonase produces little hydrolysis of polygalacturonic acid in media with a pH of 8.0 (Chatterjee and Starr, 1977) so this can be used to test for pectate transeliminase (which correlates with virulence, Section II.D.1) and media with a pH of 5.2 have been used to test for polygalacturonase (Fucikovsky *et al.*, 1978). It is probable that the wide zone of pectolysis around *E. chrysanthemi* and *E. carotovora* on most isolation media is due mainly to pectate transeliminase. The pectic enzymes of *E. carotovora* have been listed by Lund (1979).

Pectate degradation provides a rapid and easy method of recognizing virulent soft rot *Erwinia* spp. which are difficult to distinguish on media that do not incorporate pectate gels, particularly when outnumbered by saprophytic bacteria. Selective media without pectate have been described by Noble and Graham (1956), Kado and Heskett (1970), Miller and Schroth (1970), Miller (1972), Segall (1971) and Naumann and Ficke (1972). These media are not widely used and have not been comprehensively compared with media containing pectate. The media of Miller (1972) and Naumann and Ficke (1972) were found to inhibit the growth of *E. carotovora* subsp. *atroseptica* (O'Neill and Logan, 1975). *E. chrysanthemi* will not grow on Miller's medium (Miller and Schroth, 1970).

D. Erwinia carotovora

E. carotovora causes soft rots in a large number of field crops and stored vegetables, but is of only minor importance in the spoilage of most fruit, probably because of the acid pH (< 4.5) of most fruit juices. Susceptible fruit include cucumber, pepper and tomato (Arsenijevic, 1978; Lund, 1971). There are few reliable estimates of the economic losses due to *E. carotovora* – partly because the annual variation is high, being much increased when vegetables are grown and stored in wet conditions where the bacteria in the surface film of water can enter the stomata, lenticels and any mechanical wounds. During the storage and transport of potatoes, bacterial soft rot has been estimated to damage 3–5% of the crop in the UK and the USA (Church *et al.*, 1970; De Boer, 1976; Lund, 1979), possibly costing eight million pounds in the UK (Lund, 1979). Potato blackleg was estimated to have caused an annual loss of two and a quarter million dollars during the period 1951–1960 (US Department of Agriculture, 1965). *E. carotovora* is the most important cause of microbial spoilage of potatoes (Lund, 1979). The bacterial ooze from rotting potatoes rapidly infects surrounding tubers and can lead to the loss of hundreds of tons within days (Lund, 1971). Under these conditions the rotting may be greatly increased by pectinolytic clostridia (Lund, 1972, 1979). Other pectinolytic bacteria, such as fluorescent pseudomonads (e.g. *P. marginalis*) and *Flavobacterium* (Lund, 1969) are only weakly pathogenic for potatoes (stored at 5–10°C), but are important causes of soft rot in vegetables (e.g. lettuce) stored at lower temperatures (0–4°C), below the minimum growth temperature of *E. carotovora*. Peppers, tomatoes and cucumbers are normally stored at 7–10°C (Lund, 1971). *Bacillus* spp. are probably only important as the cause of rotting at relatively high temperatures > 30°C (Lund, 1979; Togashi, 1972). If vegetables are washed before retail, the chances of infection with *E. carotovora* are greatly increased.

When soil is inoculated with *E. carotovora*, the numbers drop below easily detectable levels within weeks (Graham, 1958; Graham and Harper, 1967; Logan, 1968; Thompson *et al.*, 1977), so it seems probable that *E. carotovora* can survive the winter in the soil, only if associated with plant material. Crop plants remaining in the soil can harbour the organism (McIntyre *et al.*, 1978) and it is able to survive for long periods in the rhizosphere of susceptible plants such as potato (De Boer *et al.*, 1974, 1978) chinese cabbage and other crucifers (Kikumoto and Sakumoto, 1969a,b; Kikumoto, 1974; Mew *et al.*, 1976). It has been isolated from the rhizosphere of cotton plants (Klinger *et al.*, 1971) sugarbeet, wheat and various weeds (Mendonça and Stanghellini, 1979). *E. carotovora* survives in the lenticels of stored potato tubers, so infected seed potato is a major method of spreading the disease. The introduction of disease-free potato stocks has been reviewed by Lund (1979). Insects feeding on decomposing vegetable matter have been implicated in the recontamination of *E. carotovora*-free potato stocks propagated from stem cuttings (Graham *et al.*, 1976; Harrison *et al.*, 1977). Large numbers of *E. carotovora* can be dispersed as an aerosol during rain or overhead watering (Quinn *et al.*, 1980; Graham and Harrison, 1975; Graham *et al.*, 1977) and on dust when potato haulm is pulverized before harvest (Perombelon *et al.*, 1979; Perombelon, 1978). *E. carotovora* can also be transmitted on infected seeds (McIntyre *et al.*, 1978).

E. carotovora can be isolated from within healthy plant tissue (Meneley and Stanghellini, 1972). Jones and Paton (1973) suggested that *E. carotovora* could grow intracellularly in plant tissue as an avirulent L form and, after a long period of time, revert to virulent organisms.

1. *Pathotypes*

E. carotovora strains can be divided into three groups according to their pathogenicity. Two of the pathotypes cause well characterized diseases, i.e. potato black leg caused by *E. carotovora* subsp. *atroseptica* and a root rot of sugar beet, whereas the third type, *E. carotovora* subsp. *carotovora*, causes rotting in a wide range of plants. Tests distinguishing the three pathotypes of *E. carotovora* are shown in Table VI.

The most important characteristic of *E. carotovora* subsp. *atroseptica* is its ability to cause stem rot in potato (black leg) and other plants (e.g. tomato, cauliflower and chinese cabbage) in temperate climates. This is demonstrated in the laboratory by its ability to cause blackleg when inoculated into potato stems at 19°C, whereas most other soft rot *Erwinia* are able to cause blackleg only if the temperature is above 24.5°C (Graham, 1964). This difference in the temperature required for infection of potato stems is also observed in the field as *E. carotovora* subsp. *carotovora* is reported to cause potato blackleg in Japan (Tanii and Akai, 1975), Mexico (Fucikovsky *et al.*, 1978) and southern areas of the USA (Stanghellini and Meneley, 1975), and *E. chrysanthemi* causes blackleg in tropical countries (Graham, 1971). Potato blackleg due to *E. carotovora* subsp. *carotovora* is rare in temperate climates, but has been detected in Scotland in stocks free of *E. carotovora* subsp. *atroseptica* (Graham *et al.*, 1976). *E. carotovora* subsp. *atroseptica* is also an important cause of soft rot in stored potatoes (Lund, 1979) and rots sunflower heads in Mexico (Fucikovsky *et al.*, 1978).

Strains with the biochemical and physiological characteristics of *E. carotovora* subsp. *atroseptica* isolated from stem rot of plants other than potato may fail to produce potato blackleg (Graham, 1971). The serological differences between isolates from potato and other hosts (Section III.D.2) also suggest that some host specificity may occur. Isolates from sunflower could be distinguished from potato isolates by the production of levan and inability to degrade pectate at pH 5.2 (Fucikovsky *et al.*, 1978).

TABLE VI

Tests differentiating *Erwinia chrysanthemi* and the three pathotypes of *E. carotovora*[a]

	Methods[c]	*E. carotovora* subsp. *atroseptica*	Sugar beet isolates	*E. carotovora* subsp. *carotovora*	*E. chrysanthemi*
Blackleg of potato at 19°C	A	+[b]	NT	−	−
Crown rot of sugar beet at 25–29°C	B	−	+	−	NT
Growth at 37°C	C	−	+	+	+
Acid production from					
α-Methyl D-glucoside	D	+	+	−	−
i-Inositol	D	−	+	±	±
Inulin	D	−	+	−	±
Ethanol	D	−	+	−	+
Utilization of Citrate	E	+	−	+	+
Cis-aconitate	F	+	NT	−	−*
Casein hydrolysis	G	±	−	+	±
KCN tolerance	H	+	−	+	±
Agglutination in *E. carotovora* subsp. *atroseptica* antiserum	I	+		±	−
Agglutination in sugar beet isolates antisera	J	−	+	−	NT
Utilization of palatinose	L	+	+	−	−
Sensitivity to benzyl penicillin (2 units = 1.2 μg per disc)	K	R or S	R	S	R or S

[a] Data on the reactions of the sugar beet isolates taken from Dickey (1979) and Stanghellini et al. (1977).

[b] Symbols: +, >90% of strains positive; −, >90% of strains negative; R, organism shows no inhibition zone around antibiotic disc; S, clear inhibition zone visible around antibiotic disc; NT, not tested or information not available; *, only one strain tested.

[c] Cultures were incubated at 27°C unless otherwise stated. A: stem rot of potato (cultivar Majestic) incubated in a controlled environment (Graham, 1971; Graham and Dowson, 1960); B: crown rot and root decay of sugar beet within 20 days of inoculating the wounded crown with 5×10^6 cells (Ruppell et al., 1975); C: growth in nutrient broth within seven days; D: acid production within seven days from 1% sugar in 1% peptone water (Dye, 1968); E: utilization of citrate on Simmon's citrate agar (Difco) (Dickey, 1979); F: utilization of organic acids in Slade's (1978) modification of Dye's OY medium (Dye, 1968); G: casein hydrolysis by the methods of Dye (1968; Dickey, 1979); H: KCN tolerance by the methods of Edwards and Fife (1956) without the addition of 2,3,5,-triphenyltetrazolium chloride (Dye, 1968; Dickey, 1979); I: slide agglutination in antiserum raised against *E. carotovora* subsp. *atroseptica* (Graham, 1971; Stanghellini et al., 1977); J: slide agglutination in antisera raised against strains KSB6 and CB2, two serologically distinct isolates of the sugar beet pathogen (Stanghellini et al., 1977; Section III.D.2); K: sensitivity to benzylpenicillin by the methods of Dickey (1979); L: utilization of palatinose as sole carbon source by the methods of Sands and Dickey (1978).

Another pathotype of *E. carotovora* has been reported to cause a root rot in sugar beet in western areas of the USA (Ruppell *et al.*, 1975) and appears to be a specific pathotype selected from an existing heterogeneous *E. carotovora* population by susceptible cultivars of sugar beet (Mendonça and Stanghellini, 1979; Stanghellini *et al.*, 1977). The sugar beet pathogen appears to be endemic in Californian soils and may cause severe disease in areas planted with sugar beets for the first time. The sugar beet isolates were pathogenic when inoculated into tomatoes, potatoes and chrysanthemums and were able to cause potato blackleg at 18°C but most *E. carotovora* strains isolated from potato would not cause sugar beet to rot under the same conditions as would the sugar beet pathogen. The taxonomic relationship between the sugar beet isolates and the two subspecies of *E. carotovora* has not been established.

The third pathotype includes the majority of strains of *E. carotovora* subsp. *carotovora* which cause rotting in a wide range of plants with little evidence for host specificity. However, a single strain may sometimes cause particular rots within an area. For example, a single serotype of *E. carotovora* subsp. *carotovora* is reported to cause blackleg of potato in Japan (Tanii and Akai, 1975).

2. *Serology*

In contrast to members of the amylovora group, *E. carotovora* is serologically heterogeneous. The most comprehensive serological studies were those of Goto and Okabe in Japan (only the summaries of the papers are available in English). Okabe and Goto (1955, 1956a) studied the flagellar antigens of 180 isolates and were able to distinguish thirty-three antigenic factors. The flagella of some strains were shown to exhibit antigenic phase changes (Goto and Okabe, 1957). Antigens that were not removed from the cell wall by heating at 100°C for 30 min were referred to as somatic antigens (Goto and Okabe, 1958). One somatic antigen was possessed by all the isolates, whilst 13 other strain-specific antigens were demonstrated. Cross-absorption studies revealed that there may be two strain-specific somatic antigens or antigenic factors on any one isolate. Unheated *E. carotovora* subsp. *carotovora* strains frequently failed to agglutinate with antisera to their somatic antigens which suggested that the somatic antigens were being masked by capsular components in the unheated cells (Goto and Okabe, 1958).

The serological classification developed by Goto and Okabe does not appear to have been used by other workers.

All subsequent serological studies have used what appear to be less sensitive techniques.

During the next 20 years the only significant work in this field was the development of a slide agglutination test for the identification of *E. carotovora* subsp. *atroseptica* (Novakova, 1957; Graham, 1963; Section III.D.2.a). Recently however, De Boer *et al.* (1979) published a serological classification of *E. carotovora* strains isolated from potato. They used immunodiffusion techniques to compare antigens released by the addition of one drop of phenol to 10^9-10^{10} cells suspended in 1 ml of distilled water. Reactions of partial or non-identity between the principal precipitation lines allowed the recognition of 18 serogroups with which they were able to classify 83% of 1001 isolates from potato. Extension of the classification to 29 serogroups enabled them to type about 90% of *E. carotovora* isolates from potato in British Columbia, Canada and about 70% of strains from other laboratories (S. H. De Boer, personal communication). Clearly a comprehensive classification, including strains from other hosts, would involve many more serotypes. The strains used to produce antisera have been deposited in the International Collection of Phytopathogenic Bacteria, University of California, Davis, California 95616, USA.

In immunodiffusion studies, strong cross-reactions were observed between serogroups III and XXIX, whereas both II and XVIII cross-reacted with I, but not with each other (De Boer *et al.*, 1979; S. H. De Boer, personal communication). Small numbers of weak cross-reactions were observed between strains of other serogroups. Thus it is possible that the serogroups could be subdivided by antigenic analysis using more sensitive techniques.

This serotyping scheme, using immunodiffusion, suffers from the complication that the strains possess more than one precipitating antigen (De Boer *et al.*, 1979; Slade, 1978). Immunodiffusion requires interpretation by an experienced eye, is tedious to apply to large numbers of isolates and may be affected by quantitative differences between antisera. Further work has demonstrated that serogroups I, II, III, V and XVIII can be identified using passive haemagglutination techniques (PHA), but absorbed sera are required to eliminate cross-reaction between serogroups (De Boer, 1980). For the PHA tests, sheep red blood cells were coated using either the supernatant fluid from heat-treated cells (121°C, 1 h) or using crude phenol-extracted lipopolysaccharide preparations which had been treated with dilute alkali (0.02 N NaOH, 37°C, 18 h), neutralized and dialysed.

Agglutination of bacterial cells which had been autoclaved (121°C, 2 h) was not suitable for studying the somatic antigens of *E. carotovora*, since many of these preparations auto-agglutinated (S. H. De Boer, personal communication; De Boer, 1980). However, this problem was not encountered by Goto and Okabe (1958) who used cells heated at 100°C for 30 min, but they found more complex cross-reactions than were detected by De Boer. Live cells of serogroup II were agglutinated by antibodies to the somatic antigen. Consequently, in this group the somatic antigens do not always appear to be masked as was reported by Goto and Okabe (1958). Cross-reactions have been reported between strains of *E. carotovora* and *E. chrysanthemi* which have been heated at 100°C for 2 h (Samson, 1973).

All motile strains in serogroups I, III, V and XVIII possess serologically related flagella (De Boer, 1980). Live cells of 16 other serogroups had no agglutinating antigens in common with serogroup I.

Most strains of *E. carotovora* subsp. *atroseptica* isolated from potato belong to serogroup I, with a small proportion (2.3% in British Columbia, Canada) belonging to serogroup XVIII (De Boer *et al.*, 1979). In serogroup XVIII, 2/14 of the strains had biochemical characteristics similar to those of *E. carotovora* subsp. *carotovora*. Additional, unusual serotypes of *E. carotovora* subsp. *atroseptica* isolated from potato have been placed in serogroups XX and XXII (S. H. De Boer, personal communication).

(a) *Serological identification of* E. carotovora *subsp.* atroseptica. The major interest has so far been in using slide agglutination as a rapid diagnostic test for the serologically closeknit group of *E. carotovora* subsp. *atroseptica* isolates from potato. The test usually uses antiserum diluted 1:10, and live bacteria from an isolated colony suspended in saline (Novakova, 1957; Graham, 1963). Normal serum is used as a negative control. The test can be applied to sap squeezed from rotting tissue (Graham, 1963). In the case of rotting potato tubers, tissue from the edge of the rot should be used and care taken not to confuse particles of plant tissue (e.g. starch grains) with agglutinated bacteria. Results should always be confirmed by isolation of the organism followed by biochemical and physiological characterization. The principal drawback to slide agglutination for *E. carotovora* subsp. *atroseptica* is that some strains of *E. carotovora* subsp. *carotovora* cross-react (Graham, 1963, 1971; Lazar, 1971b) and may have agglutination titres as high as strains of *E. carotovora* subsp. *atroseptica* (Vruggink and Maas Geesteranus, 1975). These cross-reactions may be due to both flagellar and somatic antigens (Section III.D.2). The identification of *E. carotovora* subsp. *atroseptica* has been improved by using antiserum to serotype I absorbed with serotype II for immunodiffusion tests (Vruggink and Maas Geesteranus, 1975). This antiserum should be reliable in the detection of serogroups I and XVIII, but immunodiffusion is more laborious than slide agglutination and need only be used for confirmation. However, De Boer *et al.* (1979) found that 2/532 (0.4%) of *E. carotovora* subsp. *carotovora* isolates gave precipitation reactions identical with 2.3% of his *E. carotovora* subsp. *atroseptica* isolates (serogroup XVIII). Agglutination cross-reactions between *E. carotovora* subsp. *atroseptica* antiserum and one strain of a *Klebsiella* sp. and two strains of *Pseudomonas fluorescens* have been reported (Graham, 1963).

These immunodiffusion techniques will identify isolates of *E. carotovora* subsp. *atroseptica* belonging to serogroups I and XVIII which represent nearly all isolates from potato. Serogroup I has also been isolated from calabrese and tomato stem (S. H. De Boer, personal communication) but other isolates of *E. carotovora* subsp. *atroseptica* from tomato, cauliflower, sunflower and iris did not cross-react with serogroup I (Graham, 1971; Allan and Kelman, 1977; Fucikovsky *et al.*, 1978; S. H. De Boer, personal communication).

The isolates of *E. carotovora* subsp. *atroseptica* examined by Vruggink and Maas Geesteranus (1975) possessed one common antigen and were variable with respect to a second antigen. These are likely to be the same as two heat stable (100°C, 30 min) antigens detected in sonicated cell suspensions of three strains of *E. carotovora* subsp. *atroseptica* by Slade (1978). These antigens were thought to be capsular, since they were found predominantly in the supernatant fluid after centrifuging (105 000 g) the LPS extracted by phenol–water from strain NCPPB 549. These antigens are probably neutral polysaccharides since they did not precipitate with cetyl trimethyl ammonium bromide (Scott, 1965). Strain 549 may possess a R-LPS since it could be extracted by the phenol–chloroform–petroleum ether method of Galanos *et al.* (1969). This R-LPS antigen could not be detected by immunodiffusion techniques in sonicated cell suspensions and did not react in passive haemagglutination tests with antisera raised against two other strains of *E. carotovora* subsp. *atroseptica*. The presence of structural similarities in the core LPS of *E. carotovora* subsp. *carotovora*, *E. herbicola*, *E. rhapontici* and *E. amylovora* was suggested since cells of these species absorbed antibodies to the R-LPS of *E. carotovora* subsp. *atroseptica*, but cells of *Salmonella abortus-bovis* and *Proteus vulgaris* did not (Slade, 1978).

Immunofluorescent techniques are reported to have been successfully used to identify *E. carotovora* subsp. *atroseptica* in plant tissues, soil insects and on isolation plates (Allan and Kelman, 1977). However, it seems probable that immunofluorescent techniques are no more specific than immunoprecipitation. Immunofluorescent techniques are particularly suitable for detecting bacteria in plant tissue (Vruggink and De Boer, 1977), but enzyme-linked immunosorbent assay (ELISA) techniques are likely to have considerable technical advantages for most purposes (Vruggink, 1978).

(b) *Isolates from sugar beet.* Two serotypes were present in isolates of *E. carotovora* from sugar beet (Stanghellini *et al.*, 1977). The two serotypes did not cross-react in slide agglutination tests using live organisms grown for two days on YDC agar and antiserum diluted 1:20 and do not cross-react with any of the 29 serogroups of *E. carotovora* isolated from potato (S. H. De Boer, personal communication).

3. *Bacteriocin typing*

Crowley and De Boer (1980) found that 8/18 strains representing 18 serogroups of *E. carotovora* produced bacteriocins that were morphologically similar to phage tails. There was a partial correlation between serological specificity and sensitivity to the bacteriocins. No bacteriocin-typing scheme was proposed, but four of the bacteriocins gave good subdivisions of serogroups III and V. The bacteriocins were able to differentiate clearly between the two principal serogroups of *E. carotovora* subsp. *atroseptica*.

E. *Erwinia chrysanthemi*

E. chrysanthemi causes vascular wilts and parenchymatal necrosis in a large number of plants (Table VII). (Dickey, 1979; Dickey and Victoria, 1980; Dinesen, 1979.) Evidence that individual strains may have a limited host range has encouraged the use of typing methods to differentiate between different pathotypes. However, research on the epidemiology is rather fragmentary because of the large number of hosts, different horticultural and agricultural practices, and no doubt, because it rarely causes diseases of field crops in temperate climates.

Using DNA hybridization techniques, Brenner *et al.* (1977) distinguished four genetic groups within *E. chrysanthemi*: isolates from *Chrysanthemum* and one isolate from *Guayule*; isolates from *Dahlia* and *Dieffenbachia*; isolates from corn and grass and a single isolate from sugar cane.

1. *Biotypes and pathotypes*

The ability of pathogens to attack several hosts is clearly important for disease control. There is evidence that *E. chrysanthemi* is differentiated into path-otypes, since individual strains are not virulent for all hosts. For example, strains from *Chrysanthemum* and *Parthenium* are only weakly pathogenic for *Philodendron* (Lelliott, 1974). However the extent of the host specificity is unclear since different methods and conditions have been used to compare different hosts. Circumstantial evidence that host specificity occurs in nature is that most strains from one specific host found in different areas generally possess the same biochemical reactions (Dickey, 1979; Table VIII) and somatic antigens (Samson, 1973; Samson and Nassan-Agha, 1978). These biotyping differences and very limited host range studies have been used to propose six pathovars of *E. chrysanthemi* (Table VII; Young *et al.*, 1978; Dickey, 1979; Dickey and Victoria, 1980). Comparative host specificity studies confirming these pathotypes is awaited.

In the case of biotypes I, II and VI, which have been isolated from a specific

TABLE VII

Hosts from which biotypes of *Erwinia chrysanthemi* have been isolated[a]

Biotype	Proposed pathovar	Representative strain	Host
I	pv. *dieffenbachiae*	NCPPB 1157	*Dieffenbachia* sp.
II	pv. *parthenii*	NCPPB 516	*Parthenium argentatum*
III	pv. *chrysanthemi*	NCPPB 402, ATCC 11663	*Chrysanthemum* sp.
			Dianthus caryophyllus
			Euphorbia pulcherrima
IV	pv. *zeae*	NCPPB 2538	*Aglaonema commutatum*
			Ananas comosus
			Chrysanthemum X *morifolium*
			Cyclamen
			Dieffenbachia sp.
			Dracaena marginata
			Ipomoea batatus
			Musa sp.
			Pelargonium capitatum
			Philodendron sp.
			Saintpaulia ionantha
			Syngonium podophyllum
			Zea mays
V	pv. *dianthicola*	NCPPB 453	*Begonia intermedia*
			Dahlia pinnata
			Daucus carota
			Dianthus sp.
			Lycopersicon esculentum
			Sedum spectabile
VI	pv. *paradisiaca*	PDDCC 2349	*Musa paradisiaca*

[a] Data from Dickey (1979) and Dickey and Victoria (1980). ATCC, American Type Culture Collection, Rockville, Maryland, USA. NCPPB, National Collection of Plant Pathogenic Bacteria, Harpenden, Hertfordshire. PDDCC, Plant Diseases Division Culture Collection, Auckland, New Zealand.

host or closely related hosts, typing can be used to recognize which crops a strain is likely to attack. However, biotype IV occurs on a diverse group of host plants, so additional characteristics to subdivide this group would be useful. Samson and Nassan-Agha (1978) have also proposed a biotyping scheme which tests for the production of arginine deaminase in addition to the tests shown in Table VIII.

2. Serology

Slide agglutination is a useful rapid diagnostic test for *E. chrysanthemi*. Graham (1971) found that a single antiserum raised against NCPPB 1385

TABLE VIII
Biotypes of *Erwinia chrysanthemi*[a]

			Biotypes[b]					
Test		Method[c]	I	II	III	IV	V	VI
Acid production from	Mannitol	A	+	+	+	+	+	−
	Sorbitol		+	+	+	+	+	−
	Arabinose		+	−	−	+	−	+
	Raffinose		−	+	+	+	±	+
	Inulin		−	−	+	−	+	−
Utilization of	Tartrate	B	+	+	−	+	+	+
Lecithinase		C	+	+	+	+	+	−
Gelatin liquefaction		D	+	+	+	+	±	−

[a] Data from Dickey (1979) and Dickey and Victoria (1980).
[b] Symbols: +, >90% strains positive; −, <10% strains positive; ±, 10–90% strains positive.
[c] A: acid production within seven days from 1% sugar in 1% peptone water (Dye, 1968); B: utilization of organic acids in Dye's OY medium (Dye, 1968); C: lecithinase detected three to seven days after spot inoculation of McClung Toabe agar base (Difco) plus 50% egg yolk (Difco) (Dickey, 1979); D: gelatin liquefaction detected in tubes of nutrient gelatin (Difco) after 14 days incubation (Dickey, 1979).

agglutinated live cells of all 623 isolates of *E. chrysanthemi* from seventeen countries. This agglutination is probably due to a heat labile antigen, since antisera raised against heat treated cells (100°C, 2 h) agglutinate only a restricted number of *E. chrysanthemi* strains (Samson, 1973). Antisera raised against live *E. chrysanthemi* will also agglutinate live cells of a few strains of *E. carotovora* subsp. *carotovora*, but not *E. carotovora* subsp. *atroseptica* (Graham, 1971; Lazar, 1971b). Samson (1973) has described an antigen responsible for low titre (1:4–1:320) agglutination of heat treated (100°C, 2 h) cells (but not live cells) of *E. chrysanthemi* and *E. carotovora*. This reaction was removed by absorbing antisera with a strain of *E. carotovora*. Slade (1978) observed a very strong cross-reaction in immunoprecipitation tests between *E. chrysanthemi* NCPPB 453 and strain AL2, which appeared to be a previously undescribed member of the *amylovora* group isolated from apple. Strain AL2 was non-pectolytic and its biochemical reactions showed little similarity with *E. chrysanthemi*.

Samson (1973) and Samson and Nassan-Agha (1978) studied the somatic antigens of *E. chrysanthemi* using antisera raised against cells heated at 100°C for 2 h and divided *E. chrysanthemi* into five groups. Table IX shows the serological relationships within Group I and the plants from which the strains were isolated. Additional strains isolated from *Aglaonema*, *Euphorbia*, *Phalaenopsis*, *Syngonium* and *Zea mays* have also been identified as Group I (Samson and Nassan-Agha, 1978). The cross-absorptions for producing

TABLE IX

Correlation between host and serotype of Serogroup I *Erwinia chrysanthemi* strains[a]

Host	Somatic antigens
Dianthus, Dahlia, Begonia	1, 2, 3, 5
Saintpaulia, Pelargonium	1, 2, 5
Dieffenbachia, Philodendron	1, 5
Chrysanthemum	1, 4
Cyclamen	1

[a] Data from Samson (1973).

specific antisera to Group I antigens are shown in Table X. Live cells of Group II were agglutinated by antiserum to strain 1236, but not by antisera to Group I strains. Group II strains have been isolated from *Parthenium* sp. and *Chrysanthemum maximum*. Cross-reacting strains from *Musa paradisiaca* form Group III. Group IV contains a single strain (1502) isolated from *Zea mays* (Samson and Nassan-Agha, 1978). The final Group V is comprised of strains from *Zea mays* and grass that do not agglutinate in antiserum raised against the other groups. The presence of isolates from *Zea mays* in several groups is the only exception to a remarkable correlation between the serotype of isolates from different countries and the hosts from which they were isolated.

3. Bacteriocin typing

Most *E. chrysanthemi* strains readily produce bacteriocins (Echandi and Moyer, 1978, 1979). The detection of bacteriocins is profoundly influenced by

TABLE X

Methods for production of antisera against *Erwinia chrysanthemi* serogroup I antigens[a]

Antigenic factor	Antiserum raised against	Absorbing strain
2, 3	795[b]	1247[b]
3	795	1361
4	1273	1247
5	1247	1273

[a] Data from Samson (1973).
[b] Strain No. in INRA Collection of Plant Pathogenic Bacteria, Station de Pathologie végétale et de Phytobactériologie, Centre de Recherches d'Angers, INRA, Beaucouze, 49000 Angers, France.

the concentration of agar, incubation temperature for growth of producer colonies and concentration of bacteria in the lawn. For optimal production of bacteriocin, producer strains are spotted onto 25 ml nutrient agar plates containing 15 **g** per litre of agar and incubated for 48 h at 20°C. The bacteria are killed with chloroform vapour. Inhibition zones in a lawn overlay (containing 2×10^6 cells of an indicator strain in 4 ml water agar) were visible after 24–48 h incubation at 40°C. For bacteriocin typing, production was enhanced by inducing nutrient broth cultures (grown for 24 h) with ultraviolet irradiation or $1\,\mu\mathrm{g\,ml}^{-1}$ mitomycin C. The culture supernatants were centrifuged and filter sterilized before spotting on to lawn plates of indicator strains.

Three high molecular weight bacteriocins have been used for typing and two of these (Ech 25 and Ech 33) were shown to resemble the tails of defective phage, consisting of an inner core and contractile sheath (Echandi and Moyer, 1979). These two bacteriocins were unstable in 0.85% NaCl or Tris-phosphate buffer (pH 7.0), but could be stored for one month after the addition of 0.02% bovine serum albumin (Table XI).

TABLE XI
Bacteriocin typing of *Erwinia chrysanthemi*[a]

Typing pattern	Bacteriocin-producing strains			Total strains per type	Host plants
	Ech 10	Ech 25	Ech 33		
A	+	+	+	9	*Dianthus, Zea, Dahlia*
B	+	−	+	5	*Dieffenbachia*
C	−	+	+	1	Tobacco
D	−	+	−	1	Tobacco
E	−	−	−	2	*Chrysanthemum, Zea*

[a] Data from Echandi and Moyer (1978, 1979).

4. Phage typing

Paulin and Nassan-Agha (1978) found that 17/35 strains of *E. chrysanthemi* contained lysogenic bacteriophage, with some strains producing four different plaque morphologies. However, it was not clear whether some of the zones of lysis observed were due to the production of bacteriocins. This is a strong possibility since many strains produce bacteriocins which morphologically resemble defective phage (Echandi and Moyer, 1978). Attempts to isolate phage lytic for *E. chrysanthemi* were unsuccessful (Paulin and Nassan-Agha, 1978).

Paulin and Nassan-Agha (1978) used the following procedure to screen for lysogenic bacteriophage. Producer strains were spotted on to NYGA agar and incubated for 48 h at 27°C. The colonies were removed with a glass rod and the remaining cells killed by ultraviolet irradiation. A second layer of half strength NYGA agar seeded with a susceptible strain was poured on to the plate and zones of lysis observed after 12 h incubation.

Five bacteriophage were used to divide 61 isolates of *E. chrysanthemi* into 11 groups (Table XII). Group I contained isolates from *Dianthus, Dahlia, Zea mays* and *Dieffenbachia*. Group II consisted only of strains from *Dieffenbachia* and all the other groups contained only strains isolated from *Zea mays*. Thus the phage types were able to subdivide isolates derived from *Zea mays* and *Dieffenbachia*. However, it is not clear how stable the phage types are, since the results may depend on the strains' complement of lysogenic phage which may be readily gained and lost. Practical difficulties of using these methods are that stock bacteriophage cannot be stored for long periods and many of the phage give faint turbid plaques.

TABLE XII
Phage typing of *Erwinia chrysanthemi*[a]

	Phages				
Phage types	28	34	3	166	24
I	+	+	+	+	+
II	+	+	+	+	−
III	−	+	−	−	−
IV	+	+	−	−	+
V	+	+	+	−	−
VI	+	−	−	−	−
VII	+	+	−	−	−
VIII	+	+	−	+	+
IX	+	−	−	−	+
X	−	−	−	−	+
XI	−	−	−	−	−

[a] Data from Paulin and Nassan-Agha (1978).

5. Conclusions

There are two distinct reasons for typing *E. chrysanthemi*. Methods distinguishing between isolates with different host specificities are needed for disease control purposes, whereas epidemiological work requires the identification of individual strains within each pathotype. Methods fulfilling these two needs are being developed, but their reliability has not yet been proven. Most typing methods have been applied only to a limited number of strains from a few host species.

F. Other members of the *carotovora* group

The other species usually placed in the *carotovora* group are *E. rhapontici* and *E. cypripedii*. *E. rhapontici* is discussed with the *herbicola* group in accordance with a recent numerical taxonomic study (Slade, 1978; Slade and Tiffin, 1980) and evidence that the ecological role of this organism is as a member of the epiphytic flora and not a virulent plant pathogen (Roberts, 1974; Slade, 1978; Sellwood and Lelliott, 1978). Evidence from DNA hybridization studies has indicated that *E. cypripedii* should also be transferred to the *herbicola* group (Section I.B; Brenner *et al.*, 1973), but unfortunately this organism has not been included in any numerical taxonomic study. *E. cypripedii* causes a brown rot of *Cypripedium* orchids in California and Natal, South Africa (Wallis *et al.*, 1975), but is non-pectinolytic and does not cause a typical soft rot disease. Little is known concerning the distribution and epidemiology of *E. cypripedii* and it is likely that, if it occurs in the epiphytic flora of other plants, it would be mistaken for *E. herbicola*.

The colonial morphology and pigmentation of *E. cypripedii* differs on different media. Colonies are colourless, round, smooth and glistening on nutrient agar at 31°C. On YDC agar (1% yeast extract, 2% glucose, 2% $CaCO_3$, 2% agar) colonies are colourless to light yellow-brown, smooth with glistening surface and a viscid consistency after 24 h, but after 72 h the colonies and surrounding areas become yellow-brown to red-brown (Sutton *et al.*, 1960). *E. cypripedii* has the characteristic ability (unusual for *Erwinia* spp.) of both producing gas from glucose and giving a positive methyl red test. *E. cypripedii* can readily be distinguished from other members of the *carotovora* group by its production of pigment and lack of pectinase. It can also be distinguished from most non-pigmented members of the *herbicola* group by its ability to hydrolyse lipids and grow at 37°C. *E. cypripedii* was reported to ferment lactose by Sutton *et al.* (1960) and Wallis *et al.* (1975), but not by Dye (1969a).

IV. The *herbicola* group

A. General description

This is a poorly defined group with a wide range of biochemical characteristics. Their ecological roles range from virulent plant pathogens to general ecological opportunists which are common on plant surfaces, but they can also be isolated from a wide range of other habitats and can cause opportunist infections of plants, animals and insects. It is not clear whether these organisms are capable of sustained growth in many of the habitats from which they have been isolated or whether they are simply transient organisms which have been shed from plant surfaces.

The following tests are of use in differentiating members of the group from members of the *carotovora* group and *Enterobacter* spp. No member of the *herbicola* group is able to decarboxylate arginine, glutamine, lysine or ornithine. They rarely produce urease, gas from glucose or grow at 42°C. It is possible that a serological test for the presence of antigen AP(H) (Section IV.B.5.a) would provide a satisfactory positive character.

B. *Erwinia herbicola*

1. *Definition*

The name *Erwinia herbicola* (Löhnis) (Dye, 1964) is derived from *Bacterium herbicola aureum* (Duggeli, 1904) and *Bacterium herbicola* (Löhnis, 1911). The term *E. herbicola* was used to encompass a group of *Erwinia* spp. which had been mistakenly described as plant pathogens (e.g. *E. lathyri*) and other strains which had been incorrectly allocated to the genus *Xanthomonas* (e.g. *X. trifolii*) (Dye, 1964, 1966). The name *E. herbicola* was first used for a collection of poorly described species and has continued as a dustbin for uncharacterized strains. All too frequently, the operative definition of *E. herbicola* has been "these isolates are not pathogenic for plants". This led to uncertainty in the species definition, since many strains have not conformed to the published descriptions of *E. herbicola*. More cautious workers reacted to the confusion by describing isolates as *Erwinia*-like organisms. These are usually members of the *herbicola* group. Some culture collection strains labelled *E. herbicola* have not been adequately characterized. *E. herbicola* was extensively characterized by Dye (1969b) and his work forms the basis of the currently accepted definition of *E. herbicola* (Lelliott, 1974). Yet it is clear that the epiphytic flora contains groups closely related to *E. herbicola* (Slade, 1978; Slade and Tiffin, 1980), but which do not entirely conform to Dye's description. Thus although the description of *E. herbicola* is now generally agreed there is no real agreement on where the limits of the species should be drawn. The taxonomy of *E. herbicola* was again thrown into confusion by the proposal of Ewing and Fife (1972) that, in order to accommodate strains of medical importance, *E. herbicola* should be renamed *Enterobacter agglomerans* (Section I.B.1). This proposal has been adopted by many medical bacteriologists, but was opposed by Cowan and Steel (1974). The plant pathologists have not accepted this change as it does not in their view encompass a natural taxonomic group. Certainly, many very diverse organisms may be identified as *E. agglomerans*, even strains outside the Enterobacteriaceae (Sakazaki *et al.*, 1976). The description of *E. agglomerans* includes many variable characteristics. Therefore a large number of isolates not conforming closely to the definitions of other species can easily be fitted into *E. agglomerans*.

Taxonomic studies of the relationship between *Erwinia herbicola*, *Enterobacter agglomerans* and other members of the genus *Enterobacter* have given equivocal results (Section I.B.1). It is clear that many *E. herbicola* strains correspond to Ewing and Fife's (1972) definition of *E. agglomerans*, although it is equally clear that the reverse is not the case. Many numerical taxonomic studies which appear to link *E. agglomerans* to the genus *Enterobacter* can be criticized because they have included few strains representing this diverse species among the strains studied and do not include representatives of the whole *herbicola* group. However, the study of Thurner and Busse (1978) is of particular interest, since it divided a large number of isolates into two clear taxa — *Erwinia* I and *Erwinia* II. *Erwinia* I, which they suggested should be called *E. agglomerans*, was closely related to *Enterobacter cloacae* and did not conform to definitions of *E. herbicola*. *Erwinia* II corresponded with the current definition (Lelliott, 1974) of *E. herbicola*. This recognition of *Enterobacter agglomerans* and *Erwinia herbicola* as two separate species has its merits, but two different definitions of *E. agglomerans* will lead to confusion, particularly as the neotype strain of *E. agglomerans* NCTC 9381 (Ewing and Fife, 1972) is undoubtedly *E. herbicola* (Goodfellow *et al.*, 1976; Slade 1978). *Erwinia* I (*E. agglomerans*) differed from *Erwinia* II (*E. herbicola*) by the fermentation of dulcitol (89%), but not *m*-inositol, production of gas from glucose (89%) and production of arginine deaminase (57%) (Thurner and Busse, 1978).

2. Occurrence

Reports of the isolation of *E. herbicola* suggest that it is virtually ubiquitous in the environment. However, due to the taxonomic confusion, the ecology of *E. herbicola* has not been adequately or systematically studied. Consequently these reports should be interpreted with caution.

E. herbicola is commonly found on the aerial surfaces of plants (Billing and Baker, 1963; Goodfellow *et al.*, 1976), on aquatic plants (Trust and Bartlett, 1976) and within healthy plant tissues (Meneley and Stanghellini, 1974; Bagley *et al.*, 1978), ovules and seeds (Mundt and Hinkle, 1976). Nitrogen-fixing strains of *E. herbicola* may be important in the rhizosphere of winter wheat and sorghum in Nebraska (Pedersen *et al.*, 1978) and they have also been found associated with fungal decay of wood (Aho *et al.*, 1974) and in paper mill process water (Neilson and Sparell, 1975; Neilson, 1979). *E. herbicola* has been isolated from water (Ursing, 1977; Thurner and Busse, 1978), soil (Graham, 1958; Miller and Schroth, 1972; Primrose, 1976) and airborne dust (Dutkiewicz, 1976). In addition to being associated with animal and plant diseases, *Erwinia* strains which are probably *E. herbicola* have caused disease in leafhoppers and spittlebugs (Whitcomb *et al.*, 1966).

(a) *Occurrence in plant diseases.*   *E. herbicola* can frequently be isolated in large numbers from damaged plant tissues and lesions (Van der Zwet, 1969). Historically, this led to the description of *E. herbicola* (under various synonyms) as a plant pathogen. The *E. herbicola* in plant lesions probably has an important role in the inhibition of pathogens by competing for nutrients, producing acidic conditions and inducing the plant's defence mechanisms (Erskine and Lopatecki, 1974; Wrather *et al.*, 1973; Goodman, 1965, 1967; McIntyre and Williams, 1972; McIntyre *et al.*, 1973; Klement, 1971; Chatterjee and Gibbins, 1969; Chatterjee *et al.*, 1969). *E. herbicola* does not generally cause disease in undamaged plant tissues, but there are a few well-documented examples of its acting as a pathogen, e.g. strains, originally described as *E. milletiae*, which produce galls on various plants in Japan (Okabe and Goto, 1956b) and produce the plant hormone β-indolyl acetic acid *in vitro* (Hirata, 1960). *E. herbicola* can cause pink disease of pineapple in which the fresh fruit appears normal, but becomes discoloured after canning (Cho *et al.*, 1978). *Erwinia* spp. (probably *E. herbicola*) also cause an internal discoloration of papaya (Nelson *et al.*, 1977) and have been associated with the internal necrosis of immature cotton (Ashworth *et al.*, 1970). A particularly interesting pathogen is *E. uredovora* which differs from *E. herbicola* only by its pathogenicity for the uredia of cereal rusts (Dye, 1969b; Pon *et al.*, 1954).

Many plant pathogens have to maintain themselves on the surface of a plant prior to gaining entry. Under these conditions, competition and antagonistic interactions among the epiphytic flora will determine whether infection occurs. Great interest is being shown in using epiphytic bacteria for biological control of plant pathogens. Partial control of *E. amylovora* has been achieved by spraying orchards with *E. herbicola* strains (Riggle and Klos, 1972; Thompson *et al.*, 1976; Paulin, 1978). Strains exist which produce bacteriocins (Thompson *et al.*, 1976) or carry phage lytic for *E. amylovora* (Erskine, 1973; Erskine and Lopatecki, 1974). *E. herbicola* is able to reduce the infection of rice by *Xanthomonas oryzae* (Hsieh and Buddenhagen, 1974).

(b) *Occurrence in animal infections.*   The first report of the isolation of yellow-pigmented anaerogenic fermentative bacilli from faecal materials was by Dresel and Stickl (1928), who called their isolate *Bacterium typhi flavum*.

After much study and controversy (Cruickshank, 1935), interest waned until a comparison of phytopathogenic strains of *Erwinia* and chromogenic isolates from animal sources by Graham and Hodgkiss (1967) indicated that *Bacterium typhi flavum* should be classified as *Erwinia herbicola*. Since Ewing and Fife (1972) suggested that the *herbicola-lathyri* bacteria should be reclassified as *Enterobacter agglomerans* there have been reports on both *Erwinia herbicola* and *Enterobacter agglomerans* without any obvious agreement on nomenclature.

Chromogenicity appears to have been the primary character used in many clinical laboratories for the selection of strains later to be identified as *Erwinia* (Bottone and Schneierson, 1972) but other clinical workers have also studied non-pigmented aerogenic strains (von Graevenitz, 1970). Ursing (1977) found that the amount of gas differed between strains and that some strains produced gas at room temperature, but not at 35°C. This reinforced the proposals of von Graevenitz (1970) and Ewing and Fife (1972) that the ability to produce gas from glucose should not be weighted taxonomically. Some workers found a negative correlation between the production of symplasmata (and biconvex bodies) and the ability to produce gas from glucose (Meyers *et al.*, 1972; Gilardi *et al.*, 1970), whereas Ursing (1977) found no such correlation.

While the majority of *E. herbicola* isolates fit neatly into the *E. agglomerans* as defined by one of the rapid multitest identification schemes (API-2OE) now commercially available, so do strains of *Erwinia carotovora* and *Erwinia amylovora*, (A. I. Tiffin, unpublished data). This would suggest that, at least in this scheme, one should rename the species *E. conglomerans*.

Although *E. herbicola* is mostly considered to be of questionable pathogenicity for man it has been isolated from almost all possible sites in the body, often in conjunction with other organisms. No very clear pattern of antibiotic susceptibility has emerged, although most authors have found strains sensitive to kanamycin and gentamycin, and resistant to erythromycin.

3. Isolation and identification

E. herbicola can readily be isolated on nutrient agar and usually forms smooth, circular, raised colonies with an entire edge within 48 h at 25°C. The colonies are 1–2 mm in diameter and can be either grey-white or yellow. Colonies frequently have a bright central spot when viewed against the light and spindle shaped, biconvex bodies may be seen within the colony under low power microscopy. In culture, strains readily give rise to variant colonial morphology.

Isolates give a fermentative reaction in Hugh and Leifson's oxidation–fermentation test and are negative, or only very weakly positive, in the methyl red test performed according to the method of Dye (1968, 1969b).

The tests listed in Section IV.A should be used to confirm that an isolate belongs to the *herbicola* group. Most *E. herbicola* strains liquefy nutrient gelatin within 30 days at 25°C and do not produce acid from adonitol. Most *E. herbicola* strains (excluding *E. herbicola* subsp. *ananas*) produce characteristic caterpillar-shaped aggregates of cells (symplasmata) in the condensation water of nutrient agar slopes after four days incubation at 25°C. Symplasmata are also produced by *Enterobacter sakazakii* (Farmer *et al.*, 1980).

Complete identification of *E. herbicola* strains, according to the criteria of Lelliott (1974), requires extensive biochemical characterization. It should be appreciated that a considerable proportion of *E. herbicola* isolates are likely to differ to a limited extent from Lelliott's description and until a comprehensive taxonomic study has been made, the limits of the species can not be defined. Particular care should be taken when assessing biochemical tests which have given different results in different laboratories (Table II). As discussed in Section IV.B.1 this apparent variability is probably due to minor differences in the performance of the tests and the assessment of the results.

4. *Subdivision*

Evidence from studies using numerical taxonomy (Goodfellow *et al.*, 1976; Bascomb *et al.*, 1971; Slade and Tiffin, 1980), DNA base ratios (Starr and Mandel, 1969) and nucleic acid segmental homology (hybridization) (Murata and Starr, 1974) shows that strains labelled *E. herbicola* constitute a very heterogeneous group of organisms. Bascomb *et al.* (1971) and Goodfellow *et al.* (1976) suggested that *E. herbicola* should be ranked as a genus, but for practical purposes it would be better to give the entire *herbicola* group generic status. If such a proposal were accepted, then the clusters defined by numerical taxonomy within *E. herbicola* would be given specific status.

Goodfellow *et al.* (1976) showed that *E. herbicola* subsp. *herbicola* and *E. herbicola* subsp. *ananas* formed two distinct clusters. Slade (1978) and Slade and Tiffin (1980) confirmed these results and demonstrated two additional clusters within *E. herbicola*: H:2 containing pigmented strains with a low maximum growth temperature and H:3 containing non-pigmented strains. Characteristics distinguishing the four clusters are shown in Table IV. Strains representing each cluster are indicated in Table XIII. Cluster H:1 contains strains from both plant and animal sources, but the low maximum growth temperatures of clusters H:2 and H:3 suggest they are unlikely to be pathogenic for mammals. With such a poorly studied, yet ecologically diverse species, it is probable that there are other types of *E. herbicola* not represented in this study. It is possible that the unclustered strains may represent additional groups.

Ewing and Fife (1972) proposed that *Enterobacter agglomerans* should be

TABLE XIII
Serotyping strains for the *herbicola* group[a]

Species	Strain	Cluster	Capsular antigens	LPS antigens	GH4	GH5
E. herbicola subsp. *herbicola*	NCPPB 1264	H3	1a	1		
	NCPPB 664	H3	1b	1		
	AL7	H3	2	1		
	NCPPB 1267	"Unclustered"	3			
	NCPPB 656	H1	4		a	
	NCPPB 1750	H1	5		a	
	NCPPB 666	H3	6			
	NCPPB 1269	H1	7		a	
	NCPPB 2035	H1	8		a	
	NCPPB 2036	"Unclustered"	9			
	NCTC 9381	H1	10	2	a	
	G150	H1	11		a	
	G152	H1	12		a	
	30	H1	13		b	
	35A	H1	14		a	
	NCPPB 2271	H2				a
	NCPPB 2272	H2				b
	NCPPB 2273	H3				b
E. herbicola subsp. *ananas*	NCPPB 1846		15	3	b	
	EA133		16			
	89		17			
E. rhapontici	NCPPB 139		18	4		
	NCPPB 1578		19			
	NCPPB 1739		20			
	AL6		21			
	NCPPB 1270					

[a] Data from Slade (1978).
All serotype strains have been deposited with the NCPPB = National Collection of Plant Pathogenic Bacteria, Hatching Green, Harpenden, Hertfordshire.

subdivided into seven biogroups (Table XIV) and *Erwinia herbicola* is found in biogroups 1, 2, 3 and 4. These biogroups do not conform to the clusters formed by numerical taxonomy (Goodfellow *et al.*, 1976; Slade, 1978). Acetoin production is unsuitable for biotyping *E. herbicola* strains since the distinction between weak positive and negative reactions is subjective.

5. Serology

(a) *Serological structure.* The antigenic structure of most strains of *E. herbicola* and *E. rhapontici* is similar, except for *E. herbicola* cluster H:2.

TABLE XIV

Biogroups of *Enterobacter agglomerans* (Ewing and Fife, 1972)

Biogroup	Nitrate reduction	Indole production	Acetoin production	*Erwinia herbicola* groups
1	+	−	+	Clusters H:1, H:3, *E. herbicola* subsp. *ananas* (Slade, 1978)
2	+	−	−	Cluster H:1, H:2 (Slade, 1978)
3	−	−	−	*E. herbicola* subsp. *ananas* (Goodfellow *et al.*, 1976)
4	−	++	+	*E. herbicola* subsp. *ananas* (Goodfellow *et al.*, 1976; Slade, 1978)
5	+	+−	−	
6	−	−	+	
7	+	+	+	

The principal antigens in most strains of *E. herbicola* are uncharged capsular polysaccharides. These antigens are not removed from the cells by heating at 100°C for 30 min and in one strain (1264) have remained after 121°C for 2 h. These capsular antigens are distinct from the lipopolysaccharide (LPS) antigens and are not pelleted by centrifugation of phenol–water extracts at 100 000 g. Many laboratory strains of *E. herbicola* contain both R and S-LPS (Slade, 1978), but it is not known whether this is also true of newly isolated strains. The presence of either or both R-LPS and S-LPS has important implications for serological studies. In immunodiffusion studies, sonicated cell suspensions of many strains do not give precipitation lines due to LPS and multiple diffuse precipitation lines may be obtained with antigen extracted from a strain containing both R- and S-LPS. LPS preparations from the same strain extracted by two different methods (phenol–water or phenol–chloroform–petroleum ether) may give different results in passive haemagglutination tests (Slade, 1978). Strains NCTC 9381 and NCPPB 2274 produce only S-LPS. It is not possible to distinguish between strains producing either S-LPS or R-LPS by colonial morphology or auto-agglutination due to the masking effect of capsular antigens. It is unknown whether the LPS is exposed on cells heated at 100°C for 30 min but the cells then do not auto-agglutinate even when it is known that the LPS is in the R form.

The LPS from *E. herbicola* strains NCPPB 1264, 2274 and NCTC 9381 contains uronic acids, but LPS from NCPPB 1846 does not (Slade, 1978). Uronic acids are relatively unusual components of LPS, but have been found in some strains of *E. coli*, *Proteus* and *Xanthomonas* (Lüderitz *et al.*, 1968, 1971; Jann *et al.*, 1970), and are reportedly present in the core region of *P. mirabilis* (Kotelko *et al.*, 1977). Uronic acids are present in the R-LPS extracted from *E. herbicola* and *E. rhapontici* by the phenol–chloroform–petroleum ether method, but this may be due to the presence of some S-LPS subunits. The uronic acids in *E. herbicola* LPS may interfere with sedimentation during centrifugation at 100 000 g as was reported for the LPS from *E. coli* (Ørskov *et al.*, 1977). The LPS samples analysed by Dutkiewicz (1976) were not centrifuged and thus were not separated from the capsular polysaccharides.

Most members of the *herbicola* group (but not *E. stewartii*) are motile, but there is no report on the antigenic structure of the flagella.

Most strains of *E. herbicola* (and *E. rhapontici*) produce antigenically similar high molecular weight acidic polysaccharides, AP(H), as an extracellular slime on 2% glycerol nutrient agar or on a mineral salts medium containing 0.5% glycerol, 0.2% asparagine and 0.05% tri-sodium citrate. Antigen AP(H) is not present as an obvious capsule, unlike the cross-reactive antigen TV of *E. amylovora*. The purification of AP(H) is discussed in Section II.A.6.d. The antigens from different strains may differ in molecular weight and have minor antigenic differences which appear to be due to alkali-labile determinants (perhaps O-acyl groups). These antigenic differences have no obvious taxonomic implication (Slade, 1978). Unpurified preparations of extracellular slime usually also contain capsular antigens.

Antigen AP(H) is not produced by members of the *carotovora* group and has not been reported from other genera. In addition to its cross-reaction with antigen TV, it cross-reacts very weakly with colanic acid produced by *Salmonella* spp., *E. coli* and *Aerobacter cloacae* (sic) (Grant *et al.*, 1969). Colanic acid contains fucose (Sutherland, 1972) in addition to the sugars found in AP(H). The *Klebsiella* capsular antigen types 8, 25 and 61 contain the same sugars as AP(H), but have not been examined for cross-reactions. The production of antisera reacting with AP(H) and their serological tests are discussed in Sections I.C.3 and II.A.6.d.

(b) *Serotyping.* Two serological schemes have been proposed for *E. her-bicola.* Muraschi *et al.* (1965) used immunodiffusion and antigens extracted by aqueous ether. They classified 55 isolates into seven serotypes, although "some cultures were mixtures of more than one serotype". The principal ether extract antigens were most likely capsular (Slade, 1978). A second scheme was established by Slade (1978) using laboratory cultures. This classification has not been used to type new isolates, a high proportion of which may possess additional antigens. This is particularly true of the LPS antigens which have not been intensively studied. In Slade's study four heat-stable antigens or groups of antigens were detected (Table XIII).

(i) *Capsular antigens.* Since live or formalin-treated cells may possess a confusing array of heat-labile antigens, the heat-stable capsular antigens are best identified by agglutination of heat-treated cells (100°C, 30 min) with suitably absorbed antisera. It is preferable to confirm the identification by immunodiffusion using 1% solutions of unknown and reference antigens with unabsorbed antiserum. The capsular antigens may be prepared by heating formalin-killed cells at 100°C for 30 min, centrifuging to remove cell debris, precipitating the polysaccharide from the supernatant fluid with three volumes of ethanol and allowing to stand at 4°C overnight.

The capsular type strains for the *herbicola* group are shown in Table XIII. Strains G150 and G152 are serotypes 1 and 3 respectively of Muraschi *et al.* (1965). The capsular antigens 1–5 found in clusters H:1 and H:3 have a complex pattern of cross-reactions, but cluster H:2 does not appear to cross-react with H:1 and H:3 (Slade, 1978). Capsular antigen 1a becomes antigenically identical with antigen 1b after being treated with 0.25 M NaOH (56°C, 2 h) or 1 M HCl (37°C, 20 h). Untreated antigen 1a can absorb all antibodies to antigen 1b.

The capsule swelling test is unsuitable for *E. herbicola* as cells within a culture may possess capsules of different sizes and some strains may have no visible capsule.

(ii) *LPS antigens.* As *E. herbicola* strains frequently produce a mixture of R- and S-LPS, it is important to distinguish between them. S-LPS can be extracted by trichloracetic acid (Sutherland and Wilkinson, 1971) which does not extract R-LPS. Antisera should be absorbed with a suitable heterologous strain to remove antibodies to R-LPS and the *Erwinia* antigen (Section I.C.1).

As the LPS may be masked by capsular material, studies on the LPS in *Erwinia* is considered to require antigen extraction, although boiling the cells may remove the capsule and be equally suitable. Slade (1978) found it most convenient to study the LPS by passive haemagglutination as the capsular antigens did not adsorb to sheep erythrocytes. Although three strains in cluster H:3 had the same LPS antigen, further work is required to show whether this would have taxonomic significance. The LPS of strain 89 cross-reacted with antiserum to antigen LPS1 (Table XIII).

(iii) *Antigen GH4.* This antigen is of taxonomic interest as it was found in all ten strains of cluster H:1 (Section IV.B.4) in *E. herbicola* subsp. *herbicola* and in only one other strain (*E. herbicola* subsp. *ananas*, 1846). It is detected by immunodiffusion of sonicated cell suspensions which have been heated to 100°C for 30 min. In many strains it gives the only distinct precipitation line other than that due to the capsular antigen (Slade, 1978). Antigen GH4 has an antigenic determinant not present in GH4b (Table XIII).

(iv) *Antigen GH5.* Strains 2272 and 2273 possessed the same heat-stable antigen GH5b which cross-reacted with GH5a produced by strain 2271. GH5b possesses a determinant not present in GH5a (Slade, 1978). These strains are in cluster H:2 and their serological structure was not studied further.

6. *Conclusions*

Improved methods of typing *E. herbicola* are required for a study of its ecology. At present, there is no distinction between *E. herbicola* pathogenic for man and epiphytic strains. It would be useful to be able to determine whether strains can move between different habitats. This is of considerable importance if epiphytic strains are to be used for the biological control of plant pathogens.

C. *Erwinia uredovora*

E. uredovora is indistinguishable from *E. herbicola* by numerical taxonomic methods (Goodfellow *et al.*, 1976). It was recognized as a separate species (Dye, 1969b) because of its unusual ability to attack the uredia of rust (*Puccinia graminis* and other species) on wheat, oats and rye (Pon *et al.*, 1954; Hevesi and Mashaal, 1975). The occurrence of *E. uredovora* on healthy plants does not appear to have been studied.

The colonies of *E. uredovora* are yellow pigmented and do not contain biconvex bodies (Graham and Hodgkiss, 1967). Symplasmata are produced. *E. uredovora* is the only *Erwinia* species reported to produce acid from adonitol in 1% peptone water (Dye, 1969b).

D. *Erwinia stewartii*

This organism causes Stewart's disease, a vascular wilt and leaf blight of maize (*Zea mays*) in which the vessels become plugged with a bright yellow slime. Affected plants are dwarfed, have pale stripes on the leaves and may have prematurely developed tassels which wither and die before the rest of the plant. If the stem of a wilting or dying sweet-corn plant is cut, small droplets of yellowish bacterial exudate appear and can be drawn into threads 2–5 mm long. The disease also occurs on teosinte (*Euchlaena mexicana*) and eastern gama grass (*Tripsacum dactyloides*); various common plants can act as symptomless hosts (Pepper, 1967). *E. stewartii* is endemic throughout the maize-growing belt of the USA, where it overwinters in the cornflea beetle (*Chaetocnema pulicaria*), and it has also been reported from Canada, China, Costa Rica, Italy, Mexico, Poland, Russia, Switzerland and Yugoslavia. Transmission is mainly by insects, but infected seeds are important in its transmission to new areas. *E. stewartii* caused heavy losses in the USA during the 1930s, but further losses were avoided by breeding maize resistant to the disease. Stewart's disease has recently become prominent again in some areas (Heichel *et al.*, 1977; Blanco *et al.*, 1979).

E. stewartii may be isolated on glucose nutrient agar where colonies are usually buff-yellow, but may range in colour from cream to orange; the colonial morphology is variable (Ivanoff *et al.*, 1938). *E. stewartii* can be distinguished from *E. herbicola* by its lack of motility and by its inability to ferment maltose, rhamnose, salicin and dextrin. Identification should be confirmed by testing its pathogenicity for seedling maize plants.

The serology of *E. stewartii* has not been studied in detail, but Garibaldi and Gibbins (1975) reported non-virulent mutants with an altered antigenic composition.

E. *Erwinia rhapontici*

E. rhapontici was initially reported to cause crown rot of rhubarb (Millard, 1924), but it has since been isolated from pink wheat grains (Roberts, 1974) and rotting hyacinth bulbs (Sellwood and Lelliott, 1978). However, *E. rhapontici* appears to have only a low pathogenicity, since rotting is induced only in damaged plant tissues (Roberts, 1974). Strains differ in their ability to rot tissues from different hosts. As it may be isolated from the surface of apple, pear and hawthorn trees (E. Billing, personal communication; Sellwood and Lelliott, 1978; Slade, 1978) it is thought likely that *E. rhapontici* is widespread amongst the epiphytic flora. There are no reports of its pathogenicity for man or animals.

   E. rhapontici may be isolated and recognized on 2% glycerol nutrient agar supplemented with 0.1% yeast extract or 5% sucrose nutrient agar. It produces mucoid colonies and a pink diffusible pigment after two days incubation at 25°C. However the frequency of non-pigment producing strains is unknown and colonies in crowded areas of the plate may not produce pigment. The pigment is not produced on nutrient agar, where the colonies are white, round and smooth. Tests suitable for differentiating *E. rhapontici* from other members of the *herbicola* group are shown in Table IV. Apart from the pink pigment, *E. rhapontici* is distinguished by its inability to liquefy nutrient gelatin within 30 days. Rotting of plant tissue is not a good diagnostic test, since some strains give only slight or erratic rotting.

   It is thought to have a serological structure similar to most strains of *E. herbicola* (Section IV.B.5.a). The known capsular serotypes are shown in Table XIII. The capsular antigens 19 and 20 cross-react.

F. Cluster E1

This small group of strains does not conform to any previously described species. Numerical taxonomy indicates that the group lies between *E. herbicola* and *E. rhapontici* (Slade, 1978). Tests distinguishing cluster E1 from other members of the herbicola group are shown in Table IV. Nothing is known of the serological structure of these organisms but in immunodiffusion studies, two strains appeared identical and did not share any heat stable antigens with the third strain. All three strains produce antigen AP(H) (Slade, 1978).

Acknowledgement

We are very grateful to Dr Eve Billing for all the help and encouragement we have received from her in connection with this work.

References

Adams, M. H. (1959). *In* "Bacteriophage", p. 454. Interscience, New York.
Adegeye, A. O. and Preece, T. F. (1978). *J. Appl. Bacteriol.* **44**, 265–277.
Aho, P. E., Seidler, R. J., Evans, H. J. and Raju, P. N. (1974). *Phytopathology* **64**, 1413–1420.
Ashworth, L. J., Hildebrand, D. C. and Schroth, M. N. (1970). *Phytopathology* **60**, 602–607.
Allan, E. and Kelman, A. (1977). *Phytopathology* **67**, 1305–1312.
Arsenijevic, M. (1978). *In* "Proc. IVth Int. Conference on Plant Pathogenic Bacteria", Vol. II, pp. 531–538. INRA, Angers.
Ayers, A. R., Ayers, S. B. and Goodman, R. N. (1979). *Appl. Environ. Microbiol.* **38**, 659–666.
Ayers, S. B., Goodman, R. N. and Stoffl, P. (1977). *Proc. Am. Phytopathol. Soc.* **4**, 107.
Bagley, S. T., Seidler, R. J., Talbot, H. W. and Morrow, J. E. (1978). *Appl. Environ. Microbiol.* **36**, 178–185.
Baker, K. F. (1971). *Hilgardia* **40**, 603–633.
Baldwin, C. H. and Goodman, R. N. (1963). *Phytopathology* **53**, 1299–1303.
Bascomb, S., Lapage, S. P., Willcox, W. R. and Curtis, M. A. (1971). *J. Gen. Microbiol.* **66**, 279–295.
Basham, H. G. and Bateman, D. F. (1975a). *Phytopathology* **65**, 141–153.
Basham, H. G. and Bateman, D. F. (1975b). *Physiol. Plant Pathol.* **5**, 249–262.
Bateman, D. F. and Millar, R. L. (1966). *Annu. Rev. Phytopathol.* **4**, 119–146.
Beer, S. V. and Aldwinckle, H. S. (1976). *Proc. Am. Phytopathol. Soc.* **3**, 300.
Bennett, R. A. (1980). *J. Gen. Microbiol.* **116**, 351–356.
Bennett, R. A. and Billing, E. (1978). *Ann. Appl. Biol.* **89**, 41–45.
Bennett, R. A. and Billing, E. (1980). *J. Gen. Microbiol.* **116**, 341–349.
Beraha, I. (1968). *Plant Dis. Rep.* **52**, 167.
Billing, E. (1960). *Nature (London)* **186**, 819–820.
Billing, E. (1976a). *In* "Microbiology of Aerial Plant Surfaces" (C. H. Dickinson and T. F. Preece, Eds), pp. 223–274. Academic Press, New York and London.
Billing, E. (1976b). *Ann. Appl. Biol.* **82**, 259–266.
Billing, E. (1980a). *Ann. Appl. Biol.* **95**, 341–364.
Billing, E. (1980b). *Ann. Appl. Biol.* **95**, 365–377.
Billing, E. and Baker, L. A. E. (1963). *J. Appl. Bacteriol.* **26**, 58–65.
Billing, E., Fletcher, J. T., Glasscock, H. H., Jones, G. E. and Lelliott, R. A. (1959). *Plant Pathol.* **8**, 152.
Billing, E., Crosse, J. E. and Garrett, C. M. E. (1960). *Plant Pathol.* **9**, 19–25.
Billing, E., Baker, L. A. E., Crosse, J. E. and Garrett, C. M. E. (1961). *J. Appl. Bacteriol.* **24**, 195–211.
Blanco, M. H., Zuber, M. S., Wallin, J. R., Loonan, D. V. and Krause, G. F. (1979). *Phytopathology* **69**, 849–853.
Bonnet, P. and Vénard, P. (1975). *Ann. Phytopathol.* **7**, 51–59.
Bottone, E. and Schneierson, S. S. (1972). *Am. J. Clin. Pathol.* **57**, 400–405.
Brenner, D. J., Fanning, G. R. and Steigerwalt, A. G. (1972). *J. Bacteriol.* **110**, 12–17.
Brenner, D. J., Steigerwalt, A. G., Miklos, G. V. and Fanning, G. R. (1973). *Int. J. Syst. Bacteriol.* **23**, 205–216.
Brenner, D. J., Fanning, G. R. and Steigerwalt, A. G. (1974). *Int. J. Syst. Bacteriol.* **24**, 197–204.
Brenner, D. J., Fanning, G. R. and Steigerwalt, A. G. (1977). *Int. J. Syst. Bacteriol.* **27**, 211–221.

Chatterjee, A. K. and Gibbins, L. N. (1969). *J. Bacteriol.* **100**, 594–600.

Chatterjee, A. K. and Starr, M. P. (1977). *J. Bacteriol.* **132**, 862–869.

Chatterjee, A. K., Gibbins, L. N. and Carpenter, J. A. (1969). *Can. J. Microbiol.* **15**, 640–642.

Chatterjee, A. K., Buss, R. F. and Starr, M. P. (1977). *Antimicrob. Agents Chemother.* **11**, 897–905.

Cho, J. J., Rohrback, K. G. and Hayward, A. C. (1978). *In* "Proc. IVth Int. Conference on Plant Pathogenic Bacteria", Vol. II, pp. 433–441. INRA, Angers.

Christofi, N., Wilson, M. I. and Old, D. C. (1979). *J. Appl. Bacteriol.* **46**, 179–183.

Church, B. M., Hampson, C. P. and Fox, W. R. (1970). *Potato Res.* **13**, 41–58.

Codner, R. C. (1971). *J. Appl. Bacteriol.* **34**, 147–160.

Cowan, S. T. and Steele, K. J. (1974). "Manual for Identification of Medical Bacteria", 2nd edn. Cambridge University Press, Cambridge.

Crosse, J. E. and Goodman, R. N. (1973). *Phytopathology* **63**, 1425–1426.

Crosse, J. E., Bennett, M. and Garrett, C. M. E. (1958). *Nature (London)* **182**, 1530.

Crowley, C. F. and De Boer, S. H. (1980). *Can. J. Microbiol.* **26**, 1023–1028.

Cruickshank, J. C. (1935). *J. Hyg.* **35**, 354.

Cuppels, D. and Kelman, A. (1974). *Phytopathology* **64**, 468–475.

Dazzo, F. B. and Hubbell, D. H. (1975). *Appl. Microbiol.* **30**, 1017–1033.

Dazzo, F. B., Napoli, C. A. and Hubbell, D. H. (1976). *Appl. Environ. Microbiol.* **32**, 166–171.

De Boer, S. H. (1976). "Ecology of *Erwinia carotovora* and Factors Affecting Tuber Susceptibility to Bacterial Soft Rot". Ph.D. Thesis, University of Wisconsin–Madison, Madison, Wisconsin.

De Boer, S. H. (1980). *Can. J. Microbiol.* **26**, 567–571.

De Boer, S. H., Cuppels, D. A. and Kelman, A. (1974). *Proc. Am. Phytopathol. Soc.* **1**, 124.

De Boer, S. H., Cuppels, D. A. and Kelman, A. (1978). *Phytopathology* **68**, 1784–1790.

De Boer, S. H., Kelman, A. and Buelow, F. H. (1975). *Proc. Am. Phytopathol. Soc.* **2**, 68.

De Boer, S. H., Copeman, R. J. and Vruggink, H. (1979). *Phytopathology* **69**, 316–319.

De Kam, M. (1976). *Antonie van Leeuwenhoek* **42**, 421–428.

Dean, M. and Wood, R. K. S. (1967). *Nature (London)* **214**, 408–410.

Dias, F. F. (1967). *Appl. Microbiol.* **15**, 1512–1513.

Dickey, R. S. (1979). *Phytopathology* **69**, 324–329.

Dickey, R. S. and Victoria, J. I. (1980). *Int. J. Syst. Bacteriol.* **30**, 129–134.

Dinesen, J. G. (1979). *Phytopathol. Z.* **95**, 59–64.

Dowson, W. J. (1957). "Plant Diseases due to Bacteria", 2nd edn. Cambridge University Press, Cambridge.

Dresel, E. G. and Stickl, O. (1928). *Deut. Med. Wochenschr.* **54**, 517–519.

Duggeli, M. (1904). *Zentralbl. Bakteriol. Parasitenkd. Infektionskr. Abt. I Orig.* **12**, 602–614.

Dutkiewicz, J. (1976). *Zentralbl. Bakteriol. Parasitenkd. Infektionskr. Abt. I Orig.* **236**, 487–508.

Dye, D. W. (1964). *N.Z. J. Sci.* **7**, 261–269.

Dye, D. W. (1966). *N.Z. J. Sci.* **9**, 843–854.

Dye, D. W. (1968). *N.Z. J. Sci.* **11**, 590–607.

Dye, D. W. (1969a). *N.Z. J. Sci.* **12**, 81–97.

Dye, D. W. (1969b). *N.Z. J. Sci.* **12**, 223–236.

Dye, D. W. (1969c). *N.Z. J. Sci.* **12**, 833–839.

Dye, D. W. (1981). *N.Z. J. Sci.* **24**, 233.

Echandi, E. and Moyer, J. W. (1978). *In* "Proc. IVth Int. Conference on Plant Pathogenic Bacteria", Vol. II, p. 639. INRA, Angers.

Echandi, E. and Moyer, J. W. (1979). *Phytopathology* **69**, 1204–1207.

Eden-Green, S. J. (1972). "Studies in Fireblight Disease of Apple, Pear and Hawthorn (*Erwinia amylovora* (Burrill) Winslow *et al.*). Ph.D. Thesis, University of London, London.

Eden-Green, S. J. and Billing, E. (1974). *Commonwealth Mycological Inst. Rev. Plant. Pathol.* **53**, 353–365.

Eden-Green, S. J. and Knee, M. (1974). *J. Gen. Microbiol.* **81**, 509–512.

Edwards, P. R. and Fife, M. A. (1956). *Appl. Microbiol.* **4**, 46–48.

Erskine, J. M. (1973). *Can. J. Microbiol.* **19**, 837–845.

Erskine, J. M. and Lopatecki, L. E. (1974). *Can. J. Microbiol.* **21**, 35–41.

Evans, G. E. and Stevenson, W. R. (1980). *Phytopathology* **69**, 540.

Ewing, W. H. and Fife, M. A. (1972). *Int. J. Syst. Bacteriol.* **22**, 4–11.

Farmer, J. J., Asbury, M. A., Hickman, F. W. and Brenner, D. J. (1980). *Int. J. Syst. Bacteriol.* **30**, 569–584.

Fonnesbech, A. (1974). *Årsskrift for den Kongelige Veterinaer-Og Landbohojskole* **24**, 176–195.

Fucikovsky, L., Rodriguez, M. and Cartin, L. (1978). *In* "Proc. IVth Int. Conference on Plant Pathogenic Bacteria", Vol. II, pp. 603–606. INRA, Angers.

Galanos, C., Lüderitz, O. and Westphal, O. (1969). *Eur. J. Biochem.* **9**, 245–249.

Gardner, J. M. and Kado, C. I. (1972). *Int. J. Syst. Bacteriol.* **22**, 201–209.

Garibaldi, A. and Gibbins, L. N. (1975). *Can. J. Microbiol.* **21**, 1282–1287.

Gavini, F., Ferragut, C., Lefebvre, B. and Leclerc, H. (1976). *Ann. Microbiol.* **127B**, 317–335.

Gilardi, G. L., Bottone, E. and Birnbaum, M. (1970). *Appl. Microbiol.* **20**, 151–155.

Goodfellow, M., Austin, B. and Dickinson, C. H. (1976). *J. Gen. Microbiol.* **97**, 219–233.

Goodman, R. N. (1965). *Phytopathology* **55**, 217–221.

Goodman, R. N. (1967). *Phytopathology* **57**, 22–24.

Goodman, R. N., Huang, J. S. and Huang, P. (1974). *Science* **183**, 1081–1082.

Goodman, R. N., Ayers, A. R. and Politis, D. J. (1978). *In* "Proc. IVth Int. Conference on Plant Pathogenic Bacteria", Vol. II, pp. 483–486. INRA, Angers.

Goto, M. (1976). *Int. J. Syst. Bacteriol.* **26**, 467–473.

Goto, M. (1979). *Plant Dis. Rep.* **63**, 100–103.

Goto, M. and Okabe, N. (1957). *Bull. Fac. Agric. Shizuoka Univ.* **7**, 11–20.

Goto, M. and Okabe, N. (1958). *Bull. Fac. Agric. Shizuoka Univ.* **8**, 1–31.

Graevenitz, von, A. (1970). *Ann. N.Y. Acad. Sci.* **174**, 436–443.

Graham, D. C. (1958). *Nature (London)* **61**, 61.

Graham, D. C. (1963). *Plant Pathol.* **12**, 142–144.

Graham, D. C. (1964). *Annu. Rev. Phytopathol.* **2**, 13–42.

Graham, D. C. (1971). *In* "Proc. IIIrd Int. Conference on Plant Pathogenic Bacteria" (H. P. Maas Geesteranus, Ed.), pp. 273–279. Wageningen.

Graham, D. C. and Dowson, W. J. (1960). *Ann. Appl. Biol.* **48**, 51–57.

Graham, D. C. and Harper, P. C. (1967). *Scot. Agric.* **46**, 68.

Graham, D. C. and Harrison, M. D. (1975). *Phytopathology* **65**, 739–741.

Graham, D. C. and Hodgkiss, W. (1967). *J. Appl. Bacteriol.* **30**, 175–189.

Graham, D. C., Quinn, C. E. and Harrison, M. D. (1976). *Potato Res.* **19**, 3–21.

Graham, D. C., Quinn, C. E. and Bradley, L. F. (1977). *J. Appl. Bacteriol.* **43**, 413–424.

Grant, D. W., Sutherland, I. W. and Wilkinson, N. F. (1969). *J. Bacteriol.* **100**, 1187–1193.

Grimont, P. A. D., Grimont, F., Dulong de Rosnay, M. L. C. and Sneath, P. H. A. (1977). *J. Gen. Microbiol.* **98**, 39–66.
Hammerling, U. and Westphal, O. (1967). *Eur. J. Biochem.* **1**, 46–50.
Harrison, M. D., Quinn, C. E., Sells, I. A. and Graham, D. C. (1977). *Potato Res.* **20**, 37–52.
Heichel, G. H., Sands, D. C. and Kring, J. B. (1977). *Plant Dis. Rep.* **61**, 149–153.
Hevesi, M. and Mashaal, S. F. (1975). *Acta Phytopathol. Acad. Sci. Hung.* **10**, 275–280.
Hildebrand, D. C. and Schroth, M. N. (1967). *Phytopathology* **57**, 250–253.
Hirata, S. (1960). *Miyazaki Deigaku Nogaku Kenkyu Jiho* **5**, 85–92.
Holding, A. J. (1960). *J. Appl. Bacteriol.* **23**, 515–525.
Hsieh, S. P. Y. and Buddenhagen, I. W. (1974). *Phytopathology* **64**, 1182–1185.
Huang, P., Huang, J. and Goodman, R. N. (1975). *Physiol. Plant Pathol.* **6**, 283–287.
Israilskij, W. P., Schklar, S. N. and Orlowa, G. I. (1966). *In* "Proc. Symp. on Host Parasite Relations in Plant Pathology", pp. 141–146. Budapest.
Ivanoff, S. S., Riker, A. J. and Dettwiler, H. A. (1938). *J. Bacteriol.* **35**, 235–253.
Jann, B., Jann, K. and Schmidt, G. (1970). *Eur. J. Biochem.* **15**, 29–39.
Jones, S. M. and Paton, A. M. (1973). *J. Appl. Bacteriol.* **36**, 729–737.
Kado, C. I. and Heskett, M. G. (1970). *Phytopathology* **60**, 969–976.
Kikumoto, T. (1974). *Bull. Inst. Agric. Res. Tohoku Univ.* **25**, 125–137.
Kikumoto, T. and Sakumoto, M. (1969a). *Ann. Phytopathol. Soc. Jpn.* **35**, 29–35.
Kikumoto, T. and Sakumoto, M. (1969b). *Ann. Phytopathol. Soc. Jpn.* **35**, 36–40.
King, E. O., Ward, M. K. and Raney, D. E. (1954). *J. Lab. Clin. Med.* **44**, 301.
Klement, Z. (1971). *In* "Proc. IIIrd Int. Conference on Plant Pathogenic Bacteria" (H. P. Maas Geesteranus, Ed.), pp. 157–164. Wageningen.
Klinger, A. E., Hildebrand, D. C. and Wilhelm, S. (1971). *Plant and Soil* **34**, 215–218.
Kloepper, J. W., Harrison, M. D. and Brewer, J. W. (1979). *Am. Potato J.* **56**, 351–361.
Komagata, K., Tamagawa, Y. and Hzuka, H. (1968). *J. Gen. Appl. Microbiol.* **14**, 39–45.
Kotelko, K., Gromska, W., Papierz, M., Sidorczyk, Z., Krajewska, D. and Szer, K. (1977). *J. Hyg. Epidemiol. Microbiol. Immunol.* **21**, 271–284.
Lazar, I. (1971a). *In* "Proc. IIIrd Int. Conference on Plant Pathogenic Bacteria" (H. P. Maas Geesteranus, Ed.), pp. 125–130. Wageningen.
Lazar, I. (1971b). *In* "Proc. IIIrd Int. Conference on Plant Pathogenic Bacteria", pp. 131–141. Wageningen.
Le Minor, L., Chalon, A. M. and Vernon, M. (1972). *Ann. Inst. Pasteur* **123**, 761–774.
Lelliott, R. A. (1967). Report of the International Conference on Fireblight, Canterbury. EPPO Publications Series A. No. 45-E.
Lelliott, R. A. (1974). Erwinia. *In* "Bergey's Manual of Determinative Bacteriology" (R. E. Buchanan and N. E. Gibbons, Eds), 8th edn, pp. 332–338. Williams & Wilkins, Baltimore, Maryland.
Lelliott, R. A. (1978). *In* "Proc. IVth Int. Conference on Plant Pathogenic Bacteria", Vol. II, p. 527. INRA, Angers.
Lippincott, B. B. and Lippincott, J. A. (1969). *J. Bacteriol.* **97**, 620–628.
Logan, C. (1963). *Nature (London)* **199**, 623.
Logan, C. (1968). *Record of Agricultural Research, Ministry of Agriculture of Northern Ireland* **17**, 115–121.
Löhnis, F. (1911). Landwirtschaftlich-bakteriologisches Praktikum. Gebrüder Borntraeger, Berlin.
Lüderitz, O., Jann, K. and Wheat, R. (1968). *In* "Comprehensive Biochemistry" (M. Florkin and E. H. Stotz, Eds), Vol. 26A, pp. 105–228. Elsevier, Amsterdam.
Lüderitz, O., Westphal, O., Staub, A. M. and Nikaido, H. (1971). *In* "Microbial

Toxins" (G. Weinbaum, S. Kadis and S. J. Ajl, Eds), pp. 145–234. Academic Press, New York and London.

Lund, B. M. (1969). *J. Appl. Bacteriol.* **32**, 60–67.

Lund, B. M. (1971). *J. Appl. Bacteriol.* **34**, 9–20.

Lund, B. M. (1972). *J. Appl. Bacteriol.* **35**, 609–614.

Lund, B. M. (1979). *In* "Plant Pathogens" (D. W. Lovelock, Ed.), pp. 19–49. Academic Press, London and New York.

Lund, B. M. and Kelman, A. (1977). *Am. Potato J.* **54**, 211–225.

McClendon, J. H. (1964). *Am. J. Bot.* **51**, 628–633.

McIntyre, J. L. and Williams, E. B. (1972). *Phytopathology* **62**, 777.

McIntyre, J. L., Kuc, J. and Williams, E. B. (1973). *Phytopathology* **63**, 872–877.

McIntyre, J. L., Sands, D. C. and Taylor, G. S. (1978). *Phytopathology* **68**, 435–440.

Mendonça, M. and Stanghellini, M. E. (1979). *Phytopathology* **69**, 1096–1099.

Meneley, J. C. and Stanghellini, M. E. (1972). *Phytopathology* **66**, 367–370.

Meneley, J. C. and Stanghellini, M. E. (1974). *J. Food Sci.* **39**, 1267–1268.

Meneley, J. C. and Stanghellini, M. E. (1976). *Phytopathology* **66**, 367–370.

Mew, T. W., Ho, W. C. and Chu, L. (1976). *Phytopathology* **66**, 1325–1327.

Meyers, B. R., Bottone, E., Hirschman, S. Z. and Schneierson, S. S. (1972). *Ann. Int. Med.* **76**, 9–14.

Miles, A. A. and Misra, S. S. (1938). *J. Hyg.* **38**, 732–749.

Millard, W. A. (1924). *Bull. University Leeds and Yorkshire Council for Agric. Education*, No. 134.

Miller, H. J. (1979). *EPPO Bull.* **9**, 7–11.

Miller, T. D. (1972). *Phytopathology* **62**, 778.

Miller, T. D. and Schroth, M. N. (1970). *Phytopathology* **60**, 1304.

Miller, T. D. and Schroth, M. N. (1972). *Phytopathology* **62**, 1175–1182.

Ministry of Agriculture, Fisheries and Food (1969). Fireblight of Apple and Pear, Advisory Leaflet 571.

Molina, J. J., Harrison, M. A. and Brewer, J. W. (1974). *Am. Potato J.* **51**, 245–250.

Møller, W. J., Thompson, S. V., Schroth, M. N., Reil, W. O. and Beutal, J. A. (1978). *Western Fruit Grower* **98**, 14–48.

Morgan, F., Nasuno, S. and Starr, M. P. (1968). *Arch. Biochem. Biophys.* **123**, 298–306.

Mount, M. S., Bateman, D. F. and Basham, H. G. (1970). *Phytopathology* **60**, 924–931.

Mundt, J. O. and Hinkle, N. F. (1976). *Appl. Environ. Microbiol.* **32**, 694–698.

Muraschi, T. F., Friend, M. and Bolles, D. (1965). *Appl. Microbiol.* **13**, 128–131.

Murata, N. and Starr, M. P. (1974). *Can. J. Microbiol.* **20**, 1545–1565.

Nasuno, S. and Starr, M. P. (1966). *J. Biol. Chem.* **241**, 5298.

Naumann, K. and Ficke, W. (1972). *Zentralbl. Bakteriol. Parasitenkd. Infektionskr. Abt. Orig.* **127**, 180–189.

Neilson, A. H. (1979). *J. Appl. Bacteriol.* **46**, 483–491.

Neilson, A. H. and Sparell, L. (1975). *Appl. Environ. Microbiol.* **32**, 197–205.

Nelson, M., Alvarez, A. and Chun, W. (1977). *Proc. Am. Phytopathol. Soc.* **4**, 106.

Noble, M. and Graham, D. C. (1956). *Nature (London)* **178**, 1479–1480.

Novakova, J. (1957). *Phytopathology Z.* **29**, 72–74.

O'Neill, R. and Logan, C. (1975). *J. Appl. Bacteriol.* **39**, 139–146.

Okabe, N. and Goto, M. (1955). *Bull. Fac. Agric. Shizuoka Univ.* **5**, 72–86.

Okabe, N. and Goto, M. (1956a). *Bull. Fac. Agric. Shizuoka Univ.* **6**, 16–32.

Okabe, N. and Goto, M. (1956b). *Bull. Fac. Agric. Shizuoka Univ.* **6**, 14–15.

Ørskov, I., Ørskov, F., Jann, B. and Jann, K. (1977). *Bacteriol. Rev.* **41**, 667–710.

Paton, A. M. (1959). *Nature (London)* **183**, 1812–1813.

Panopoulos, N. J. (1978). *In* "Proc. IVth Int. Conference on Plant Pathogenic Bacteria", Vol. II, pp. 467–470. INRA, Angers.
Paulin, J. P. (1978). *In* "Proc. IVth Int. Conference on Plant Pathogenic Bacteria", Vol. II, pp. 525–526. INRA, Angers.
Paulin, J. P. and Lachaud, G. (1978). *In* "Proc. IVth Int. Conference on Plant Pathogenic Bacteria", Vol. II, pp. 519–522. INRA, Angers.
Paulin, J. P. and Nassan-Agha, N. A. (1978). *In* "Proc. IVth Int. Conference on Plant Pathogenic Bacteria", Vol. II, pp. 539–546. INRA, Angers.
Pedersen, W. L., Chakrabarty, K., Klucas, R. V. and Vidaver, A. K. (1978). *Appl. Environ. Microbiol.* **35**, 129–135.
Pepper, E. H. (1967). Monograph 4. American Phytopathological Society. Heffernan Press, Worcester, Massachusetts.
Perombelon, M. C. M. (1971a). *J. Appl. Bacteriol.* **34**, 793–799.
Perombelon, M. C. M. (1971b). *Potato Res.* **14**, 158–160.
Perombelon, M. C. M. (1972). *Ann. Appl. Biol.* **71**, 111–117.
Perombelon, M. C. M. (1978). *In* "Proc. IVth Int. Conference on Plant Pathogenic Bacteria", Vol. II, pp. 563–566. INRA, Angers.
Perombelon, M. C. M. (1979). *Potato Res.* **22**, 63–68.
Perombelon, M. C. M. and Lowe, R. (1971). Report of the Scottish Horticultural Research Institute for 1970, pp. 32–33.
Perombelon, M. C. M., Fox, R. A. and Lowe, R. (1979). *Phytopathology Z.* **94**, 249–260.
Petrescu, P. M. (1974). *Eur. J. Pathol.* **4**, 222–227.
Pon, D. S., Townsend, C. E., Wessman, G. E., Schmitt, C. G. and Kingsolver, C. H. (1954). *Phytopathology* **44**, 707–710.
Preece, T. F. (1977). "Watermark Disease of the Cricket Bat Willow." Forestry Commission Leaflet No. 20, HMSO, London.
Preece, T. F., Wong, W. C. and Adegeye, A. O. (1979). *In* "Plant Pathogens" (D. W. Lovelock, Ed.), pp. 1–17. Academic Press, London and New York.
Primrose, S. B. (1976). *J. Gen. Microbiol.* **97**, 343–346.
Quinn, C. E., Sells, I. A. and Graham, D. C. (1980). *J. Appl. Bacteriol.* **49**, 175–181.
Rapin, A. M. C. and Mayer, H. (1966). *Ann. N.Y. Acad. Sci.* **133**, 425–432.
Ries, S. M. and Otterbacher, A. G. (1977). *Plant Dis. Rep.* **61**, 232–235.
Riggle, J. H. and Klos, E. J. (1972). *Can. J. Bot.* **50**, 1077–1083.
Ritchie, D. F. (1978). "Bacteriophages of *Erwinia amylovora:* Their Isolation, Distribution, Characterization and Possible Involvement in the Etiology and Epidemiology of Fire Blight". Ph.D. Thesis, Michigan State University, Michigan.
Ritchie, D. F. and Klos, E. J. (1977). *Phytopathology* **67**, 101–104.
Ritchie, D. F. and Klos, E. J. (1978). *Plant Dis. Rep.* **62**, 167–169.
Ritchie, D. F. and Klos, E. J. (1979). *Phytopathology* **69**, 1076–1083.
Roberts, P. (1974). *J. Appl. Bacteriol.* **37**, 353–358.
Roberts, W. P. and Kerr, A. (1974). *Physiol. Plant Pathol.* **4**, 81–91.
Ruppell, E. G., Harrison, M. D. and Kent Nielson, A. (1975). *Plant Dis. Rep.* **59**, 837–840.
Sakazaki, R., Tamura, K., Johnson, R. and Colwell, R. R. (1976). *Int. J. Syst. Bacteriol.* **26**, 158–179.
Samson, R. (1972). *Ann. Phytopathol.* **4**, 157–163.
Samson, R. (1973). *Ann. Phytopathol.* **5**, 377–388.
Samson, R. and Nassan-Agha, N. (1978). *In* "Proc. IVth Int. Conference on Plant Pathogenic Bacteria", Vol. II, pp. 547–554, INRA, Angers.

Sands, D. C. and Dickey, R. S. (1978). *In* "Proc. IVth Int. Conference on Plant Pathogenic Bacteria", Vol. II, pp. 555–560. INRA, Angers.

Schroth, M. N., Thomson, S. V., Hildebrand, D. C. and Moller, W. J. (1974). *Annu. Rev. Phytopathol.* **12**, 389–412.

Schroth, M. N., Thomson, S. V. and Moller, W. J. (1978). *In* "Proc. IVth Int. Conference on Plant Pathogenic Bacteria", Vol. II, p. 527. INRA, Angers.

Schroth, M. N., Thomson, S. V. and Moller, W. J. (1979). *Phytopathology* **69**, 565–568.

Scott, J. E. (1965). *In* "Methods in Carbohydrate Chemistry" (R. L. Whistler, J. N. Bemiller and M. L. Wolfrom, Eds), Vol. V, pp. 39–44. Academic Press, New York and London.

Seemuller, E. A. and Beer, S. V. (1976). *Phytopathology* **66**, 433–436.

Segall, R. H. (1971). *Phytopathology* **61**, 425–426.

Sellwood, J. E. and Lelliott, R. A. (1978). *Plant Pathol.* **27**, 120–124.

Shinde, P. A. and Lukezic, F. L. (1974a). *Phytopathology* **64**, 865–871.

Shinde, P. A. and Lukezic, F. L. (1974b). *Phytopathology* **64**, 871–876.

Sjulin, T. M. and Beer, S. V. (1978). *Phytopathology* **68**, 89–94.

Slade, M. B. (1978). "Serological Studies on the Genus *Erwinia*." Ph.D. Thesis, University of Reading, Reading.

Slade, M. B. and Tiffin, A. I. (1978). *In* "Proc. IVth Int. Conference on Plant Pathogenic Bacteria", Vol. I, pp. 289–294. INRA, Angers.

Slade, M. B. and Tiffin, A. I. (1980). *J. Appl. Bacteriol.* **49**, xiii.

Stanghellini, M. E. and Meneley, J. C. (1975). *Phytopathology* **65**, 86–87.

Stanghellini, M. E., Sands, D. C., Kronland, W. C. and Mendonca, M. M. (1977). *Phytopathology* **67**, 1178–1182.

Starr, M. P. (1959). *Annu. Rev. Microbiol.* **13**, 211–238.

Starr, M. P. and Chatterjee, A. K. (1972). *Annu. Rev. Microbiol.* **26**, 389–426.

Starr, M. P. and Mandel, M. (1969). *J. Gen. Microbiol.* **56**, 113–123.

Starr, M. P., Cardona, C. and Folsom, D. (1951). *Phytopathology* **41**, 915–919.

Starr, M. P., Cosens, G. and Knackmiss, H. J. (1966). *Appl. Microbiol.* **14**, 870–872.

Starr, M. P., Chatterjee, A. K., Starr, P. B. and Buchanan, G. E. (1977). *J. Clin. Microbiol.* **6**, 379–386.

Stephens, G. J. and Wood, R. K. S. (1975). *Physiol. Plant Pathol.* **5**, 165–181.

Stewart, D. J. (1962). *Nature (London)* **195**, 1023.

Sutherland, F. W. (1972). *In* "Advances in Microbial Physiology" (A. H. Rose and D. W. Tempest, Eds), Vol. 8, pp. 143–213. Academic Press, New York and London.

Sutherland, I. W. and Wilkinson, J. F. (1971). *In* "Methods in Microbiology" (J. R. Norris and D. W. Ribbons, Eds), Vol. 5B, pp. 343–384. Academic Press, London and New York.

Sutton, D. D., Ark, P. A. and Starr, M. P. (1960). *Phytopathology* **20**, 182–186.

Tanii, A. and Akai, J. (1975). *Ann. Phytopathol. Soc. Jpn.* **41**, 513–517.

Teviotdale, B. L. (1979). *Phytopathology* **69**, 921.

Thompson, S. V., Schroth, M. N., Moller, W. J. and Reil, W. O. (1976). *Phytopathology* **66**, 1457–1459.

Thompson, S. V., Schroth, M. N., Hills, F. J., Whitney, E. D. and Hildebrand, D. C. (1977). *Phytopathology* **67**, 1183–1189.

Thurner, K. and Busse, M. (1978). *Zentralbl. Bakteriol. Parasitenkd. Infektionskr. Abt. I. Orig.* **167**, 262–271.

Togashi, J. (1972). *Rep. Inst. Agric. Res. Tohoku Univ.* **23**, 17–52.

Trust, T. J. and Bartlett, K. H. (1976). *Appl. Environ. Microbiol.* **31**, 992–994.

Tseng, T. C. and Mount, M. S. (1974). *Phytopathology* **64**, 229–236.

US Department of Agriculture (1965). "Losses in Agriculture", Handbook 291. US Dept of Agriculture, Washington, DC.
Ursing, J. (1977). *Acta Path. Microbiol. Scand. Sect. B* **85**, 61–66.
Van der Zwet, T. (1969). *Phytopathology* **59**, 607–613.
Van der Zwet, T. (1970). *F.A.O. Plant Protection Bulletin* **18**, 83–88.
Von Riesen, V. L. (1976). *Int. J. Syst. Bacteriol.* **26**, 143–145.
Vruggink, H. (1978). *In* "Proc. IVth Int. Conference on Plant Pathogenic Bacteria", Vol. I, pp. 307–310, INRA, Angers.
Vruggink, H. and De Boer, S. H. (1977). *Potato Res.* **20**, 268.
Vruggink, H. and Maas Geesteranus, H. P. (1975). *Potato Res.* **18**, 546–555.
Waldee, E. L. (1945). *Iowa State College J. Sci.* **19**, 435–484.
Wallis, F. M., Joubert, J. J. and Schlosser, I. F. (1975). *Phytophylactica* **7**, 125–130.
Webb, L. E. (1974). *Phytopathology Z.* **80**, 267–278.
Wheeler, H. (1975). "Plant Pathogenesis", p. 31. Springer-Verlag, New York.
Whitcomb, R. F., Shapiro, M. and Richardson, J. (1966). *J. Invertebr. Pathol.* **8**, 299–307.
Wilson, E. E., Zeitoun, F. M. and Fredrickson, D. L. (1967). *Phytopathology* **57**, 618–621.
Winslow, C. E. A., Broadhurst, J., Buchanan, R. E., Krumviede, C., Rogers, L. A. and Smith, G. H. (1917). *J. Bacteriol.* **2**, 505–566.
Wolpert, J. S. and Albersheim, P. (1976). *Biochem. Biophys. Res. Comm.* **70**, 729–737.
Wong, W. C. (1974). "Watermark Disease of Cricket Bat Willow: Epidemiology and the Nature of the Symptoms in the Wood". Ph.D. Thesis. The University of Leeds, Leeds.
Wong, W. C. and Preece, T. F. (1973). *Plant Pathol.* **22**, 96–97.
Wong, W. C., Nash, T. H. and Preece, T. F. (1974). *Plant Pathol.* **23**, 25–29.
Wrather, J. A., Kuc, J. and Williams, E. B. (1973). *Phytopathology* **63**, 1073–1076.
Young, J. M., Dye, D. W., Bradbury, J. F., Panagopoulos, C. G. and Robbs, C. F. (1978). *N.Z. J. Agric. Res.* **21**, 153–177.
Zeitoun, F. M. and Wilson, E. E. (1966). *Phytopathology* **56**, 1381–1385.
Zeitoun, F. M. and Wilson, E. E. (1969). *Phytopathology* **59**, 756–761.

Appendix

Recognized names and type species taken from the Approved Lists of Bacterial Names (V. B. D. Skerman, V. McGowan and P. H. A. Sneath, Eds) (1980), American Society for Microbiology, Washington, D.C.

E. amylovora	Type strain ATCC 15580
E. ananas (*E. herbicola* subsp. *ananas*, Bergey 8th ed)	Type strain NCPPB 1846
E. carotovora subsp. *atroseptica*	Type strain NCPPB 549
E. carotovora subsp. *carotovora*	Type strain ATCC 15713 = NCPPB 312
E. chrysanthemi	Type strain ATCC 11663 = NCPPB 402
E. cyprepedii	Type strain PDDCC 1591
E. herbicola (*E. herbicola* subsp. *herbicola*, Bergey 8th edn	Type strain NCPPB 2971
E. mallstivora	Type strain ATCC 29573
E. milletiae	Type strain NCPPB 2519
E. nigrifluens	Type strain ATCC 13028 = NCPPB 564
E. quercina	Type strain ATCC 29281
E. rhapontici	Type strain ATCC 29283
E. rubrifaciens	Type strain ATCC 29291 = NCPPB 2020
E. salicis	Type strain ATCC 15712 = NCPPB 447
E. stewartii	Type strain ATCC 8199 = NCPPB 2295
E. tracheiphila	Type strain NCPPB 2452
E. uredovora	Type strain ATCC 19321 = NCPPB 800

NCPPB   National Collection of Plant Pathogenic Bacteria, Plant Pathology Laboratory, Harpenden, Hertfordshire.

ATCC   American Type Culture Collection, Rockville, Maryland 20852, USA.

PDDCC   Culture Collection of Plant Diseases Division, New Zealand Department of Scientific & Industrial Research, Auckland, New Zealand.

6

Biochemical and Serological Characterization of *Yersinia enterocolitica*

G. KAPPERUD AND T. BERGAN

Institute of Zoology and Department of Microbiology, Institute of Pharmacy, University of Oslo, Oslo, Norway

METHODS IN MICROBIOLOGY
VOLUME 15 ISBN 0–12–521515–0

I. General introduction

Infections due to *Yersinia enterocolitica* have been the focus of increasing interest during the last few decades. The first recognized description of the bacterium presently known as *Y. enterocolitica* appeared in 1939 after earlier outbreaks of human gastro-enteritis in the USA (Gilbert, 1933, 1939). Frederiksen (1964; Wauters, 1970) on the basis of strain collection deposits at the Danish Statens Seruminstitut managed to trace one strain back to the period 1926–1932 when it was isolated by M. Kristensen. Some 15 strains were isolated in the USA between 1923 and 1957 (Gilbert, 1933; Wauters, 1970), and one from France in 1949 (Hässig *et al.*, 1949).

The current phase of interest in *Y. enterocolitica* started with the involvement of the bacterium in epizootic outbreaks among chinchillas, hares and swine in Europe and Africa during the period 1958–1963 (Wauters, 1970) and with the establishment of a causative relationship with abscedizing lymphadenitis in man (Knapp and Masshoff, 1954; Knapp and Thal, 1963; Masshoff and Dolle, 1953). The species has attracted considerable interest because of a rising frequency of isolation from human infections (Winblad, 1973). Moreover, *Y. enterocolitica* and *Y. enterocolitica*-like bacteria are commonly encountered in the environment and among healthy carriers from both the terrestrial and freshwater animal faunas (Mollaret *et al.*, 1979). The species, consequently, is ubiquitous and may have both human and animal reservoirs. The latter is diverse and appears, in part, to include strains other than those isolated from man.

Three methodological approaches have been employed for routine sub-division of *Y. enterocolitica*: serogrouping, biotyping and phage typing (Bergan, 1978; Niléhn, 1969; Wauters, 1970; Winblad, 1978). Serological characterization on the basis of O-antigens has become the most useful typing method, which is a valuable and convenient tool of considerable versatility in epidemiological tracing.

It is the purpose of this chapter to review the methodology and applications of serogrouping of *Y. enterocolitica*.

II. Taxonomy

A. Previous designations

What is now known to us as *Y. enterocolitica* has been labelled differently as *Bacterium enterocoliticum* (Schleifstein and Coleman, 1943), *Pasteurella pseudotuberculosis rodentium* (Hässig *et al.*, 1949), *Pasteurella pseudotuberculosis* type b (Dickinson and Mocquot, 1961), *Pasteurella X* (Daniels and Goudzwaard, 1963), *Pasteurella pseudotuberculosis* atypique (by depositor 1963 to collection of the Pasteur Institute, Paris, cited by Wauters, 1970), germe X (Mollaret and Destombes, 1964) and *Yersinia* X (Smith and Thal, 1965). The name *Y. enterocolitica* was formally proposed by Frederiksen (1964).

B. Genus affiliation

The genus *Yersinia* was proposed in 1944 by Van Loghem for bacteria which were then related to the genus *Pasteurella*, and were pathogenic, but non-septicaemic. Thal (1954) pointed out evidence relating *Yersinia* to the Enterobacteriaceae. A general numerical taxonomic study from 1958 placed *Yersinia* between *Klebsiella* and *Escherichia* (Sneath and Cowan, 1958). The allocation of *Yersinia* to the enterobacteriae was supported further by the findings of Frederiksen in 1964. Numerical taxonomy confirmed the allocation of the oxidase-negative entities known as *Pasteurella pestis*, *P. pseudotuberculosis* and bacterium X into *Yersinia*, as they were distinct from the oxidase-positive taxa remaining in the genus *Pasteurella* (Smith and Thal, 1965; Talbot and Sneath, 1960). Moore and Brubaker (1975) stated that the relationship of *Y. enterocolitica* to the genus *Yersinia* required further evaluation, e.g. by DNA hybridization. This was carried out by Brenner *et al.* (1976) and Brenner (1979). The results confirmed the genus relationship of the species.

It is now generally accepted that *Y. enterocolitica* belongs to the family Enterobacteriaceae. This has been further elucidated in a large series of studies involving both biochemical–cultural methods and DNA hybridization (Brenner *et al.*, 1980a; Bercovier *et al.*, 1980a). Accordingly, *Y. enterocolitica* shares many of the characteristics defining the Enterobacteriaceae. It is thus rod-shaped (sometimes cocco-bacillary), facultatively aerobic, Gram-negative, oxidase-negative, catalase-positive and has nitrate reductase. It is non-encapsulated (after growth on artificial media, capsule may be formed in peritoneal exudate), asporogenous and possesses relatively few peritrichous flagella (Mollaret and Thal, 1974). The % (G + C) contents of DNA in *Y. enterocolitica* ranges from 48.5 to ± 1.5 moles (Brenner *et al.*, 1980a). This characteristic is of major taxonomic significance and differentiates *Y. enterocolitica* from other enterobacteria. There is ample margin to species with lower %(G + C) ratios; the proteae have values in the range 38–42% for four species and about 50% for *P. morganii*. All other genera have % (G + C) above 50% (Cowan, 1974). Another distinguishing characteristic of *Y. enterocolitica* is motility at 22–28°C, but not at 35–37°C.

C. Differentiation between *Yersinia enterocolitica* and *enterocolitica*-like organisms

The biochemical and cultural determinative properties of *Y. enterocolitica* have been given in detail by Bottone (1977) and Bercovier *et al.* (1980a,b). The classification as reflected by the description in the last edition of "Bergey's Manual of Determinative Bacteriology" (Mollaret and Thal, 1974) recognized *Y. enterocolitica* as phenotypically heterogeneous. The need for a taxonomic revision has been recognized by several authors (Brenner *et al.*, 1976, 1980a). Atypical variants have tentatively been designated as *Y. enterocolitica*-like organisms. The taxonomy of the genus *Yersinia* has recently been subject to thorough analysis. Hence, a summary of the present taxonomic status within this group of bacteria is needed here.

Accordingly, several proposals revising the genus *Yersinia* have been raised after an examination of 175 strains of *Y. enterocolitica* and *Y. enterocolitica*-like organisms by biochemical–cultural characteristics supplemented by DNA hybridization (Bercovier *et al.*, 1980a,b; Brenner *et al.*, 1980a,b; Ursing *et al.*, 1980a,b). These data were supplemented by extensive biochemical–cultural characterization of considerably more strains (7000 *Y. enterocolitica*, 637 *Y. enterocolitica*-like organisms and 100 strains of *Y. pestis* and *Y. pseudotuberculosis*).

They defined a *Y. enterocolitica* (*sensu stricto*) and proposed three new species, *Y. frederiksenii*, *Y. intermedia* and *Y. kristensenii*, which were derived from the preceding broader concept of *Y. enterocolitica*. A detailed description of the properties of these and other *Yersinia* species appear in Table I. For routine purposes, the diagnosis of *Y. frederiksenii* may be applied to strains producing acid from rhamnose (Rh$^+$), *Y. intermedia* from rhamnose and melibiose and *Y. kristensenii* from trehalose but not from sucrose (S$^-$).

The amount of new data was formidable, but there may be problems recognizing some of the new species. The distinction between *Y. enterocolitica*, *Y. frederiksenii*, *Y. intermedia* and *Y. kristensenii* was based on the choice of radioactively labelled reference DNA from only two to six strains for each

TABLE I

Biochemical–cultural characteristic of *Yersinia* species

Characteristic	*Y. enterocolitica*	*Y. frederiksenii*	*Y. intermedia*	*Y. kristensenii*	*Y. pseudotuberculosis*	*Y. pestis*	*Y. ruckeri*
Acetoin (Voges–Proskauer)							
(37°C)	−(0)[a]	−(0)	−(0)	−(0)	−(0)	−(0)	−(0)
(28°C)	+(90)[b]	+(98)	+(100)	−(0)	−(0)	−(0)	−(3)
Arginine dihydrolase	−(0)	−(0)	−(0)	−(0)	−(0)	−(0)	−(3)
Catalase	+(100)	+(100)	+(100)	+(100)	+(100)	+(100)	
Citrate							
Simmon's							
(37°C)	−(0)	−(0)	−(0)	−(0)	−(0)	−(0)	−(0)
(28°C)	−(0)	−(0)	−(0)	−(0)	−(1)	−(0)	−(3)
Christensen's	V(65)	V(51)	V(70)	V(40)	V(85)	−(0)	V(76)
Deoxyribonuclease	V(40)	V(53)	−(2)	−(1)	−(0)	+(91)	−(0)
β-Galactosidase (ONPG, 37°C)	+(96)[b]	+(100)	+(100)	+(97)	+(100)	+(52)	+(100)
Gelatin (film)	−(0)	−(0)	−(0)	−(0)	−(0)	−(0)	−(0)
H_2S (Kliegler's)	−(0)	−(0)	−(0)	−(0)	−(0)	−(0)	−(0)
Indole	V(27)	+(99)	+(100)	V(61)	−(0)	−(0)	−(0)
KCN	−(0)	−(0)	−(0)	−(0)	−(0)	−(0)	V(27)
Lipase (Tween 80)	V(21)	V(70)	V(68)	V(74)	−(0)	−(0)	V(55)
Lysine decarboxylase	−(0)	−(0)	−(0)	−(0)	−(0)	−(0)	+(88)
Malonate	−(0)	−(0)	−(0)	−(0)	V[c]	−(0)	−(0)
Methyl red							
(37°C)	+(100)	+(100)	+(100)	+(100)	+(100)	+(100)	+(95)
(28°C)	V(60)	+(100)	+(100)	+(100)	+(100)	+(100)	+(97)
Motility							
(37°C)	−(0)	−(0)	−(0)	−(0)	−(0)	−(0)	−(0)
(28°C)	+(88)	+(100)	+(100)	+(100)	+(98)	−(0)	V(82)
Mucate	−(0)	−(0)	V(12)	−(0)	−(0)	−(0)	−(0)
NO_3 reduction to NO_2	+(97)[b]	+(100)	+(100)	+(100)	+(99)	V[d]	V(85)
Oxidase	−(0)	−(0)	−(0)	−(0)	−(0)	−(0)	−(0)
Ornithine decarboxylase	+(97)[b]	+(99)	+(100)	+(100)	−(0)	−(0)	+(100)
Phenylalanine	−(0)	−(0)	−(0)	−(0)	−(0)	−(0)	−(0)
Polypectate	+(100)[e]	+(100)[e]	+(100)[e]	+(100)[e]	+(100)	+(100)	−(0)

TABLE I – *continued*

Characteristic	Y. enterocolitica	Y. frederiksenii	Y. intermedia	Y. kristensenii	Y. pseudotuberculosis	Y. pestis	Y. ruckeri
Tetrathionate reductase	V(35)	+(96)	+(93)	V(83)	-(0)	V(15)	-(0)
Tryptophan deaminase	-(0)	-(0)	-(0)	-(0)	-(0)	-(0)	-(0)
Urease	+(99)	+(99)	+(99)	+(100)	+(100)	-(0)	-(0)
β-Xylosidase (PNPX, 37°C)	-(0)	V(47)	-(0)	-(0)	+(100)	+(100)	
(G+C) mole % DNA	48.5±0.5	48	48.5±0.5	48.5±0.5	46.5	46.0	48.5±0.5
Type strain[f]	ATCC 9610	CIP 80-29	ATCC 29909	CIP 80-30	ATCC 29833	ATCC 19428	ATCC 29473
Acid production from:							
N-acetylglucosamine	+(100)	+(100)	+(100)	+(100)	+(100)	+(100)	-(0)
Adonitol	-(0)	-(0)	-(0)	-(0)	-(5)	-(0)	-(0)
Amygdalin	V(15)	+(100)	+(100)	-(0)	-(0)	-(0)	-(0)
Amylose	-(0)	-(0)	-(0)	-(0)	-(0)	-(0)	
D-Arabinose	-(u1)	-(0)	-(0)	-(0)	-(0)	-(0)	-(0)
L-Arabinose	+(99)	+(100)	+(100)	+(100)	+(100)	+(98)	-(0)
Arbutin	V(60)	+(100)	+(100)	+(100)	V(88)	+(100)	-(0)
D-Cellobiose	+(99)	+(100)	+(100)	+(100)	-(0)	-(0)	-(0)
Dextrin	-(5)	+(100)	+(100)	+(100)	-(5)	V(62)	
Dulcitol	-(0)	-(0)	-(0)	-(0)	-(0)	-(0)	-(0)
Erythritol	-(0)	-(0)	-(0)	-(0)	-(0)	-(0)	-(0)
Esculin	V(31)	+(100)	+(100)	-(0)	+(100)	+(100)	-(0)
D-Fructose	+(100)	+(100)	+(100)	+(100)	+(100)	+(100)	
Galactose	+(99)	+(100)	+(100)	+(100)	+(100)	+(100)	
D-Glucose	+(100)	+(100)	+(100)	+(100)	+(100)	+(100)	+(100)
Glycerol	+(100)	+(100)	+(100)	+(100)	+(99)	V[a]	V(70)
Glycogen	-(0)	-(0)	-(0)	-(0)	-(0)	-(0)	
i-Inositol	+(70)[b]	+(90)	+(81)	V(28)	-(0)	-(0)	-(0)
Inulin	-(0)	-(0)	-(0)	-(0)	-(0)	-(0)	
Lactose	-(8)	V(10)	-(2)	-(0)	-(0)	-(0)	
Maltose	+(99)	+(100)	+(100)	+(100)	+(100)	V(84)	-(0)
D-Mannitol	+(100)	+(100)	+(100)	+(100)	+(100)	+(100)	+(100)
D-Mannose	+(100)	+(100)	+(100)	+(99)	+(97)	+(100)	+(97)

D-Melibiose	-(u1)	-(0)	+(99)	-(0)	+(99)	V^d	-(0)
D-Melizitose	-(0)	-(0)	-(0)	-(0)	-(0)	-(2)	-(0)
α-Methyl-D-glucoside	-(0)	-(6)	+(93)	-(0)	-(0)	-(0)	-(0)
α-Methyl-D-mannoside	-(0)	-(0)	-(0)	-(0)	-(0)	-(0)	
α-Methyl-xyloside	-(0)	-(0)	-(0)	-(0)	-(0)	-(0)	
Raffinose	-(u1)	-(3)	+(95)	-(0)	V(11)	-(2)	-(0)
L-Rhamnose	-(0)	+(100)	+(95)	-(0)	+(99)	-(3)	-(0)
Ribose	+(100)	+(100)	+(100)	+(100)	+(100)	+(100)	
Salicin	V(15)	+(100)	+(100)	-(0)	V(65)	V(88)	
D-Sorbitol	+(97)^b	+(100)	+(100)	+(100)	-(2)	-(0)	-(0)
L-Sorbose	+(90)^b	+(100)	+(98)	+(97)	-(0)	-(0)	-(0)
Starch	V(22)^e	+(100)^e	+(100)^e	+(100)^e	+(100)	+(100)	
Sucrose	+(98)^b	+(100)	+(100)	-(0)	-(0)	-(0)	
D-Trehalose	V(97)^b	+(100)	+(100)	+(100)	+(100)	+(100)	+(97)
D-Xylose	V(26)	+(100)	+(100)	+(100)	+(100)	+(100)	-(0)
L-Xylose	-(0)	-(0)	-(0)	-(0)	-(0)	-(0)	-(0)
Gas from glucose	-(0)	-(0)	-(0)	-(0)	-(0)	-(9)	-(9)

Adapted from Bercovier et al. (1980a, b, c), Brenner et al. (1980b), Ewing et al. (1978) and Ursing et al. (1980a, b).

Incubation at 28°C for all strains, except Y. ruckeri to which 22–25°C applies.

^{a} +, more than 90% of strains positive reaction; −, less than 10% positive reaction; V, between 10% and 90% of strains positive reactions. Numbers in parentheses refer to percentage of strains with positive reactions.

^{b} Most negative strains are biotype 5.

^{c} Serogroup IV strains of Y. pseudotuberculosis are malonate-positive.

^{d} Reaction varies in different biotypes of P. pestis.

^{e} Within four to seven days.

^{f} ATCC, American Type Culture Collection; CIP, Collection of the Pasteur Institute, Paris; NCTC, National Type Culture Collection. Conferring with the Approved List of Bacterial Names the listed strains have been given type strain status. ATCC 29833 = NCTC 10275; ATCC 19428 = NCTC 5923.

taxon. The hybridization data showed considerable heterogeneity within each of the nomenspecies; many hybridization percentages for *Y. enterocolitica* (*sensu stricto*) were in the range 40–60% in spite of the limited number of reference DNAs. Consequently, further elucidation based on a broader spectrum of reference and radioactively labelled DNA may be required to more fully assess the degree of genotypic overlapping between the proposed species. The problems of the new speciation are reflected by our own studies.

In a numerical taxonomic study comparing 332 strains, we found a marked phenetic continuum between *Y. enterocolitica* (*sensu stricto*) on the one hand and *Y. frederiksenii* and *Y. intermedia* on the other (Kapperud *et al.*, 1981). The overlapping was considerably less pronounced between *Y. enterocolitica* (*sensu stricto*) and *Y. kristensenii*. The overlapping was reflected further by antigenic properties, habitat preferences and pathogenicity characteristics. The same applies to similarities in fatty acid or phospholipid composition as demonstrated by gas–liquid chromatography (g.l.c.) (Bercovier and Carlier, 1979; Jantzen and Lassen, 1980; Sandhu *et al.*, 1980).

There are several probable explanations. One aspect to keep in mind is that the chance of having outliers or interphasing strains between taxa increase with the number of strains in the study. However, the two materials were probably basically different.

Thus, our strains were isolated from wild animals and environmental sources representing numerous habitats within the same ecosystems, whereas the strains from man were more predominant in the collections of Brenner *et al.* (1980a,b), Bercovier *et al.* (1980a,b) and Ursing *et al.* (1980a). A broad range of habitats increase the chance of including a spectrum of intermediate strains. This circumstance is reflected by a difference between the percentage of positive reactions for several characteristics in the groups of strains included in each of the two studies (Table II) and different distributions of O-antigens among the strains of each species (*vide infra*).

The taxonomic allocation of such strains may pose a problem to the diagnostic microbiologist. Our data were based on incubation at 37°C, in accordance with standard methods in medical diagnostic microbiology. It is possible that the phenotypic data obtained at 28°C, as employed by Brenner and co-workers, will result in a better separation between taxa and, consequently, increase the ability to distinguish between *Y. enterocolitica* (*sensu strictu*), *Y. frederiksenii*, *Y. intermedia* and *Y. kristensenii*.

Because of the phenetic diversity of *Y. enterocolitica*, biotyping has been employed. The biotyping scheme of Niléhn (1969) was modified by Wauters (1970), Winblad (1978) and Bercovier *et al.* (1980a). As seen in Table III the original biotypic procedure included several properties which had an identical pattern of + and − and would thus be redundant in a scheme of differentiating characteristics. Thus, production of acetoin, *β*-galactosidase,

TABLE II

Biochemical–cultural characteristics of *Yersinia* species as determined by different investigators

Characteristic	Y. enterocolitica		Y. frederiksenii		Y. intermedia		Y. kristensenii	
	K^a	Ba^b	K	U^c	K	B^d	K	Bb^e
Adonitol	18	0	2	0	12	0	14	0
Dextrin	29	30	14	100	28	100	36	100
Dulcitol	6	0	0	0	12	0	5	0
Glycerol	83	100	64	100	72	100	84	100
Inositol	53	91	44	100	64	100	27	88
Inulin	14	0	7	0	4	0	5	0
Raffinose	7	0	12	3	36	99	0	0
Starch	44	100	31	100	28	100	68	100
Xylose	41	0	20	0	12	100	23	100

[a] K, Kapperud *et al.* (1981), incubation at 37°C.
[b] Ba, Bercovier *et al.* (1980a), incubation at 28°C.
[c] U, Ursing *et al.* (1980a), incubation at 28°C.
[d] B, Brenner *et al.* (1980a), incubation at 28°C.
[e] Bb, Bercovier *et al.* (1980b), incubation at 28°C.

nitratase, ornithine decarboxylase, and acid formation from four of the seven carbohydrates included by Niléhn (1969) showed the pattern $+, +, +, +$ and $-$. Therefore, Wauters (1970) deleted acetoin, and three of the sugars. He also removed esculin and salicin, which develop positive reactions slowly and which were less reproducible, and instead introduced lecithinase, which exhibited the same pattern of positive and negative reactions. Winblad (1978) presented an intermediate modification between the schemes of Niléhn and Wauters. Bercovier *et al.* (1980a) found none of these alternatives, nor the variant of Knapp and Thal (1973), completely acceptable. Bercovier *et al.* therefore presented a new biotyping alternative, which deleted five of the carbohydrates employed by Niléhn, changed from lecithinase to lipase, and introduced a few other modifications. Since the new scheme did not employ acetoin – by many considered a major differential character for the definition of biotype 5, which is otherwise negative in all properties, nitrate reductase is an important and corresponding property. Instead they suggested deoxyribonuclease, although it develops late in two of the five types.

There has been general agreement that the biotypes 1–4 of Niléhn and Wauters belong to *Y. enterocolitica* (*sensu stricto*), but the question has been raised as to the taxonomical status of the biotype 5 strains. The hybridization data indicate that biotype 5 also constitutes a genetically distinct group, but within *Y. enterocolitica*. The heterogeneity within *Y. intermedia* has resulted in the biotyping scheme presented in Table IV.

TABLE III

Biotypes of *Y. enterocolitica*

Characteristic	Niléhn (1969)					Wauters (1970)					Knapp and Thal (1973)				Winblad (1978)					Bercovier *et al.* (1980a)				
	1	2	3	4	5	1	2	3	4	5	1	2	3	4	1	2	3	4	5	1	2	3	4	5
Acetoin production (Voges–Proskauer)	+	+	+	+	−	Not included					Not included				Not included					Not included				
Deoxyribonuclease	Not included					Not included					Not included				Not included					−	−[a]	−[a]	−	−
Esculin hydrolysis	+	−[a]	−[a]	+	−	Not included					−	−	+	V	+	−	−	−	−	Not included				
β-Galactosidase	+	+	+	+	−	+(L)	+	+	−	−	Not included				+	+	+	+	−	Not included				
Indole production	+	+	−	+	−	+	+	−	−	−	−	+	+	V	+	+	+	−	−	+	−	−	+	−
Lecithinase	Not included					−	−	−	−	−	Not included				Not included					Not included				
Lipase	Not included					+	+	+	+	−	Not included				+	+	−	−	−	+	−	+	+	−
Nitrate reduction	+	+	+	+	−	+	+	+	+	−	Not included				+	+	+	+	−	+	+	+	+	−
Ornithine decarboxylase	+	+	+	+	−	+	+	+	+	−	Not included				+	+	+	+	−	Not included				
Acid formation in OF-medium (48 h) from:																								
Lactose	+	+	+	−[a]	−[a]	Not included					Not included				+	+	+	−	−	Not included				
Salicin (acid)	+	−[a]	−[a]	−[a]	−[a]	Not included					−	−	+	V	+	−	−	+	−	Not included				
Sorbitol	+	+	+	+	−	Not included					Not included				+	+	+	+	−	Not included				
Sorbose	+	+	+	+	−	Not included					Not included				Not included					Not included				
Sucrose	+	+	+	+	−	Not included					Not included				Not included					Not included				
Trehalose	+	+	+	+	−	+	+	+	+	−	Not included				+	+	+	+	−	+	+	+	+	−
D-Xylose	+	+	+	−[a]	−[a]	+	+	+	−	−	V	+	+	V	+	+	+	−	−	−	+	+	−	−[a]

Symbols: +, more than 90% of strains positive reaction; −, less than 10% positive reaction; V, between 10% and 90% of strains positive reactions. Numbers in parentheses refer to percentage of strains with positive reactions; (L), late; a, delayed positive reaction after 72 h exhibited by some strains.

TABLE IV

Biotypes of *Y. intermedia*

Characteristic	Biotype							
	1	2	3	4	5	6	7	8
Melibiose	+	+	+	+	+	+	+	−
Raffinose	+	+	+	+	+	−	−	+
Rhamnose	+	+	+	−	−	+	+	+
Simmon's citrate	+	−	+	+	+or−	−	−	+
α-Methyl-D-glucoside	+	+	−	+	−	+	−	+
Distribution (%) upon biotypes	69	20	4	3	2	1	1	1

Adapted from Brenner *et al.* (1980b).

Symbols: +, positive reactions in more than 90% of the strains; −, positive reactions in less than 10% of the strains.

D. Distinction between *Yersinia pestis* and *pseudotuberculosis*

DNA hybridization data and biochemical determinative properties have shown that *Y. pestis* and *Y. pseudotuberculosis* clearly belong to the family Enterobacteriaceae (Bercovier *et al.*, 1980c; Brenner *et al.*, 1976) and that their interrelationship is consistent with recognition at subspecies status within the same entity, *Y. pseudotuberculosis*, as *Y. pseudotuberculosis* subsp. *pseudotuberculosis* and *Y. pseudotuberculosis* subsp. *pestis*. The interrelatedness appeared well substantiated, although somewhat lower heterospecies ranges were observed with one of the two *Y. pseudotuberculosis*-labelled DNA against *Y. pestis*. The only labelled DNA from *Y. pestis* was not tested against any strain of *Y. pseudotuberculosis*. Consequently, it would appear that a cluster relationship between the species is needed before a final conclusion is drawn regarding the subspecies status of the two taxons. *Y. pestis* was first described and was first suggested as a *Yersinia* species before *Y. pseudotuberculosis* (Mollaret and Thal, 1974). *Y. pestis* was suggested as belonging to *Yersinia* in 1944 when the genus was first named by Van Loghem (1944) and *Y. pseudotuberculosis* followed in 1965 (Smith and Thal, 1965). The species *Y. pestis* was described as *Bacterium pestis* by Lehmann and Neumann in 1896. In comparison, *Bacillus pseudotubekulosis* (Sic!) was described by Pfeiffer in 1899. The epithet *pestis* was thus proposed before the epithet *pseudotuberculosis*. The name *Y. pestis*, consequently, has priority and is the recognized type species of the genus *Yersinia* in both "Bergey's Manual of Bacteriology" and in the Approved List of Bacterial Names (Skerman *et al.*, 1980).

These circumstances would appear to be in contrast to the recommendation that *Y. pseudotuberculosis* subsp. *pseudotuberculosis* be the type species of the genus *Yersinia*, with the motivation that *P. pseudotuberculosis* was described before *Y. pestis* and therefore has taxonomic priority (Bercovier *et al.*, 1980c). This, however, refers to dates of probable isolation, but both papers reporting that a bacterium causes plague (Yersin, 1894) and pseudotuberculosis (Malassez and Vignal, 1883) were without a name proposal. The former was described briefly as chain-forming rods, Gram-negative, with polar staining and capsules. The latter bacterium was only characterized as micrococcal and with zoogloea formation in liquid culture medium making the reidentification criteria limited. The Approved List of Bacterial Names states unequivocally that the list represents the reference deciding the priority of bacterial names to be recognized after 1 January, 1980. The date of publication of this list nearly coincided with the time when the publications with the new proposals of 1980 were submitted for publication, but when priority is the question, the date of publication is decisive. It follows from the above that later proposals of a revised genus *Yersinia* must adhere to the priorities defined by this list. Accordingly, *Y. pestis* is the type species of this genus, and not *Y. pseudotuberculosis*. If scientific data are to be interpreted to the effect that both *Y. pestis* and *Y. pseudotuberculosis* were only to be recognized as subspecies of the same species priority of *Y. pestis* as type species warrants the designations *Y. pestis* subsp. *pestis* and *Y. pestis* subsp. *pseudotuberculosis*. These are, consequently, proposed as legitimate and with priority over *Y. pseudotuber- culosis* subsp. *pestis* and *Y. pseudotuberculosis* subsp. *pseudotuberculosis*.

From a pragmatic point of view, considering the differences in disease panorama, reservoir, mode of spread of infections and that the two taxa are biochemically distinctly differentiable (rhamnose, urease, motility at 28°C), we have kept them separate as in the last edition of "Bergey's Manual of Determinative Bacteriology" (Mollaret and Thal, 1974). This is also con- sistent with the view of Bercovier *et al.* (1980c) to the effect that the designations *Y. pestis* and *Y. pseudotuberculosis* should continue to be used for medical purposes: "It would still be perfectly acceptable for clinical bacteri- ologists and public health officers to refer to *Y. pseudotuberculosis* and *Y. pestis*. In fact, we strongly recommend this approach. Thus, the taxonomy of these organisms would be consistent with available scientific knowledge and the need of the medical community for practical designations would be met."

E. Status of *Yersinia ruckeri*

The relationship between *Y. ruckeri* and recognized members of the En- terobacteriaceae is reflected by a DNA relatedness of about 30% to *Y. enterocolitica*, *Y. pseudotuberculosis* and *Serratia* species. The % (G+C) of

DNA is about 48%, i.e. within the range of accepted *Yersinia* species (Brenner, 1979). A survey of the cultural–biochemical properties of *Yersinia* species appear in Table I.

F. Status of *Yersinia philomiragia*

The proposed species *Y. philomiragia* has been considered to be genotypically distinct from *Pasteurella*, *Yersinia* and the Enterobacteriaceae. Inclusion in these taxa is therefore not warranted (Ursing *et al.*, 1980b).

G. Status of X1 and X2

Two additional taxa designated X1 and X2 belong to the *Y. enterocolitica*-like groups of bacteria (Bercovier *et al.*, 1980a). X1 is characterized as sucrose- and ornithine decarboxylase-negative. It differs from *Y. enterocolitica* biotype 5 by positive-nitrate and trehalase reactions and the other species, e.g. *Y. kristensenii*, in several properties. The X2 shares with *Y. frederiksenii* and *Y. intermedia* the ability to ferment rhamnose but not sucrose, as in *Y. kristensenii* (Table V). The X1 and X2 are also differentiable from each other by DNA hybridization; although both are sufficiently related to *Y. enterocolitica* to allow them to be included in the genus *Yersinia* but clearly distinct from *Y. enterocolitica* (*sensu stricto*), *Y. frederiksenii*, *Y. intermedia* and *Y. kristensenii*.

III. Identification

A. Isolation

Y. enterocolitica has been reported to cause diarrhoeas as frequently as *Salmonella* and *Campylobacter* in some countries (Van Noyen *et al.*, 1980). The yersiniae occur less frequently as pathogens of pyogenic infections than other enterobacteria. However, they are often difficult to isolate because they grow more slowly than companion species in mixed cultures, e.g. from faecal samples.

To improve the rate of isolation, selective procedures directed specifically towards isolation of this group of agents are worthwhile. Enrichment at low temperatures (4–10°C) has been employed. Cold enrichment in selenite broth has been recommended (Wauters, 1970). Further studies have shown the success of a combination of direct plating on SS agar and two days enrichment in a modified Rappaport broth with carbenicillin (Wauters, 1973) at room temperature (Van Noyen *et al.*, 1980). This procedure results in improved

TABLE V

Key biochemical–cultural characteristics of *Yersinia* species of *Y. enterocolitica* and *Y. enterocolitica*-like organisms

Characteristic	Y. enterocolitica	Y. frederiksenii	Y. intermedia	Y. kristensenii	X1	X2	Y. pseudotuberculosis	Y. pestis	Y. ruckeri
Melibiose	−(0)[a]	−(0)	+(99)	−(0)	−	−	+(99)	v[b]	−(0)
Raffinose	−(1)	−(3)	+(95)	−(0)	−	−	−(11)	−(2)	−(0)
Rhamnose	−(0)	+(100)	+(95)	−(0)	−	+	+(99)	−(3)	−(0)
Sucrose	+(98)	+(100)	+(100)	−(0)	−	−	−(0)	−(0)	−(0)
Ornithine decarboxylase	+(97)	+(99)	+(98)	+(100)	−	+	−(0)	−(0)	+(100)
Sorbose	+(90)	+(100)	+(98)	+(97)	−	−	−(2)	−(0)	−(0)
Cellobiose	+(99)	+(100)	+(100)	+(100)	+	−	−(0)	−(0)	−(0)
Acetoin (VP, 28°C)	+(90)	+(98)	+(100)	−(0)	−	+	−(0)	−(0)	−(3)
α-D-Methylglucoside	−(0)	−(6)	+(93)	−(0)	−	−	−(0)	−(0)	−(0)

Adapted from Brenner (1979), Brenner *et al.* (1980a, b), Bercovier *et al.* (1980a, b), Ursing *et al.* (1980a).
[a] Symbols: +, more than 90% of strains positive reaction; −, less than 10% positive reaction; v, between 10% and 90% of strains positive reactions. Numbers in parentheses refer to percentage of strains with positive reactions. Incubation at 28°C.
[b] Reaction depends on *Y. pestis* biotype.

isolation of serotypes O:3 and O:9 from human faeces with the recovery of *Y. enterocolitica* from 3.7% of the faecal samples. For the recovery of other serotypes, particularly environmental in origin, the recommended method is cold enrichment at 4–10°C in 1/15 M phosphate buffer at pH 7.6 subcultured weekly for four weeks. Cold enrichment in buffer increases the isolation rate, but additional isolates have belonged to biotype 1, which is considered apathogenic for man. No strain was obtained after selenite cold enrichment which was not also isolated after buffer cold enrichment, and the buffer method did not enhance the isolation of O:3- and O:9-antigens.

   Y. enterocolitica grows on standard media, but somewhat more slowly than most enterobacterial strains and is, consequently, easily overlooked also on the usual enteric media. The colony appearance resembles that of other enterobacteria, although *Y. enterocolitica* develops more slowly. After incubation at 22 or 37°C overnight, the colonies are pinpoints of 0.5 mm in diameter. After 48 h, the size increases to 0.5–2.5 mm. This pertains to clinical strains. Environmental strains may show colonies which are larger. Some environmental strains, e.g. sucrose-negative variants, have a characteristic earthy odour resembling cauliflower, potato or apple. Incubation for 48 h at room temperature (22–25°C) enhances the recovery. On lactose bromothymol blue agar, *Y. enterocolitica* is usually lactose-negative, although it produces β-galactosidase (ONPG-positive). Clinical strains remain lactose-negative even after storage at room temperature for days. Some environmental strains, in contrast, actually exhibit as distinctly lactose-positive colonies, which particularly develop after a couple of days at room temperature. On deoxycholate or SS agar the colonies are translucent and colourless and these media are recommended by Weissfeld and Sonnenwirth (1981). Growth on LSU agar (Juhlin and Ericson, 1961) at 22–37°C shows pinpoint translucent colourless colonies. After 48 h or longer clinical strains become whitish-green to grey. Environmental strains are more white. The margin of the colonies may be undulant, particularly from environmental strains, although an even smooth surface and margin is most common.

B. Identification

It must be ascertained that the species is selected for serological characterization. For routine purposes, *Y. enterocolitica* and *Y. enterocolitica*-like organisms may be distinguished on the basis of eight key cultural–biochemical characteristics as presented in Table V.

   The biochemical–cultural tests used for identification may be composed of media and methods described by Cowan and Steel (1970) or by Martin and Washington (1980). Strains which are urease-positive, H$_2$S-negative, citrate-negative and exhibit more motility at 22°C than at 37°C are candidates for the

TABLE VI

Serogrouping scheme of *Y. enterocolitica* as developed essentially by Winblad[a]

Serogroup[b]	Reference strain for immunization[c]	Synonym strain designations	Original source of strain	Minor O-antigen	Strains for absorption to obtain factor sera	O-antigen groups which may occur with major antigen in new isolates
1	64	348 (Winblad), 57/6 (Becht)	Chinchilla	2, 3	14	2, 3
2	14	338 (Winblad), 404 (Lucas)	Hare	3	134	3,
3	134	XWbd (Winblad), My 0 (Winblad)	Human			
4	96	357 (Winblad)	Chinchilla	32	64	32, 33, 13, 16
5	124					27
6	102	219 (Winblad)	Human	30	64+134	30, 31
7	106		Guinea-pig	8	161	8, 13
8	161		Human			19
9	383		Human			
10[d]	500		Ice cream			
11	105		Human	23		23, 24
12	490	63 (Lucas)	Hare	25	474	25, 26
13	553		Human	7	106	7, 8, 4
14	480		Human		490	
15	614	188/69 (Esseveld)	Human		64	4, 18, 21, 22, 29
16	1475		Water			
17	955	333B (Lassen)	Water		134	
18	846	4542 (Weaver)	Human		474	
19	842		Human	8	161	8
20	845	4403 (Weaver)	Human		64+474	
21	1110		Human			4, 16, 18
22	1367		Human			

23	105	Human	11	841	11
24	841	Human	11	105	11
25	490	Hare	12	103	12
26	103	Sheep	12	490	12
27	885	Dog	5	124	5
28	1474	Water			
29	867	Human	16	1475	16
30	102	Human	6	1477	6
31	1477	Water	6	102	6
32	96	Chinchilla	4	1476	4
33	1476	Water	4	96	4
34	1501	Human			

[a] Adapted from Wauters (1970), Wauters et al. (1972), Winblad (1978). See Note added in proof for the serogrouping scheme of Y. enterocolitica, based on the O-antigen system developed primarily by Wauters.

[b] All reference strains possess a major antigen corresponding to the respective serogroup. In addition, several reference strains carry minor O-antigens.

[c] Strain Nos. refer to IP, Collection of the Pasteur Institute, Paris (at National Yersinia Center).

[d] CIP 500 carries only O:10, and not K:1; CIP 551 carries O:10; K:1.

genus *Yersinia*. Brenner *et al.* (1980a,b) and Bercovier *et al.* (1980a,b) point out that these species lack gas formation from glucose, but here the amount of gas is probably decisive, as we observe very small amounts of gas more often and record this as positive (Kapperud *et al.*, 1981).

Commercial rapid methods like API-20E, Enterotube with 12 reactions, Micro-ID or the Minitek system are all suitable for the identification of the *Y. enterocolitica* and *Y. enterocolitica*-like bacteria. The identification is improved if the test systems are incubated for 48 h at 22°C before they are read (Le Gendre and Johnson, 1981; Restaino *et al.*, 1979). Incubation at 28°C would probably also function well.

IV. Serotyping schemes

A. O-antigen

The serogrouping scheme of *Y. enterocolitica* is based upon the polysaccharide O-antigens of the cell wall. Winblad (1967, 1968) originally recognized eight serogroup antigens defined by cross-reactions of eight reference strains. Ahvonen and Jansson (1968) and Niléhn (1969) described a ninth antigen. Supplement to 17 O-groups evolved after the work of Wauters (1970) and Wauters *et al.* (1971). Winblad in 1978 recognized 26 serogroups. This has since been extended to 34 serogroups (Wauters *et al.*, 1972). Serogroup reference strains and cross-reactions appear in Table VI. An allusion to serotype (Sic!) O:54 has been made (Van Noyen *et al.*, 1980) and indicates that further antigens have been identified.

One problem of *Y. enterocolitica* O-antigen grouping is a tendency for the strains to become rough upon repeated subcultivation. If a strain throws off rough colonies, these should be avoided as they will agglutinate spontaneously. It is important to always employ blank negative serum controls to discover whether a cell preparation is constituted wholly or perhaps partially of rough colony type cells. Polyagglutinable strains should be examined particularly carefully with this in mind. The maintenance of the smooth colony form is reportedly better at 22°C than at 37°C (Wauters, 1970).

The composition of the polysaccharides of the somatic group antigens is shown in Table VII. The distribution of serogroups among *Y. enterocolitica* and *Y. enterocolitica*-like species is shown in Table VIII. There is a marked difference between our collection and the characteristics presented in the recent studies of Bercovier *et al.* (1980a,b,c), Brenner *et al.* (1980a,b) and Ursing *et al.* (1980a,b). In comparison to Table VIII, Winblad (1978) observed among 915 *Y. enterocolitica* strains 66.5% O:3, 5.6% O:2, 6.8% O:1, 1.1% O:11 and O:12, 8.4% O:9, 4.6% O:8, 1.6% O:17 and the remainder distributed

TABLE VII

Composition of lipopolysaccharide components of somatic antigen of *Y. enterocolitica*

Sero-group	KDO	L-Glycero-manno-heptose	D-Glycero-manno-heptose	Hexos-amines[a] (a)	Glucose	Galactose	Rhamnose	Desoxy-hexose
1	+	+	+	+	+	+	+	+
2	+	+	+	+	+	+	+	+
3	+	+	+	+	+	+/−	−	+
4	+	+	+	+	+	+	+/−	−/+
5	+	+	+	+	+	+	+/−	−/+
6	+	+	+	+	+	+	+/−	−/+
7	+	+	+	+	+	+	−	+
8		+	−	+	+	+	+	−
9	+	+	+	+	+	+	−	−

Adapted from Rische *et al.* (1973), Wartenberg *et al.* (1975).

[a] All strains contain both galactosamine and glucosamine, except one O:1 strain (CIP 39) with only glucosamine; this also lacked galactose. Mannose and fructose were both present in one of several O:6 and in the one O:7 strain tested.

between a few of the other serogroups (from O:4 to O:20). The great majority of his strains were of human origin. Among human strains, the frequency of O:3 and O:9 dominate with O:8 and O:17 clearly less frequent; this is based upon data without differentiation between *Y. enterocolitica* and *Y. enterocolitica*-like organisms.

Winblad (1978) and Wauters (1970) subdivided O:5 according to H-antigens. The O:5A carry a variety of H-antigens and are biotype 1, and O:5B (O:5 to O:27) have H:a, b, c and are biotype 2.

It is notable that *Y. enterocolitica*, *Y. frederiksenii* and *Y. intermedia* may carry the antigens of *Y. pseudotuberculosis* according to the scheme described by Thal and Knapp (1971).

B. H-antigen

Heat-labile antigens have been described (Wauters *et al.*, 1971, 1972). These are presumably flagellar, since the H-antigens of *Y. enterocolitica* are absent when grown at 37°C where the strains emerge as non-motile. This class of antigens, though, has only a modest application to routine typing. A total of 19 H-antigens has been designated, a–t exempting j. Several of the O-antigen groups have not been further differentiated by H-antigens; subdivision by means of H-antigens has been demonstrated for groups O:1, 2, 3, 5, 6, 7, 8 and 10. The H-antigen type reference strains and recommended absorption

TABLE VIII

Percentage distribution upon serogroups within *Yersinia* species

Serogroup	*Y. enterocolitica*		*Y. frederiksenii*		*Y. intermedia*		*Y. kristensenii*	
	Our	Ba[b]	Our	U[c]	Our	B[d]	Our	Bb[e]
No. strains tested	148	7000	55	201	25	321	51	115
1	2.8						6.7	15
2	1.1			6				
3	4.0		1.3	6				
4	10.2		25.6		14.3	14		
5	3.4							1.6
6	14.2				16.7		1.7	
7	8.5							
8	3.4							
9	1.7							
10	0.6			4				
11							15.0	1.6
12	1.1						16.7	12
13	2.8		9.0		7.1			26
14	2.3		7.7					
15	0.6		1.3					
16	5.7		7.7	24	7.1		6.7	7
17	1.7		5.1		2.4	8		
18	2.3		10.3		7.1			
19	0.6							
20	0.6							
21	0.6		11.5		7.1			0.8
22	1.7							
23	1.1						1.7	

Y. pseudotuberculosis O-antigens[a]

Serogroup	1	2	3	4	5	6	7	8	9
24								1.7	
25								1.7	
26								1.7	
27	1.1							20.0	
28			1.3						
29									
30	0.6								
31	0.6								
32			1.3						
33	0.6								
34	0.6								
IA	1.1		1.3						
IB	1.1		2.6						
IIA	0.6		3.8		2.4				
IIB			2.6						
IV									
Percentage typable	77.3	82	92.3	49	64.3	58	63.3	76	
No. of serogroups	29	84	15	20	8	27	10	8	12

[a] From Thal and Knapp (1979).
[b] Ba, Bercovier et al. (1980a), incubation at 28°C.
[c] U, Ursing et al. (1980), incubation at 28°C.
[d] B, Brenner et al. (1980a), incubation at 28°C.
[e] Bb, Bercovier et al. (1980b), incubation at 28°C.
[f] Also K:1.

TABLE IX
Required absorption to produce H-antigen factor sera for *Y. enterocolitica*

H-antigen	Reference strain[a] used as antigen	Reference strains[a] used for absorption
a	211	178
b	178	614
c	64	211
d	123	64 + 96
e	123	64 + 102
f	96	123
g	102	123 + 211
h	106	96 + 102
i	123	64 + 106
k	551	96
l	105	—
m	480	—
n	553	—
o	490	—
p	867	—
q	955	—
r	841	—
s	1474	—
t	1494	—

Adapted from Wauters (1981) and Wauters *et al.* (1971).
[a]Collection reference number, Pasteur Institute, National Yersinia Center, Paris.

TABLE X
H-antigens and O-antigens produced by respective H-antigen reference strains for *Y. enterocolitica*

Reference strain[a]	O-antigens	H-antigens
64	1, 2a, 3	a, b, c
178	2a, 2b, 3	b, c
96	4, 32	b, e, f, i
123	5	b, c, d, e, i
102	6, 30	a, b, d, g, i
106	7, 8	d, e, f, g, h
161	8	b, e, f, i
211	9	a, b
551	10	b, (f), k
105	11, 23	l
841	11, 24	r
490	12, 25	o
1494	12, 26	t
553	13, 7	n
480	14	m
614	15	(b), c
867	16	p
955	17	q
1474	28	s

Adapted from Wauters (1981) and Wauters *et al.* (1971).
[a]Collection reference number, Pasteur Institute, National Yersinia Center, Paris.

appears in Table IX. Their antigen mosaics are shown in Table X. Before H-factor absorption, the crude sera are diluted 1:10 and absorbed by heat-treated homologous antigen cells to remove O-antibody.

The most commonly encountered combination of O- and H-antigen is presented in Table XI.

TABLE XI
The most common combinations of O- and H-antigens

O-antigen	H-antigen
1	a, b, c
2	b, c
3	a, b, c
4	b, e, f, i
5	b, c, d, e, i
	a, b, c, d, g
	a, b, c,
6	a, b, d, g, i
	b, d, e, g, i
	d, e, f, g, h
7	d, e, f, g, h
	b, d, e, g, i
	b, c, d, k
8	b, e, f, i
9	a, b
10	b, (f), k
	b, c, d, e, i
	b, c, e, f, i
11	l
12	l
	o
13	n
14	m
15	c
16	p

C. K-antigen

Surface K-antigen has been detected in strains of O:10 (Wauters, 1970). The envelope antigen is heat-labile and masks the O-agglutinin to the extent that inhibits O-agglutinability. Only one antigen of this class has been characterized, K:1. This is present in the majority, but not all O:10 strains and in occasional O:21 strains. Very rarely a more luxurious and distinct capsule is detectable by India ink preparations. The capsular substance does not cross-react with K:1.

The envelope K:1-antigen is related to fimbriae sized 1.7–2.5 nm in diameter (Aleksic *et al.*, 1976; MacLagan and Old, 1980). Fimbriated cells of this species cannot be identified by pellicle formation. The antigen is proteinaceous with 17 amino acids and has a pI of 3.9 (isoelectric focusing) (Aleksic and Aleksic, 1979). It is moderately heat-labile (Aleksic *et al.*, 1976). Boiling at 100°C up to 3 h does not eliminate, but reduces, agglutinability in fimbriae-specific antiserum. The antigenicity disappears after autoclaving at 120°C for at least 30 min. Cell treatment with alcohol or HCl does not eliminate the K-agglutinability.

The K-antigen is present on fimbriae described as 4–4.5 nm in diameter, which may be the same as the fimbriae described by others as 1.7–2.5 nm (*vide supra*). It represents a mannose-resistant adhesin agglutinating fowl erythrocytes (MacLagan and Old, 1980). In addition, *Y. enterocolitica* may have more coarse fimbriae of 8 nm in diameter, which is a broad spectrum mannose-resistant haemagglutinin labelled MR/Y-adhesin unrelated to biotype or serotype (MacLagan and Old, 1980), and is lost after growth at 37°C.

V. Preparation of antigens: immunization

A. O-antigens

1. *Antigen preparation*

S-form colonies of the reference strain are selected for plating on nutrient agar (1.4% agar*) and incubated overnight at 22–25°C. Slow-growing strains may be incubated for 36–48 h (Wauters, 1970; Winblad, 1978). Some strains may form their specific antigens only when grown at 29°C (Wauters, 1970). Fresh suspensions are examined for spontaneous agglutination before use.

The cells are harvested and suspended in physiological saline, autoclaved for 1 h at 120°C, and washed once in saline. The practice of autoclaving instead of boiling, which is preferred for O-antigen preparation of many other species, follows the experience of Winblad (1967) that lower titres were obtained with cells which were simply boiled. The same has been observed with other Gram-negative rods, e.g. *Pseudomonas aeruginosa* (T. Bergan, unpublished data). The antigenic specificity of boiled cells and autoclaved cells is in principle identical. Consequently, boiled cells should be fully acceptable. Autoclaved cells tend to exhibit spontaneous auto-agglutination

* Beef-extract 0.5%, NaCl 0.3%, Na-phosphate 0.2% (Winblad, 1978). Growth proceeds in Roux bottles.

consistent with the Enterobacteriaceae. Autoclaving induces R-form behaviour by stripping the outer cell wall layers. This is in accordance with Wauters (1970), but contrary to Winblad (1967), who found less spontaneous agglutination for heated cells than with suspensions of live cells.

The agglutination results of live cultures is analogous to that of boiled or autoclaved cells as regards antigenic specificity, although the titres are lower and weaker cross-reactions may disappear. Live cultures may be inactivated by suspension in saline containing 0.01% merthiolate to protect laboratory personnel.

2. Immunization schedule

Immunization proceeds according to Wauters (1970). Cell suspensions in saline of 15×10^8 cells per millilitre are injected intravenously on five occasions in volumes of 1, 1, 1.5, 1.5 and 2 ml at intervals of three to four days. The animals are bled six days after the final injection. Homologous titres of 1.600–12.800 are usual. Winblad (1978) uses a lower density of 10^5 cells per millilitre, starts with 0.5 ml and injects 1.0 ml on days 7, 14 and 21.

B. H-antigens

1. Antigen preparation

The strains are grown in Petri dishes with nutrient medium* with 0.3% agar at 22–25°C. The inoculum is deposited at the periphery of the plate. Plating is repeated to enhance motility and hence formation of flagellae until the culture develops zones of 3–5 cm per 18 h. A few strains are equipped with a small number of flagella per cell, for instance groups O:2 and 3 (Niléhn, 1969).

Cultures with 3–5 cm overnight migratory zones are picked from the growth margin, suspended in 5 ml peptone broth and kept at room temperature for 5–6 h. Subsequent plating is performed on semi-solid 1% nutrient agar in a 10-cm diameter Petri dish topped up with 2 ml of peptone water and further incubated for 12–18 h at 22–25°C. The ensuing culture is mixed with formalin to a final concentration of 0.5%.

2. Immunization schedule

Immunization follows the protocol used to produce O-sera. The titres are usually in the range of 6.400–51.200

C. K-antigens

K-antigen sera are produced with non-heated cells grown at 37°C (to suppress H-antigen formation) (Wauters, 1970). O-antigen reference strains No. IP

474, IP 551 or IP 679 (Pasteur Institute, National Yersinia Center, Paris) (of O:10) are suitable for the production of K-antisera. The crude antisera are absorbed with cells of the homologous strain which have been autoclaved to remove all but the O-antigens. Consequently, only K-antibodies remain after absorption. Such antisera do not agglutinate heated cells.

VI. Agglutination technique

Identification of both O- and H-antigens proceeds as tube agglutination in saline. The H-agglutination is read after 2 h at 37°C, whereas O-agglutination is recorded after further incubation for 18–20 h at 22–25°C.

As antigens for O-grouping are preferred cells boiled for 2 h. This is more practical than autoclaving, although the titres are higher with autoclaved antigens. Live antigen may be employed. Cultures incubated at 37°C to suppress development of the H-antigen are in that case preferred. If grown at room temperature, suitable H-antigen is obtained. The live cells are suspended in saline with 0.01% merthiolate. The appearance of H-agglutinates are more fluffy and easily disruptable, whereas the O-agglutinates are more granular and hard. Hence, the same cell suspension, if derived from growth at 22–25°C may be used for both determinations, but this is not recommended as the results may be difficult to differentiate for the inexperienced.

The observations of Wauters (1970) serve well to assess the suitability of different methods of antigen treatment and incubation temperatures (Table XII). Reading the results is improved with good light, holding the tubes against a dark background and, with the aid of a magnifying glass or concave mirror to reflect the bottom of the tubes.

VII. Cross-reactions

The fimbrial K:1-antigen of O:10 strains of *Y. enterocolitica* is shared by salmonellae of the subgenera I, II and III (Arizona). The K:1-antigen of *Y. enterocolitica* is identical to antigen in *Salmonella* subgenus I (Aleksic *et al.*, 1978).

All strains of the family Enterobacteriaceae share a common enterobacterial antigen (CEA) (Mäkelä and Mayer, 1976). This was demonstrated in all 63 strains of *Y. pestis*, *Y. pseudotuberculosis* and *Y. enterocolitica* and in 236 strains of various other Enterobacteriaceae, but not in 69 strains of non-enterobacterial species (Le Minor, 1979). The antigen resides in the outer membrane of the bacterial cell wall, but inside the O-antigen layer (Acker *et al.*, 1981). Absorption removing the common antigen renders sera which can

TABLE XII

Comparison of methods for tube agglutination: titres obtained with live versus heat-treated cells, and agglutination at temperature of 37°C or 50°C

Serogroup	Patient identification	Live cells		Heat-treated cells	
		37°C	50°C	37°C	50°C
3	1	3.200	1.600	200	100
	2	3.200	3.200	800	1.600
	3	12.800	12.800	3.200	1.600
	4	6.400	6.400	800	800
	5	800	400	200	0
	6	800	800	0	0
9	7	1.600	800	400	0
	8	3.200	400	800	0
	9	800	200	0	0
	10	400	200	0	0
	11	1.600	800	0	0

Adapted from Wauters (1970).

be used diagnostically for the genus *Yersinia* (Winblad, 1978), similar to common practice with salmonellas and shigellas.

Cross-reactions have been found between the somatic antigens of *Y. enterocolitica* O:9 and O:3 and *Brucella* spp. and *Salmonella* (Winblad, 1978; Corbel, 1975). In particular, a cross-reactivity has been observed between *Yersinia* O:12 and *Salmonella* O:47 (Wauters *et al.*, 1971). These bacteria did not share protective antigens (Corbel, 1979). Cross-reacting antigens have been found in *Y. enterocolitica* and *Pseudomonas aeruginosa* (Winblad, 1978). No cross-reaction has been observed between *Y. enterocolitica* and *V. cholerae* (Wauters *et al.*, 1971).

It is of pathogenetic significance that *Y. enterocolitica* injected intravenously has stimulated the production of antibody against both lipopolysaccharide and a protein substance, whereas oral infection resulted only in immunoglobulin against the protein component (Ogata *et al.*, 1979).

The plasmid determined plague virulence antigens (V- and W-antigens) of *Y. pestis* and *Y. pseudotuberculosis* have both been detected in *Y. enterocolitica* O:8 biotype 2 (Ben-Gurion and Shafferman, 1981; Carter *et al.*, 1980; Straley and Brubaker, 1981).

VIII. Other typing systems

Phage typing of *Y. enterocolitica* has been developed by Niléhn (1969, 1973), Mollaret and Nicolle (1965) and Nicolle *et al.* (1973) as described by Bergan (1978). Bacteriocin typing has been investigated by Hamon *et al.* (1966), but with negative results. Bottone *et al.* (1979) have found that *Y. enterocolitica* and related species have a temperature-dependent production of bacteriocins. The system was not developed to a typing procedure. Biotyping of *Y. enterocolitica* and *Y. intermedia* appear in Tables III and IV.

IX. Susceptibility to antibiotics

The most common isolates of human *Y. enterocolitica* are usually resistant to penicillins such as ampicillin, benzylpenicillin, carbenicillin, phenoxymethyl-penicillin and cephalosporins because of β-lactamase production (Bottone, 1977; Hausnerova, 1973; Winblad, 1978). Serotypes O:3 and O:9 produce two distinct β-lactamases. *Y. enterocolitica* is mostly sensitive to chloramphenicol, colistin, aminoglycosides such as gentamicin, tetracyclines and sulpho-namides (Frederiksen, 1964; Winblad, 1978), although naturally occurring strains may be resistant of R-factors (Bottone, 1977). *Y. enterocolitica* strains are resistant to erythromycin, bacitracin, novobiocin and fucidin (Frederik-sen, 1964).

X. Interrelationship between serogroups and pathogenicity

Y. enterocolitica is engaged in a range of clinical entities affecting several host species (Bottone, 1977; Mollaret *et al.*, 1979). The clinical manifestations of yersiniosis in man or animals have mainly been associated with only a restricted number of serogroups. This permits reliable identification of the typical agents of yersiniosis using O-antigens. In fact, two different groups of strains have been distinguished according to their clinical significance: host-adapted pathogens and environmental strains.

    Host-adapted pathogens giving rise to well-defined clinical syndromes (*vide infra*) belong to a few serogroups, each of which exhibits host specificity (Mollaret *et al.*, 1979; Wauters, 1970). Thus, O:3, O:8 and O:9 are the most important causative agents in man. Likewise, O:1 and O:2 have been associated with disease in chinchillas and hares respectively.

    Environmental strains constitute a spectrum of phenotypic variants which display a variety of antigenic factors. As a group, these bacteria usually lack clinical significance and are ubiquitous. Occasionally, however, such strains have been associated with atypical clinical syndromes mainly affecting patients with compromised host defense.

XI. Serogroups, biotypes and phage types

The serogroups commonly involved in pathological processes in man or animals belong to distinct serogroup–biotype–phage type combinations (Bottone, 1977; Mollaret *et al.*, 1979; Wauters, 1970) (Table XIII). Thus, serogroup 1, which has caused yersiniosis in chinchillas, belongs to Wauters' biotype 3, phage type II. Likewise, the strains reported from epizootic outbreaks among hares, are serogroup 2, biotype 5 and phage types XI or II. Serogroups 3 and 9, the most common human pathogens, invariably belong to biotypes 4 and 2 respectively. O:9 is predominantly phage type X_3, whereas O:3 belongs to phage types VIII, IX_A or IX_B, according to the geographic origin of the strains (Table XIII).

In contrast, the environmental serogroups are biochemically as well as antigenically diverse. A majority belongs to biotype 1. Many isolates, however, cannot be ascribed to any presently defined biotype and constitute a spectrum of intermediate phenotypes. The phage types observed among these strains are X_o or X_z.

XII. Clinical manifestations

Y. enterocolitica is associated with a spectrum of clinical syndromes in man. The range of manifestations has been reviewed by Bottone (1977) and Mollaret *et al.* (1979).

A. Acute abdominal forms

Gastro-enteritis is by far the most frequently encountered manifestation. Acute non-complicated enteritis is usually observed in children below seven years of age, and the frequency of such cases decreases with age. Occasionally, an acute infection is confined to the right iliac fossa in the form of acute terminal ileitis or acute mesenteric lymphadenitis, giving rise to symptoms resembling appendicitis.

In most cases, a single serogroup of *Y. enterocolitica* is recovered from the intestinal contents. However, isolations of multiple concomitant serogroups or biotypes have been reported (Bottone, 1977). Enteric infections with *Y. enterocolitica* have primarily been elicited by strains belonging to sero-group O:3, biotype 4; by O:9, biotype 2; or by O:8, biotype 1. The

TABLE XIII

Interrelationship between serogroup, biotype, phage type, habitat, pathogenicity and geographic distribution of *Y. enterocolitica*

Serogroup	Biotype[a]	Phage type	Habitat	Clinical significance	Geographic distribution
1	3	II	Chinchilla	Epizootics (1958–1963)	Europe, USA
2	5	XI or II	Hare, goat		Europe
3		VIII	Human	Gastro-intestinal infections Reactive arthritis Erythema nodosum	Europe, Japan
			Swine	Healthy carrier	
	4	IX$_a$	Human	Gastro-intestinal infections	South Africa
			Swine	Healthy carrier	
		IX$_b$	Human	Gastro-intestinal infections	Canada
			Swine	Healthy carrier	
9	2	X$_3$	Human	Gastro-intestinal infections Reactive arthritis Erythema nodosum	Europe
			Swine	Healthy carrier	
8	1	X	Human	Gastro-intestinal infections	USA, Canada

	Human		Miscellaneous
4		Healthy carrier	
5		Gastro-intestinal infections	
6		Extra-intestinal infections	
16	X_z	Septicaemia	
Miscellaneous	X_0		
NAG[b]			
1		Terrestrial fauna	Healthy carrier
2		Aquatic fauna	—
3		Food	—
NT[c]		Environment	

Adapted from Brenner et al. (1980a) and Mollaret et al. (1979).

[a] According to Wauters (1970).

[b] NAG, non-agglutinable.

[c] NT, non-typable.

symptoms may be closely similar to those associated with other enteric pathogens, i.e. *Salmonella*, *Shigella* and enteropathogenic *E. coli*. Therefore, the diagnosis depends on the isolation of the causative agent. Serogrouping of isolates is recommended to distinguish pathogens from commensals. Isolation of environmental serogroups might reflect only chance contamination or a harmless carrier state without any causal connection to the disease.

Patients develop specific antibodies with rising titres during the acute course of the disease. The titres usually decline within two months (Bottone, 1977). Hence, demonstration of specific serum agglutinins may be diagnostically helpful.

B. Erythema nodosum and arthritis

Cutaneous manifestations, usually in the form of erythema nodosum, may follow infections with *Y. enterocolitica*. It has been reported that 80% of the cases are women, most frequently over forty years of age; 40% of the cases show no history of gastro-intestinal symptoms (WHO Scientific Working Group, 1980). Likewise, post-infectious arthritis or poly-arthritis, usually preceded by fever and gastro-intestinal manifestations, have been observed. This form has occurred most commonly in young adults and has an equal sex distribution.

In Europe, there is a high incidence of both reactive arthritis and erythema nodosum. The causative agents belong to O:3, biotype 4, phage type VIII, or O:9, biotype 2. These manifestations have been virtually absent in the USA, Canada and South Africa (Bottone, 1977). This observation has been related to the geographic distribution of serogroups and phage types. Thus, O:3 has only exceptionally been isolated in the USA. O:3, biotype 4, has frequently been detected in Canada, but the strains differ from the European isolates by belonging to phage type IX_B. Likewise, the South African isolates have belonged to O:3, biotype 4, phage type IX_A. Serogroup O:9 is rare in both the USA, Canada and South Africa.

C. Septicaemia

This manifestation has primarily occurred in patients with compromised host defense due to underlying illness or immunosuppressive therapy (Bottone, 1977). In France, 40% of the human clinical isolates of *Y. enterocolitica* have been derived from septicaemia, in contrast to the rare occurrence of this syndrome in other countries (Mollaret *et al.*, 1976).

D. Atypical forms

Y. enterocolitica and *Y. enterocolitica*-like bacteria have been associated with a broad range of atypical clinical manifestations, mostly affecting immuno-compromised patients (Bottone, 1977; Mollaret *et al.*, 1979). Rare cases of myocarditis, subacute hepatitis, abscesses in different organs, conjunctivitis, ophthalmitis, meningitis, urethritis and so on have been described. The strains involved have occasionally belonged to serogroups different from those associated with the classic syndromes, but similar to the strains prevailing in nature.

XIII. Pathogenicity factors

A. Invasiveness

Autopsies from human yersiniosis have revealed an invasive form of enteritis (Bradford *et al.*, 1974). Likewise, cell culture models as well as studies on experimentally infected animals have indicated that penetration into the epithelial cells of the intestinal mucosa is an essential factor in the pathogenesis of *Y. enterocolitica* infections (Lee *et al.*, 1977; Maruyama, 1973; Pai *et al.*, 1980; Une, 1977b).

Virulence in *Y. enterocolitica* has been associated with a family of related plasmids, 40–48 Mdaltons in size (Gemski *et al.*, 1980; Portnoy *et al.*, 1981; Vesikari *et al.*, 1981; Zink *et al.*, 1980). Strains harbouring these plasmids are also calcium-dependent when grown at 37°C (Gemski *et al.*, 1980). In *Y. pestis* and *Y. pseudotuberculosis*, such a temperature-dependent calcium dependency is correlated with production of a VW-antigen complex. VW-antigens have been demonstrated in virulent but not in avirulent *Y. enterocolitica* strains (Carter *et al.*, 1980). The existence of a common virulence plasmid in genus *Yersinia* has been suggested (Ben-Gurion and Shafferman, 1981).

Human pathogenic strains of serogroups O:3, O:8 and O:9 contain plasmids with a size range characteristic for each serogroup (Vesikari *et al.*, 1981). O:3 and O:9 harbour plasmids sized approximately 47 and 44 Mdaltons respectively (Vesikari *et al.*, 1981). The DNA of these plasmids is similar but not identical to the plasmid of O:8, which approximates 40–42 Mdaltons. The plasmids share a high degree of DNA sequence homology and may represent the divergence of an ancestral plasmid (Portnoy *et al.*, 1981). Thus, a family of different sized plasmids exists in *Y. enterocolitica*, each of which contributing to the pathogenicity of a separate serogroup.

The presence of virulence plasmids is correlated with the ability to auto-agglutinate in tissue culture media (Laird and Cavanaugh, 1980; Vesikari *et al.*, 1981). Consequently, a rapid auto-agglutination test has been proposed for screening of potential virulence (Laird and Cavanaugh, 1980). However, negative results should be interpreted with caution since the plasmids are easily lost when subcultured. The same statement pertains to a proposed calcium dependency test (Gemski *et al.*, 1980). The auto-agglutination test has suggested that virulence characteristics may be present in serogroups rarely associated with disease (Schiemann *et al.*, 1981). However, the reliability of this test may require further evaluation.

It has been demonstrated that the ability to invade cultured cells (HeLa, HEp-2) *in vitro* is not associated with the presence of virulence plasmids (Portnoy *et al.*, 1981; Vesikari *et al.*, 1981). Strains which have lost these factors are still able to invade host cells. In fact, strains of O:3 and O:9 cured of the plasmids show increased invasiveness to HEp-2 cells (Vesikari *et al.*, 1981). In contrast, the ability of O:8 to cause keratoconjunctivitis in the guinea-pig eye (Serény test) is lost with curing of the plasmid (Zink *et al.*, 1980). The Serény test is unsuitable for the study of virulence in O:3 and O:9 since these serogroups are frequently negative in this assay (Vesikari *et al.*, 1981).

Portnoy *et al.* (1981), working mainly with O:8, presumed that the contribution of the virulence plasmid to pathogenesis occurs subsequent to penetration of the cell membrane. On the other hand, it has been suggested that the virulence plasmids of O:3 and O:9 primarily determine epithelial cell adherence (Vesikari *et al.*, 1981). These observations might reflect different pathogenic strategies among *Y. enterocolitica* serogroups.

The ability to invade cultured cells (HeLa, HEp-2) *in vitro* is a stable property exhibited by virtually all strains belonging to the pathogenic serogroups (Lee *et al.*, 1977; Une *et al.*, 1977). This characteristic is unaffected by the loss of the virulence plasmid. However, *in vitro* invasiveness has also been demonstrated among serogroups with uncertain clinical significance (Kapperud, 1980b; Pedersen *et al.*, 1979). Obviously, cell culture invasiveness reflects the possession of attributes which are necessary but not sufficient to determine virulence. A negative result in this assay is still a reliable criterion, however, for assessing the lack of virulence (Lee *et al.*, 1981).

The adherence of *Y. enterocolitica* to host cells may be related to the production of fimbriae. Production of a mannose-resistant adhesin which agglutinates a broad spectrum of erythrocytes is associated with the presence of fimbriae of 8 nm in diameter (MacLagan and Old, 1980). These fimbriae have been detected in both pathogenic and apathogenic serogroups. HeLa cell invasiveness is not related to the presence of this factor (Rahimian and Evans, 1981). Production of fimbriae is lost by culturing at 37°C. This is analogous to the temperature dependency of virulence plasmid and accompanying auto-agglutination and calcium dependency, flagellae, adherence to HEp-2 cells, and enterotoxin production in *Y. enterocolitica*.

A distinct type of fimbriae, identified as the K:1-antigen (*vide supra*), has been recognized in O:10 and O:21 (Aleksic *et al.*, 1976; MacLagan and Old, 1980).

B. Enterotoxin production

Another possible pathogenicity factor in *Y. enterocolitica* is enterotoxin production. Production of heat-stable enterotoxin (YEST) after cultivation between 20°C and 30°C *in vitro* is widespread among *Y. enterocolitica* and *Y. enterocolitica*-like bacteria (Kapperud, 1980a, 1982; Pai *et al.*, 1978). There is no correlation with the presence of plasmids (Vesikari *et al.*, 1981). YEST production appears to be a stable property mediated by chromosomal genes (Zink *et al.*, 1980). The physiochemical, biological and antigenic characteristics of YEST closely resemble those of the heat-stable enterotoxin of *E. coli* (Boyce *et al.*, 1979; Okamoto *et al.*, 1981; Robins-Browne *et al.*, 1979).

The highest prevalence of enterotoxin production (80–100%) has been detected among the human clinical isolates belonging to serogroups O:3, O:5–27, O:8 and O:9 (Kapperud, 1980a, 1982; Olsson *et al.*, 1980; Pai *et al.*, 1978). This property is also prevalent among serogroups recovered from animals and environmental sources (approximately 50% and 20% respectively) (Kapperud, 1980a, 1982; Pai *et al.*, 1978). YEST is widespread in all serogroups, but is sparsely represented among non-agglutinable strains. Enterotoxin production has been detected among 22% of O:6 and 11% of O:4, which are two of the most frequently encountered serogroups in Scandinavian ecosystems (Kapperud, 1980a). The clinical significance of YEST is uncertain for four reasons: (1) YEST is common among strains without any recognized virulence; (2) both animals and man may be healthy carriers of enterotoxigenic strains (Kapperud, 1980d); (3) none of the clinical isolates has so far been shown to produce YEST at the human body temperature; (4) YEST production has not been demonstrated *in vivo*.

On the other hand, enterotoxin production at 37°C has been reported to be prevalent in *Y. kristensenii*. According to Kapperud (1980a, 1982), 49% of the strains belonging to this nomenspecies produced YEST at both 37°C and 22°C (serogroups O:1, O:11 and O:28 and non-agglutinable strains). All isolates of O:28 and 50% of O:11 produced enterotoxin at these temperatures. Only 1% of the environmental isolates of *Y. enterocolitica* (*sensu stricto*) showed YEST production at 37°C. *Y. kristensenii* have occasionally been associated with human disease (Bottone and Robin, 1979). The pathogenicity of this nomenspecies may deserve further evaluation.

Likewise, *Y. kristensenii* seems to be unique with respect to enterotoxin production at 4°C. Kapperud (1982) found that 47% of the strains belonging to *Y. kristensenii* produced YEST at 4°C (serogroups O:1, O:11, O:12 and

O:28 and non-agglutinable strains). This property was also detected among environmental isolates of *Y. enterocolitica* (*sensu stricto*) (O:6 and O:16), but the prevalence was relatively low (4%). This observation may give *Y. enterocolitica* and *Y. enterocolitica*-like bacteria a new significance in food hygiene as possible agents of food intoxination (Boyce *et al.*, 1979; Kapperud and Langeland, 1981). Refrigeration is currently used to prolong the time of storage of perishable food. *Y. enterocolitica* is one of the few human pathogens which is able to grow at this temperature. The occurrence of these bacteria in various kinds of food has been amply documented (Table XIV). Boyce *et al.* (1979) showed that YEST is acid-stable and is not inactivated by exposure to 121°C for 30 min or by storage at 4°C for at least five months. They suggested this substance to be able to withstand food processing and storage, and gastric acidity. Hence, preformed YEST may be capable of causing food intoxination.

In conclusion, the prevalence of YEST production at 20–30°C reaches a maximum among clinical isolates of O:3, O:5–27, O:8 and O:9, with a much lower prevalence among the other serogroups of environmental origin. YEST production at 37°C and 4°C is common in serogroups O:11, O:12 and O:28, and has also been detected among serogroups O:1, O:6 and O:16 and non-agglutinable strains.

XIV. Frequency of serogroups

A. Human clinical specimens

In Europe, the serogroups of *Y. enterocolitica* most frequently causing human disease are O:3 (54.6%–99.1%) followed by O:9 (0.4%–31.6%) (Table XIV). Other serogroups are rare in this geographical region. O:3 predominates in all countries. The prevalence of O:9 is higher in the Netherlands than elsewhere (31.6%) (Oosterom, 1979). According to Esseveld and Goudzwaard (1973), O:9 composed approximately 50% of the isolates originating from the Rotterdam region. This observation urged Vandepitte and Wauters (1979) to survey the prevalence of this serogroup in Belgium. It was evident that the prevalence was significantly lower in this country (7.2%–9.5%, average 8.6%), even in those provinces having a common border with the Netherlands. A great majority of the O:3 and O:9 isolates is derived from faeces of patients with gastro-intestinal disease (Mollaret *et al.*, 1979) (*vide supra*).

A high prevalence of O:3 has also been reported from Japan (46.6%) (Kanazawa and Ikemura, 1979) and from Canada (80.1%–100.0%) (Lafleur *et al.*, 1979; Toma *et al.*, 1979). In Canada, O:3 dominated in Ontario, Quebec and in the four Eastern provinces. In the Western provinces, the predominat-

TABLE XIV

Prevalence of *Y. enterocolitica* serogroups in clinical and non-clinical human specimens

Distribution according to serogroups (%):

Source of strains	Country	Ref.	No. of strains	1	2	3	4	5	6	7	8	9	10	11	12	13	14	15	16	17	18	19	20	21	22	28	34	MI	SA	NA
Clinical specimens	USA	a	60	—	3.3	—	3.3	6.7	8.3	1.7	33.3	3.3	—	—	—	1.7	—	—	—	—	—	3.3	5.0	—	—	—	—	8.3	—	22.0
Clinical specimens	USA	b	33	12.1	—	3.0	—	24.2	9.1	—	18.2	—	6.1	6.1	—	9.1	—	—	3.0	—	—	3.0	—	—	—	—	—	—	—	6.1
Atypical syndromes	USA	c	22	—	—	—	—	—	—	—	—	—	—	—	4.5	—	—	—	4.5	45.5	—	—	—	—	—	—	—	—	—	45.5
Patients and carriers	USA	d	52	—	—	1.9	—	3.8	5.8	—	71.2	—	—	—	—	—	—	—	—	—	—	—	—	—	—	—	—	—	—	17.3
Clinical specimens	Canada	e	977	—	—	80.1	1.0	5.1	3.0	0.2	3.3	0.1	—	—	0.1	0.5	—	—	0.6	—	—	—	0.1	0.4	—	—	0.3	—	—	5.0
Children, clin. specimens	Canada	f	146	—	—	100.0	—	—	—	—	—	—	—	—	—	—	—	—	—	—	—	—	—	—	—	—	—	—	—	—
Clinical specimens	Belgium	g	1761	0.1	—	89.2	0.1	0.7	0.8	0.1	—	8.6	0.2	—	—	0.1	0.1	—	0.1	—	—	—	—	—	—	—	—	0.1	—	0.1
Clinical specimens	Belgium	h	120	—	—	86.7	—	—	—	0.7	—	13.3	—	—	0.2	0.9	—	0.7	—	—	—	—	—	—	—	—	—	—	—	—
Clinical specimens	The Netherlands	i	588	0.3	—	54.6	0.2	2.0	3.7	—	—	31.6	1.2	—	—	0.6	—	—	1.4	—	—	0.2	—	—	0.2	—	0.2	—	—	2.0
Clinical specimens	France	j	168	—	—	61.9	—	0.6	6.5	1.8	0.6	7.1	3.6	0.6	0.6	0.6	7.1	—	1.2	—	—	0.2	—	—	0.2	—	—	—	—	7.7
Clinical specimens	Japan	k	58	—	—	46.6	—	10.3	22.4	6.9	—	—	1.7	—	1.7	—	—	6.9	—	—	—	—	·	·	·	·	·	·	·	3.4
Patients and carriers	Hungary	l	1355	0.1	—	99.1	—	0.1	0.1	15.2	—	0.4	0.1	—	8.8	3.2	5.1	0.1	—	1.4	—	—	·	·	·	·	·	·	·	—
Healthy carriers	Japan	m	217	—	—	0.5	2.8	9.7	15.7	11.1	—	—	1.4	—	—	3.2	22.2	0.5	—	—	—	—	·	·	·	·	·	·	·	35.9
Healthy carriers	Japan	n	18	—	—	16.7	5.6	22.2	22.2	—	—	—	—	—	—	—	—	—	—	—	—	—	·	·	·	·	·	·	·	—
Healthy carriers	Norway	o	20	—	—	—	5.0	10.0	25.0	5.0	—	—	—	—	—	20.0	—	—	5.0	—	—	—	—	—	—	—	—	—	—	30.0

SA, self-agglutinable; NA, non-agglutinable; MI, miscellaneous; various serogroups or serogroup combinations not compiled in this table. . . not tested.
[a] Quan (1979); [b] Bissett (1979); [c] Bottone and Robin (1979); [d] Shayegani et al. (1979); [e] Toma et al. (1979); [f] Lafleur et al. (1979); [g] Vanderpitte and Wauters (1979); [h] Van Noyen et al. (1979); [i] Oosterom (1979); [j] Alonso et al. (1979); [k] Kanazawa and Ikemura (1979); [l] Szita and Svidró (1976); [m] Kanazawa and Ikemura (1979); [n] Asakawa et al. (1979); [o] Kapperud (1980c).

ing serogroups were O:8, O:5–27 and O:4–32, analogous to the pattern observed in the USA (Toma *et al.*, 1979).

In the USA, isolates of O:3 and O:9 occur sparsely. The most frequent serogroups in the USA are O:8 (18.2%–71.2%) followed by O:5–27 (3.8%–24.2%). The reason why the USA differs from the rest of the world is unknown. The rare isolates of O:3 reported from the USA might have been contracted outside the country (Bisset, 1979; Shayegani *et al.*, 1979). With few exceptions, O:9 is conspicuously absent in the USA (Quan, 1979) and is rare in Canada (Toma *et al.*, 1979). The dominance of O:8 in the USA may be more apparent than real because of the involvement of this serogroup in outbreaks of yersiniosis (Shayegani *et al.*, 1979).

A spectrum of serogroups similar to those isolated from animals and from environmental sources has been recovered from human clinical specimens. These serogroups combined represent only a few per cent of all strains isolated. It would appear that the clinical significance is related to only a few cases, mainly patients with underlying conditions, indicating that *Y. enterocolitica* may be opportunistic. Most strains are probably harmless commensals. Furthermore, the reported frequency of environmental serogroups may, at least in part, reflect the skill of the laboratory staff to recognize strains with aberrant colony morphology.

Bottone and Robin (1979) isolated 22 strains of serogroups which are rare in human disease (O:17, O:16 and O:12) in New York. Twenty-one of these isolates belonged to biochemical entities that are now recognized as *Y. intermedia* (19 strains) or *Y. kristensenii* (2 strains).

B. Humans, asymptomatic carriers

A variety of serogroups have been isolated from the stools of symptomless persons (Table XIV). The pattern of serogroups parallels the serogroup profile recovered from healthy carriers among the terrestrial fauna. Vandepitte and Wauters (1979) felt that this observation supports the suggestion that such strains possess a lower, if any, human pathogenicity.

C. Domestic and synanthropic animals

Swine may be healthy carriers of O:3 and O:9 (Table XV), belonging to the same biotypes and phage types as those engaged in human disease (Mollaret *et al.*, 1979). This may account for a potential important role of swine as a reservoir of human infection. O:3 is the most frequently encountered serogroup in porcine faecal specimens in Europe, Canada, Japan and South Africa (26.4%–100%). In contrast, O:8, the predominant human pathogen in the USA, appears to be rare in the faeces of swine.

TABLE XV

Prevalence of *Y. enterocolitica* serogroups in domestic and synanthropic animals

Source of strains	Country	Ref.	No. of strains	1	2	3	4	5	6	7	8	9	10	11	12	13	14	15	16	17	18	19	20	21	22	28	34	MI	SA	NA
Swine, cecal contents	Sweden	a	18	—	—	100.0	—	—	—	—	—	—	—	—	—	—	—	—	—	—	—	—	—	—	—	—	—	—	—	—
Swine, colon contents	Denmark	b	78	—	—	34.6	3.8	15.4	16.7	9.0	—	—	—	1.3	5.1	—	—	1.3	—	1.3	—	5.1	—	—	—	—	—	—	—	6.4
Swine, throat swabs	Denmark	c	84	—	—	100.0	—	—	—	—	—	—	—	—	—	—	—	—	—	—	—	—	—	—	—	—	—	—	—	—
Swine	Canada	d	34	—	—	38.2	—	23.5	2.9	—	—	—	—	—	3.8	2.9	2.9	—	1.9	9.4	—	—	—	—	—	—	—	5.9	—	23.5
Swine	Japan	e	53	—	—	26.4	—	41.5	11.3	3.8	—	—	1.9	—	22.1	—	—	—	—	—	—	—	—	—	—	—	—	—	—	3.5
Swine	Japan	f	231	—	—	59.3	—	11.7	2.2	1.3	—	—	—	—	—	—	—	—	—	—	—	—	—	—	—	—	—	—	—	—
Rats	Japan	g	76	—	—	—	—	13.2	32.9	22.4	—	—	—	—	—	—	—	—	—	—	—	—	—	—	—	—	—	—	—	17.1
Rats	Czechoslovakia	h	26	—	—	61.5	—	—	3.8	—	—	1.3	3.8	—	6.6	3.8	3.9	—	2.6	—	—	—	—	—	—	—	3.8	—	—	19.2
Dogs	Japan	i	395	—	—	71.4	—	15.7	1.0	—	—	—	—	—	3.8	—	—	—	—	2.0	—	—	—	—	—	—	—	—	—	—
Flies	Japan	j	50	—	—	6.0	—	12.0	28.0	—	—	11.9	6.0	—	—	16.0	10.0	—	—	—	—	—	—	—	—	—	—	—	—	20.0

For abbreviations see Table XIV.

[a] Hurvell *et al.* (1979); [b] Pedersen and Winblad (1979); [c] Pedersen (1979); [d] Toma *et al.* (1979); [e] Asakawa *et al.* (1979); [f] Zen-Yoji *et al.* (1974); [g] Zen-Yoji *et al.* (1974); [h] Aldová *et al.* (1977); [i] Kaneko *et al.* (1977); [j] Fukushima *et al.* (1979).

In addition to a function as faecal commensals, O:3 and O:9 also inhabit the oral cavity of swine (Christensen, 1980; Pedersen, 1979; Wauters, 1979). The organisms remain longer on the tonsils than in stools (Wauters and Pohl, 1972). O:8 has recently been isolated from porcine tongues in the USA (Doyle et al., 1981).

The prevalence of O:9 among swine reflects the importance of this serogroup in human illness. High percentages of O:9 have been reported from swine in the Netherlands, corresponding to the importance of this serogroup in human yersiniosis (Esseveld and Goudzwaard, 1973). Analogously, O:9 has not been isolated from the buccal cavity of swine in Denmark, where human infections with this serogroup are rare (Pedersen and Winblad, 1979).

Dogs may also be faecal carriers of O:3 and O:9. The relatively intimate contact between dogs and man suggests a potential reciprocal transmission between man and pets, although such an epidemiological link has not yet been clearly confirmed. The clinical significance of Y. enterocolitica in canine infection also remains uncertain (Kaneko et al., 1977).

Cross-contamination between swine, rats and flies has been suggested. Aldova and Laznickova (1979) detected strains belonging to O:3 only among rats captured in pig houses, but not among rats lacking in porcine contact. Likewise, flies collected from a piggery carried strains belonging to serogroup O:3 (Fukushima et al., 1979).

Both domestic and synanthropic animals may be symptomless carriers of serogroups similar to those prevailing among wild animals and in the environment. These strains seem to constitute a normal element in their intestinal flora.

D. Foods

The occurrence of Y. enterocolitica and Y. enterocolitica-like bacteria in various foods has been amply documented (Lee, 1977). The presence of serogroups O:3, O:8 and O:9 in meat products has been reported by several authors (Table XVI). The occurrence of O:3 and O:9 in meats seems to be mainly related to products of porcine origin. Pork contaminated either primarily or secondarily with Y. enterocolitica has been proposed as the most important source of human yersiniosis (Asakawa et al., 1979). Clinically relevant serogroups have rarely been isolated from bovine or caprine milk. However, milk products were implicated as the source of infection in large outbreaks of yersiniosis in the USA (Black et al., 1978; Lee, 1977). With these exceptions, there is no firm evidence linking Y. enterocolitica in food to human infection.

Both meat and milk products may contain environmental serogroups, suggestive of contamination from a wide spectrum of sources.

TABLE XVI

Prevalence of *Y. enterocolitica* serogroups in foods

Source of strains	Country	Ref	No. of strains	Distribution according to serogroups (%)																									
				1	2	3	4	5	6	7	8	9	10	11	12	13	14	15	16	17	18	19	20	21	22	28	34	MI	SA	NA
Milk	Canada	a	114	—	—	—	10,5	7,0	8,8	1,0	—	—	—	—	—	1,8	1,8	—	—	1,8	—	—	—	—	—	—	—	6,1	—	62,3
Raw milk	Canada	b	98	—	—	—	10,2	7,1	4,1	—	—	—	—	—	—	2,0	2,0	2,0	—	2,0	1,0	—	—	1,0	—	—	—	—	—	67,3
Raw milk	France	c	40	—	—	—	5,0	57,5	10,0	27,5	—	—	—	—	—	—	—	—	—	—	—	—	—	—	—	—	—	—	—	—
Raw goat's milk	Australia	d	35	—	—	—	—	14,3	—	—	—	—	14,3	—	—	—	2,9	2,9	51,4	—	—	—	—	—	2,9	—	—	8,6	—	2,9
Pig's tongues	Belgium	e	168	—	—	98,2	—	—	—	—	—	1,8	—	—	—	—	—	—	—	—	—	—	—	—	—	—	—	—	—	—
Minced meat	The Netherlands	f	27	—	—	3,7	—	63,0	7,4	—	—	3,7	—	—	—	—	—	—	11,1	—	—	—	—	—	—	—	—	11,1	—	—
Pork products	Canada	g	80	—	—	15,0	—	13,8	—	—	1,3	—	—	—	—	3,8	—	—	1,3	3,8	—	—	—	1,3	—	—	—	—	—	60,0
Food and chopping boards	Japan	h	39	—	—	10,3	—	30,8	5,1	2,6	—	—	2,6	5,1	30,8	10,3	2,6	—	—	—	—	—	2,6	—	—	—	—	—	—	—
Foods	Czechoslovakia	i	76	—	—	1,3	—	18,4	9,2	13,2	—	—	2,6	2,6	2,6	7,9	1,3	—	1,3	—	—	1,3	2,6	—	—	—	—	15,8	14,5	—
Foods	Canada	j	21	—	—	—	28,6	4,8	14,3	—	5,3	—	—	—	—	4,8	4,8	—	—	—	—	—	—	—	—	—	—	—	—	42,9

For abbreviations see Table XIV.

[a] Toma *et al.* (1979); [b] Schiemann (1979); [c] Vidon and Delmas (1981); [d] Hughes and Jensen (1981); [e] Wauters (1979); [f] Oosterom (1979); [g] Schiemann (1980); [h] Asakawa (1979); [i] Aldová and Lazníčková (1979); [j] Toma *et al.* (1979).

E. Wild animals

Y. enterocolitica and *Y. enterocolitica*-like bacteria recovered from the intestinal contents of healthy carriers among terrestrial and freshwater animals are antigenically diverse (Table XVII). Clinically significant serogroups have only exceptionally been isolated from these sources (Alonso *et al.*, 1979; Kapperud, 1981). Kapperud (1975, 1977) reported that 4.4% of the yersiniae from Scandinavian wildlife mammals belonged to serogroup O:3. All of these isolates differed in biochemical properties from strains relevant to human medicine.

Kapperud (1981) found that the serogroups recovered from terrestrial animals were partly different from those obtained from freshwater fish, although there was a definite overlap. This may be attributable to the discernible habitat preferences of the nomenspecies proposed by Brenner *et. al.* (1980a). Thus, *Y. frederiksenii* and *Y. intermedia* seem to be more prevalent in aquatic ecosystems than *Y. enterocolitica* (*sensu stricto*), which has been isolated from both terrestrial and aquatic habitats. The highest prevalence of *Y. kristensenii* has been detected among terrestrial vertebrates and in soil (Bercovier *et al.*, 1978, 1980b; Kapperud, 1981; Ursing *et al.*, 1980a). These nomespecies are characterized by different, although partially overlapping, serogroup profiles (Table VIII). Hence, serogroups O:11, O:12 and O:28 have frequently been isolated in terrestrial ecosystems. This reflects the habitat preference of *Y. kristensenii*. Likewise, O:4 and non-agglutinable strains have occurred relatively more often in fish, reflecting the predominance of *Y. frederiksenii* and *Y. intermedia* in this group (Kapperud and Jonsson, 1978; Kapperud, 1981).

F. Water and sewage

Water is the recipient of bacterial contamination from both terrestrial and aquatic sources. Consequently, a broad diversity of serogroups has been isolated from water (Table XVIII). Clinically significant serogroups have only seldom been recovered from surface or well water not subjected to any apparant pollution. However, sewage water may contain a high proportion of strains belonging to serogroup O:3 (Oosterom, 1979).

XV. Epidemiology

The correlation observed between O-antigens and pathogenetic properties makes serological characterization an indispensable tool in epidemiological investigations. Despite considerable efforts, however, the epidemiology of *Y.*

TABLE XVII

Prevalence of *Y. enterocolitica* serogroups in wild-living animals

Source of strains	Country	Ref.	No. of strains	Distribution according to serogroups (%)																										
				1	2	3	4	5	6	7	8	9	10	11	12	13	14	15	16	17	18	19	20	21	22	28	34	MI	SA	NA
Small rodents	Czechoslovakia	a	30	3.3	26.7	—	10.0	—	—	—	6.7	—	3.3	—	23.3	—	—	—	36.7	—	—	—	—	—	—	3.3	—	—	—	10.0
Small mammals	France	b	73	—	2.7	—	4.1	6.8	12.3	5.5	—	—	6.8	6.8	7.0	1.0	—	—	8.2	—	6.8	—	—	—	—	—	—	5.5	—	11.0
Small mammals	France	c	199	18.6	—	1.0	5.5	.0	2.5	2.5	—	—	9.5	6.5	5.9	—	—	0.5	14.1	—	—	—	—	—	—	—	—	—	—	26.1
Small mammals, fox	Norway	d	136	5.1	—	4.4	5.9	0.7	23.5	3.7	—	—	—	4.4	—	—	—	—	5.9	0.7	—	—	—	—	—	6.6	—	8.1	—	25.0
Birds	France	e	53	—	—	—	3.8	—	15.1	7.5	1.9	—	9.4	7.5	13.2	—	—	1.9	1.9	—	5.7	—	—	—	—	—	—	13.2	—	18.9
Earthworms	France	f	24	—	—	—	4.2	8.3	—	—	4.2	—	4.2	—	12.5	—	—	—	4.2	—	4.2	—	4.2	—	—	—	4.2	29.2	—	20.8
Freshwater fish	Norway	g	129	1.6	—	—	17.1	—	0.8	—	—	—	—	—	—	3.9	5.4	—	3.1	3.9	—	—	—	—	—	—	—	10.9	—	53.5

For abbreviations see Table XIV.

[a] Aldová and Lím (1974); [b] Bercovier et al. (1978); [c] Alonso et al. (1979); [d] Kapperud (1975/77); [e] Bercovier et al. (1978); [f] Bercovier et al. (1978); [g] Kapperud and Jonsson (1978).

TABLE XVIII

Prevalence of *Y. enterocolitica* serogroups in water, sewage, sludge and soil

Source of strains	Country	Ref.	No. of strains	Distribution according to serogroups (%)																										
				1	2	3	4	5	6	7	8	9	10	11	12	13	14	15	16	17	18	19	20	21	22	28	34	MI	SA	NA
Lakes, rivers, brooks	Norway	a	57	—	—	1.8	8.8	1.8	10.5	—	—	—	—	3.5	5.3	—	3.5	1.8	—	3.5	—	—	—	—	—	5.3	—	12.3	—	42.1
Water	Canada	b	45	—	—	—	11.1	4.4	20.0	—	—	—	—	—	—	—	4.4	—	—	4.4	—	—	—	—	—	—	—	6.7	—	48.9
Lakes, rivers, wells	USA	c	140	—	9.3	0.7	12.9	0.7	2.1	—	1.4	—	—	—	—	—	1.4	—	4.3	3.6	2.9	1.4	1.4	2.1	2.1	—	—	11.5	—	42.9
Drinking and surface water	Czechoslovakia	d	139	4.3	—	—	12.9	3.6	11.5	—	9.4	0.7	1.4	—	—	1.4	4.3	4.3	0.7	5.0	—	1.4	0.7	5.0	—	0.7	0.7	—	2.2	29.5
Drinking water	Norway	e	11	27.2	—	—	18.2	—	9.1	9.1	—	—	—	—	—	9.1	—	—	9.1	9.1	—	—	—	—	—	9.1	—	—	—	—
Water, sewage sludge	Norway	f	116	1.7	—	—	4.3	1.7	3.4	10.3	0.9	—	1.7	—	5.2	0.9	1.7	—	6.9	12.1	2.6	—	—	—	—	2.6	—	3.4	—	43.1
Water, sewage	Canada	g	152	—	—	—	3.3	1.3	0.7	1.3	—	—	—	3.3	0.7	—	5.3	5.3	13.2	5.9	—	—	—	0.7	—	—	—	—	—	61.8
Sewage water	The Netherlands	h	33	—	—	12.1	—	6.1	18.2	6.1	—	—	—	—	—	—	—	—	12.1	12.1	—	—	—	—	—	—	—	—	—	45.5
Water, sludge, sediments	Denmark	i	185	—	—	0.5	9.7	3.8	7.0	—	2.7	0.5	2.7	0.5	2.2	4.3	0.5	1.6	2.7	3.2	—	—	—	—	—	—	—	0.5	—	57.3
Soil	France	j	125	—	—	—	2.4	—	2.4	4.0	0.8	—	0.8	1.6	30.4	—	—	—	6.4	—	7.2	—	—	—	—	—	0.8	27.2	—	16.0

For abbreviations see Table XIV.

[a] Kapperud (1977); [b] Toma *et al.* (1979); [c] Saari and Jansen (1979); [d] Aldová and Lázničková (1979); [e] Lassen (1972); [f] Langeland (1981); [g] Lautenschlager (1980); [h] Oosterom (1970); [i] Krøngaard Kristensen (1977); [j] Bercovier *et al.* (1978).

enterocolitica infections is still not completely known. Three modes of human contamination have been proposed.

1. Ingestion of contaminated food or water.
2. Contact with animals carrying *Y. enterocolitica*.
3. Direct or indirect transmission between man.

Numerous studies have indicated swine to be important reservoirs of O:3- and O:9-strains (Table XV). The porcine strains also exhibit the same biotype and phage type characteristics as those responsible for human infection. On this background, it has been suggested that swine is the major source of human contamination. In Norway, however, practically no cases of human yersiniosis have been traced to contamination from swine (J. Lassen, personal communication). In accordance with this statement, Mollaret *et al.* (1979) expressed the opinion that "no proof, direct or indirect, has been brought up to show human infection originating from animals". Such observations have inspired questions as to the stability of serological features within *Y. enterocolitica*, and have directed attention towards a possible clinical significance of serogroups prevailing in the environment (Mollaret *et al.*, 1979). It is possible that virulent strains may occasionally be derived from avirulent forms through uptake of the virulence plasmid and acquisition of other virulence determinants.

References

Acker, G., Knapp, W., Wartenberg, K. and Mayer, H. (1981). *J. Bacteriol.* **147**, 602–611.

Ahvonen, P. and Jansson, E. (1968). *Scand. J. Clin. Invest. Suppl. 101* **21**, 57.

Aldova, E. and Laznickova, K. (1979). *Contr. Microbiol. Immunol.* **5**, 122–131, 1979.

Aldova, E. and Lim, D. (1974). *Zentralbl. Bakteriol. Parasitenkd. Infekionskr. Abt. Orig. A* **226**, 491–496.

Aldova, E., Cerny, J. and Chmela, J. (1977). *Zentralbl. Bakteriol. Parasitenkd. Infekionskr. Abt. Orig. A* **239**, 208–212.

Aleksic, S. and Aleksic, V. (1979). *Zentralbl. Bakteriol. Parasitenkd. Infekionskr. Abt. Orig. A* **243**, 177–196.

Aleksic, S., Rohde, R., Müller, G. and Wohlers, B. (1976). *Zentralbl. Bakteriol. Parasitenkd. Infekionskr. Abt. Orig. A* **234**, 513–520, 1976.

Aleksic, S., Rohde, R. and Aleksic, A. (1978). *Zentralbl. Bakteriol. Parasitenkd. Infekionskr. Abt. Orig. A* **241**, 418–426, 1978.

Alonso, J. M., Bercovier, H., Servan, J. and Mollaret, H. H. (1979). *Contr. Microbiol. Immunol.* **5**, 132–143.

Anonymous (1977). *New York Morbid. Mortal. Weekly Rep.* **26**, 7.

Asakawa, Y., Akahane, S., Shiozawa, K. and Honma, T. (1979). *Contr. Microbiol. Immunol.* **5**, 115–121.

Ben-Gurion, R. and Shafferman, A. (1981). *Plasmid* **5**, 183–187.

Bercovier, H. and Carlier, J. P. (1979). *Ann. Microbiol. (Paris)* **130A**, 37–46.
Bercovier, H., Brault, J., Barre, N., Treignier, M., Alonso, J. M. and Mollaret, H. H. (1978). *Curr. Microbiol.* **4**, 201–206.
Bercovier, H., Brenner, D. J., Ursing, J., Steigerwalt, A. G., Fanning, G. R., Alonso, J. M., Carter, G. P. and Mollaret, H. H. (1980a). *Curr. Microbiol.* **4**, 201–206.
Bercovier, H., Mollaret, H. H., Alonso, J. M., Brault, J., Fanning, G. R., Steigerwalt, A. G. and Brenner, D. J. (1980b). *Curr. Microbiol.* **4**, 225–229.
Bercovier, H., Mollaret, H. H., Alonso, J. M., Brault, J., Fanning, G. R., Steigerwalt, A. G. and Brenner, D. J. (1980c). *Curr. Microbiol.* **4**, 225–229.
Bergan, T. (1978). *Methods Microbiol.* **12**, 25–36.
Bisset, M. L. (1979). *Contr. Microbiol. Immunol.* **5**, 159–168.
Black, R. E., Jackson, R. J., Trai, T., Medvesky, M., Shayegani, M., Feeley, J. C., MacLeod, K. I. E. and Wakelee, A. M. (1978). *N. Engl. J. Med.* **298**, 76–79.
Bottone, E. J. (1977). *CRC Crit. Rev. Microbiol.* **5**, 211–241.
Bottone, E. J. and Robin, T. (1979). *Contr. Microbiol. Immunol.* **5**, 95–105.
Bottone, E. J., Sandhu, K. K. and Pisano, M. A. (1979). *J. Clin. Microbiol.* **10**, 433–436.
Boyce, J. M., Evans, D. J., Evans, D. G. and DuPont, H. L. (1979). *Infect. Immunol.* **25**, 532–537.
Bradford, W. D., Noce, P. S. and Gutman, L. T. (1974). *Arch. Pathol.* **98**, 17–22.
Brenner, D. J. (1979). *Contr. Microbiol. Immunol.* **5**, 33–43.
Brenner, D. J., Steigerwalt, A. G., Falcão, D. P., Weaver, R. E. and Fanning, G. R. (1976). *Int. J. Syst. Bacteriol.* **26**, 180–194.
Brenner, D. J., Ursing, J., Bercovier, H., Steigerwalt, A. G., Fanning, G. F., Alonso, J. M. and Mollaret, H. H. (1980a). *Curr. Microbiol.* **4**, 195–200.
Brenner, D. J., Bercovier, H., Ursing, J., Alonso, J. M., Steigerwalt, A. G., Fanning, G. R., Carter, G. P. and Mollaret, H. H. (1980b). *Curr. Microbiol.* **4**, 207-12.
Carter, P. B., Zahorchak, R. J. and Brubaker, R. R. (1980). *Infect. Immun.* **28**, 638–640.
Christensen, S. G. (1980). *J. Appl. Bacteriol.* **48**, 377–382.
Corbel, M. J. (1975). *J. Hyg.* **75**, 151–171.
Corbel, M. J. (1979). *Contr. Microbiol. Immunol.* **5**, 50–63.
Cowan, S. T. (1974). *In* "Bergey's Manual of Determinative Bacteriology" (R. E. Buchanan and N. E. Gibbons, Eds), pp. 290–293. Williams & Wilkins, Baltimore, Maryland.
Cowan, S. T. and Steel, K. J. (1970). "Manual for the Identification of Medical Bacteria". Cambridge University Press, Cambridge.
Daniels, J. J. H. M. and Goudzwaard, C. (1963). *Tijdschr. Diergeneesk.* **88**, 96–102.
Dickinson, A. and Mocquot, G. (1961). *J. Appl. Bacteriol.* **24**, 252–284.
Doyle, M. P., Hugdahl, M. B. and Taylor, S. L. (1981). *Appl. Environ. Microbiol.* **42**, 661–666.
Dudley, M. V. and Shotts, E. B. (1979). *J. Clin. Microbiol.* **10**, 180–183.
Esseveld, H. and Goudzwaard, C. (1973). *Contr. Microbiol. Immunol.* **2**, 99–101.
Frederiksen, W. (1964). Proc. XIV Scand. Congr. Path. Microbiol. Oslo 25–27, 1964. Universitetsforlagets Trykningssentral. 103–104.
Fukushima, H., Ito, Y., Saito, K., Tsubokura, M. and Otsuki, K. (1979). *Appl. Environ. Microbiol.* **38**, 1009–1010.
Gemski, P., Lazere, J. R. and Casey, T. (1980). *Infect. Immun.* **27**, 682–685.
Gilbert, R. (1933). "Annual Report of the Division of Laboratories and Research", pp. 57–58. New York Department of Health, Albany.
Gilbert, R. (1939). "Annual Report of the Division of Laboratories and Research", pp. 45–46. New York State Department of Health, Albany.

Hamon, Y., Nicolle, P., Vieu, J. F. and Mollaret, H. H. (1966). *Ann. Inst. Pasteur* **111**, 368–372.

Hässig, A., Karrer, J. and Pusterla, A. F. (1949). *Schweiz. Med. Wschr.* **79**, 971–973.

Hausnerova, S., Hausner, O. and Pauckova, V. (1973). *Contr. Microbiol. Immunol.* **2**, 76–80.

Hughes, D. and Jensen, N. (1981). *Appl. Environ. Microbiol.* **41**, 309–310.

Hurvell, B., Glatthard, V. and Thal, E. (1979). *Contr. Microbiol. Immunol.* **5**, 243–248.

Jantzen, E. and Lassen, J. (1980). *Int. J. Syst. Bacteriol.* **30**, 421–428.

Juhlin, I. and Ericson, C. (1961). *Acta Pathol. Microbiol. Scand.* **52**, 185–200.

Kanazawa, Y. and Ikemura, K. (1979). *Contr. Microbiol. Immunol.* **5**, 106–114.

Kaneko, K., Hamada, S. and Kato, E. (1977). *Jpn. J. Vet. Sci.* **39**, 407–414.

Kapperud, G. (1975). *Acta Pathol. Microbiol. Scand. Sect. B* **83**, 335–342.

Kapperud, G. (1977). *Acta Pathol. Microbiol. Scand. Sect. B* **85**, 129–135.

Kapperud, G. (1980a). *Acta Pathol. Microbiol. Scand. Sect. B* **88**, 287–291.

Kapperud, G. (1980b). *Acta Pathol. Microbiol. Scand. Sect. B* **88**, 293–297.

Kapperud, G. (1980c). *Acta Pathol. Microbiol. Scand. Sect. B* **88**, 303–306.

Kapperud, G. (1980d). "*Yersinia enterocolitica* and *Yersinia enterocolitica*-like Bacteria: Aspects of Ecology, Pathogenicity and Taxonomy". Ph.D. Thesis, University of Oslo.

Kapperud, G. (1981). *Acta Pathol. Microbiol. Scand. Sect. B* **89**, 29–35.

Kapperud, G. (1982). *Acta Pathol. Microbiol. Scand. Sect. B* **90**, 185–189.

Kapperud, G. and Jonsson, B. (1978). *Med. Malad. Infect.* **8**, 500–506.

Kapperud, G. and Langeland, G. (1981). *Curr. Microbiol.* **5**, 119–122.

Kapperud, G., Bergan, T. and Lassen, J. (1981). *Int. J. Syst. Bacteriol.* **31**, 401–419.

Knapp, W. and Masshoff, W. (1954). *Dtsch. Med. Wschr.* **79**, 1266–1271.

Knapp, W. and Thal, E. (1963). *Zentralbl. Bakteriol. Parisitenkd. Infekionskr. Abt. Orig.* **190**, 472–484.

Knapp, W. and Thal, E. (1973). *Contr. Microbiol. Immunol.* **2**, 10–16.

Lafleur, L., Hammerberg, O., Delage, G. and Pai, C. H. (1979). *Contr. Microbiol. Immunol.* **5**, 298–303.

Laird, W. J. and Cavanaugh, D. C. (1980). *J. Clin. Microbiol.* **11**, 430–432.

Lassen, J. (1972). *Scand. J. Infect. Dis.* **4**, 125–127.

Lee, W. H. (1977). *Food Protect.* **40**, 486–489.

Lee, W. H., McGrath, P. P., Carter, P. H. and Eide, E. L. (1977). *Can. J. Microbiol.* **23**, 1714–1722.

Lee, W. H., Smith, R. E., Damere, J. M., Harris, M. E. and Johnston, R. W. (1981). *J. Appl. Bacteriol.* **50**, 529–539.

Le Gendre, G. and Johnson, R. (1981). Abstract 213, Annual Meeting, American Society for Microbiology.

Lehmann, K. B. and Neumann, R. (1896). "Atlas und Grundriss der Bakteriologie und Lehrbuch der Speciellen Bacteriologischen Diagnostik". J. F. Lehmann, München.

Le Minor, L. (1979). *Contr. Microbiol. Immunol.* **5**, 1–7.

MacLagan, R. M. and Old, D. C. (1980). *J. Appl. Bacteriol.* **49**, 353–360.

Mäkelä, P. H. and Mayer, H. (1976). *Bacteriol. Rev.* **40**, 591–632.

Malassez, L. and Vignal, W. (1883). *Arch. Physiol. Normale Pathol. Sect. 3* **2**, 369–412.

Martin, W. J. and Washington, J. A. (1980). *In* "Manual of Clinical Microbiology" (E. H. Lennette, Ed.), pp. 195–219. American Society for Microbiology, Washington, D.C.

Maruyama, T. (1973). *Jpn. J. Bacteriol.* **28**, 413–422.

Masshoff, W. and Dolle, W. (1953). *Virchow, Arch. Pathol. Anat.* **323**, 664–694.

Mollaret, H. H. (1976). *Med. Mal. Infect.* **6**, 442–448.

Mollaret, H. H. and Destombes, P. (1964). *Presse Med.* **92**, 2913–2915.

Mollaret, H. H. and Nicolle, P. (1965). *C.R. Acad. Sci.* **260**, 1027–1029.

Mollaret, H. H. and Thal, E. (1974). *In* "Bergey's Manual of Determinative Bacteriology" (R. E. Buchanan and N. E. Gibbons, Eds), pp. 330–332. Williams & Wilkins, Baltimore, Maryland.

Mollaret, H. H., Alonso, J. M. and Bercovier, H. (1976). *Med. Mal. Infect.* **6**, 102–107.

Mollaret, H. H., Bercovier, H. and Alonso, J. M. (1979). *Contr. Microbiol. Immun.* **5**, 174–184.

Moore, R. L. and Brubaker, R. R. (1975). *Int. J. Syst. Bacteriol.* **25**, 336–339.

Morris, G. K. and Feeley, J. C. (1976). *Bull. W.H.O.* **54**, 79–85.

Nicolle, P., Mollaret, H. and Brault, J. (1973). *Microbiol. Immunol.* **2**, 54–58.

Niléhn, B. (1969). *Acta Pathol. Microbiol. Scand. Suppl.* **206**, 1–48.

Niléhn, B. (1973). *Microbiol. Immunol.* **2**, 59–67.

Ogata, S., Kanamori, M. and Miyashita, K. (1979). *Contr. Microbiol. Immunol.* **5**, 64–72.

Okamoto, K., Inoue, T., Ichikawa, H., Kawamoto, Y. and Miyama, A. (1981). *Infect. Immun.* **31**, 554–559.

Olsson, E., Krovacek, K., Hurvell, B. and Wadström, T. (1980). *Curr. Microbiol.* **3**, 267–271.

Oosterom, J. (1979). *Antonie van Leeuwenhoek* **45**, 630–633.

Pai, C. H., Mors, V. and Seemayer, T. A. (1980). *Infect. Immun.* **28**, 238–244.

Pai, C. H., Mors, V. and Toma, S. (1978). *Infect. Immun.* **22**, 334–338.

Pedersen, K. B. (1979). *Contr. Microbiol. Immunol.* **5**, 253–256.

Pedersen, K. B. and Winblad, S. (1979). *Acta Pathol. Microbiol. Scand. Sect. B* **87**, 137–140.

Pedersen, K. B., Winblad, S. and Bitsch, V. (1979). *Acta Pathol. Microbiol. Scand. Sect. B* **87**, 141–145.

Pfeiffer, A. (1889). "Uber die bacilläre Pseudotuberculose bei Nagetieren", pp. 1–42. Thieme, Leipzig.

Portnoy, D. A., Moseley, S. L. and Falkow, S. (1981). *Infect. Immun.* **31**, 775–782.

Quan, T. J. (1979). *Contr. Microbiol. Immunol.* **5**, 83–87.

Rahimian, F. and Evans, Z. A. (1981). Abstracts of the Annual Meeting of the American Society for Microbiology B53, p. 23.

Restaino, L., Grauman, G. S., McCall, W. A. and Hill, W. M. (1979). *J. Food Protect.* **42**, 120–123.

Rische, H., Beer, W., Seltmann, G., Thal, E. and Horn, G. (1973). *Contr. Microbiol. Immunol.* **2**, 23–26.

Robins-Browne, R. M., Still, C. S., Miliotis, M. D. and Koornhof, H. J. (1979). *Infect. Immun.* **25**, 680–684.

Saari, T. N. and Jansen, G. P. (1979). *Contr. Microbiol. Immunol.* **5**, 185–196.

Sandhu, *et al.* (1980). Annual Meeting of the American Society of Microbiology, Abstract No. 154.

Schiemann, D. A. (1979). *Contr. Microbiol. Immunol.* **5**, 212–227.

Schiemann, D. A. (1980). *J. Food Protect.* **43**, 360–365.

Schiemann, D. A., Devenish, J. A. and Toma, S. (1981). *Infect. Immun.* **32**, 400–403.

Schleifstein, J. and Coleman, M. (1943). Annual Report of the Division of Laboratory and Research. New York State Department of Health, p. 56.

Shayegani, M., Menegio, E. J., McGlynn, D. M. and Gaafar, H. A. (1979). *Contr. Microbiol. Immunol.* **5**, 196–205.

Skerman, V. B. D., McGowan, V. and Sneath, P. H. A. (1980). *Int. J. Syst. Bacteriol.* **30**, 225–420.

Smith, J. E. and Thal, E. (1965). *Acta Pathol. Microbiol. Scand.* **64**, 213–223.

Sneath, P. H. A. and Cowan, S. T. (1958). *J. Gen. Microbiol.* **19**, 551.
Straley, S. C. and Brubaker, R. R. (1981). *Proc. Natl. Acad. Sci. U.S.A.* **78**, 1224–1228.
Szita, J. and Svidro, A. (1976). *Acta Microbiol. Acad. Sci. Hung.* **23**, 191–203.
Talbot, J. M. and Sneath, P. H. A. (1960). *J. Gen. Microbiol.* **22**, 303–311.
Thal, E. (1954). "Untersuchungen über *Pasteurella pseudotuberculosis*". Thesis, University of Lund.
Thal, E. and Knapp, W. (1971). *Symp. Ser. Immunobiol. Standard* **15**, 219–226.
Toma, S., Lafleur, L. and Deidrick, V. R. (1979). *Contr. Microbiol. Immunol.* **5**, 144–149.
Une, T. (1977a). *Microbiol. Immunol.* **21**, 349–363.
Une, T. (1977b). *Microbiol. Immunol.* **21**, 365–377.
Une, T. (1977c). *Microbiol. Immunol.* **21**, 727–729.
Une, T., Zen-Yoji, H., Maruyama, T. and Yanagawa, Y. (1977). *Microbiol. Immunol.* **21**, 727–729.
Ursing, J., Brenner, D. J., Bercovier, H., Fanning, G. R., Steigerwalt, A. G., Brault, J. and Mollaret, H. H. (1980a). *Curr. Microbiol.* **4**, 213–217.
Ursing, J., Steigerwalt, A. G. and Brenner, D. J. (1980b). *Curr. Microbiol.* **4**, 231–233.
Vandepitte, J. and Wauters, G. (1979). *Contr. Microbiol. Immunol.* **5**, 150–158.
Van Loghem, J. J. (1944). *Antonie van Leeuwenhoek.* **10**, 15–15.
Van Noyen, R., Vandepitte, J. and Selderslaghs, R. (1979). *Contr. Microbiol. Immunol.* **5**, 283–291.
Van Noyen, R., Vandepitte, J. and Wauters, G. (1980). *J. Clin. Microbiol.* **11**, 127–131.
Vesikari, T., Nurmi, T., Mäki, M., Skurnik, M., Sundqvist, C., Granfors, K. and Grönroos, P. (1981). *Infect. Immun.* **33**, 870–876.
Vidon, D. J. M. and Delmas, C. L. (1981). *Appl. Environ. Microbiol.* **41**, 355–359.
Wartenberg, K., Lysy, J. and Knapp, W. (1975). *Zentralbl. Bakteriol. Parasitenkd. Infekionskr. Abt. Orig. A* **230**, 361–366.
Wauters, G. (1970). "Contribution a l'Étude de *Yersinia enterocolitica*", pp. 1–165. Vander, Louvain.
Wauters, G. (1973). *Contr. Microbiol. Immunol.* **2**, 68–70.
Wauters, G. (1979). *Contr. Microbiol. Immunol.* **5**, 249–252.
Wauters, G. (1981). *In* "*Yersinia enterocolitica*" (E. J. Bottone, Ed.), pp. 41–53. CRC Press, Boca Raton, Florida.
Wauters, G. and Pohl, P. (1972). *Rev. Ferment. Ind. Aliment.* **7**, 16.
Wauters, L., Le Minor, L. and Chalon, A. M. (1971). *Ann. Inst. Pasteur* **120**, 631–642.
Wauters, L., Le Minor, L., Chalon, A. M. and Lassen, J. (1972). *Ann. Inst. Pasteur* **122**, 951–956.
Weissfeld, A. S. and Sonnenwirth, A. C. (1981). *J. Pediatr.* **98**, 504–505.
Winblad, S. (1967). *Acta Pathol. Microbiol. Scand. Suppl.* **187**, 1–115.
Winblad, S. (1968). The epidemiology of human infections with *Yersinia pseudotuberculosis* and *Yersinia enterocolitica* in Scandinavia. Rend. Symp. Internat. Pseudo-tuberculose, pp. 133–136. Karger, Basel.
Winblad, S. (Ed.) (1973). Proceedings of the International Symposium held in Malmö, Sweden, April 10–12, 1973.
Winblad, S. (1978). *Methods in Microbiol.* **12**, 37–50.
WHO Scientific Working Group (1980). *Bull. W.H.O. Org.* **58**, 519–537.
Yersin, A. (1894). *Ann. Inst. Pasteur (Paris)* **8**, 662–667.
Zen-Yoji, H., Sakai, S., Maruyama, T. and Yanagawa, Y. (1974). *Jpn. J. Microbiol.* **18**, 103–105.
Zink, D. L., Feeley, J. C., Wells, J. G., Vanderzant, C., Vickery, J. C., Roof, W. D. and O'Donovan, A. (1980). *Nature (London)* **283**, 224–226.

Note added in proof

Sergrouping scheme for *Y. enterocolitica* as developed primarily by Wauters

Reference strain for immunization		Reference strains for absorption (Strain No.[a])	Factor sera obtained by absorption
O-antigens	Strain No.[a]		
1, 2a, 3	135	178, 134	1/1, 2a[b]
2a, 2b, 3	178	134, 134, 178	2a/2b/2a, 2b/3[b]
3	134	—	—
4, 32	96	1476	32
4, 33	1476	96	33
5	124	—	—
5, 27	885	124	27
6, 30	102	1477	30
6, 31	1477	102	31
7, 8	106	161	7
8	161	—	—
9	383	—	—
10	500	—	—
11, 23	105	841	23
11, 24	841	105	24
12, 25	490	103	25
12, 26	103	490	26
13, 7	553	106	13
14	480	—	—
15	614	—	—
16	1475	—	—
16, 29	867	1475	29
17	955	—	—
18	846	—	—
19,8	842	161	19
20	845	—	—
21	1110	—	—
22	1367	—	—
28	1474	—	—
35	3842	—	—
36	2222	—	—
37	7224	—	—
38	7175	—	—
39	7142	—	—
40	2677	—	—
41, 42	2223	3235	42
41, 43	3235	2223	43
44	7146	—	—
44, 45	7210	7146	45
46	7320	—	—
47	7184	—	—
48	3960	—	—
49, 51	7231	7229	49
50, 51	7229	7231	50
52	7209	—	—
52, 53	2842	7209	53
52, (54)	2835	7209	(54)
55	—	—	—
57	—	—	—

Adapted from Wauters (1970, 1981) and Wauters *et al.* (1972).

[a] Strain designations refer to collection reference numbers, Pasteur Institute, National Yersinia Center, Paris.

[b] The phenotypic expression of the O-antigen determinants 1, 2a, 2b and 3 (reference strains 135 and 178) is influenced by growth temperature. Factors 1, 2a and 2b are dominant at 22°C whereas factor 3 prevails above 28°C. This circumstance allows production of antisera with different specificities according to the growth temperature of the strains used for immunization and absorption.

7

Gas–Liquid Chromatography for the Assay of Fatty Acid Composition in Gram-negative Bacilli as an Aid to Classification

T. BERGAN AND K. SØRHEIM

Department of Microbiology, Institute of Pharmacy, University of Oslo, Oslo, Norway

METHODS IN MICROBIOLOGY
VOLUME 15 ISBN 0–12–521515–0

I. Introduction

Gas-liquid chromatography (GLC) has been used successfully for the characterization fatty acid (FA) and carbohydrate composition of bacteria and significant correlation with classification has been shown between FA composition and classification (Mitruka, 1975; Shaw, 1974) in Gram-positive (Jantzen et al., 1974c) as well as Gram-negative strains (Jantzen et al., 1974b, 1975). Shaw (1974) presented a composite list of the lipid composition as a guide to bacterial identification. GLC is a method with considerable potential for rapid diagnosis and is now employed routinely for classification of anaerobic bacteria in respect to metabolites derived from glucose (Sutter et al., 1980).

Since fast-growing non-fastidious Gram-negative organisms constitute the major diagnostic load of medical bacteriology, and strains which are biotypically intermediate between recognized species regularly pose diagnostic problems for the practising microbiologist, we decided to probe the usefulness of fatty acids for the classification of Gram-negative aerobic bacilli. This chapter describes a method for the analysis of bacterial fatty acids and presents the relationship between fatty acid profiles of individual strains assessed by numerical clustering procedures.

II. Materials and methods

A. Bacterial strains

The bacterial species and number of the 167 strains of Gram-negative bacteria are shown in Table I. The strains were mostly fresh isolates from specimens of routine medical microbiology classified by biochemical–cultural determinative methods as presented by Lassen (1975) and supplemented by additional tests where further differentiation was necessary as shown by Finegold and Martin (1982). Bacteria like *Salmonella*, *Shigella* or *Vibrio cholerae* were excluded from the study for reasons of laboratory safety.

B. Cultivation

Standardization of growth conditions is a crucial point if the data are to be employed in classification. In slow-growing pediococci, for instance, the relative contribution of the cellular FAs reach a stable niveau only after three days of incubation (Uchida and Iogi, 1972). The composition of the medium used is an important determinative factor. But by employing a strictly standardized growth medium and growth conditions, the reproducibility is

TABLE I
Bacterial strains studied

Species	Strain designation
Pseudomonas aeruginosa	41, 92, 93, 123
Alcaligenes faecalis	153
Escherichia coli	4, 5, 6, 11, 18, 19, 21, 25
	26, 27, 29, 31, 33, 34, 35,
	40, 49, 50, 52, 54, 55, 59,
	62, 87, 97, 98, 107, 109, 111,
	112, 119, 120, 126, 127, 132,
	140, 141, 142, 143, 147, 149,
	151, 152, 158, 161, 164, 166,
	188
Citrobacter freundii	61, 75, 94, 125, 157, 176,
	198, 199, 200
Citrobacter sp.	84
Klebsiella aerogenes	8, 10, 14, 36, 51, 57, 60,
	65, 66, 113, 122, 129, 195
K. oxytoca	9, 17, 32, 47, 53, 67, 146,
	196
K. ozaenae	15, 68, 104, 116, 160
K. rhinoscleromatis	193, 194
Enterobacter cloacae	43, 63, 124, 133, 139, 148,
	159, 162
E. aerogenes	7, 30, 48, 96, 150, 155, 156
E. hafniae	117, 189, 190
Serratia liquefaciens	64, 100
S. rubidae	85, 197
Proteus vulgaris	20, 38, 102, 110, 115, 165
P. mirabilis	16, 24, 39, 42, 45, 46, 58,
	105, 106, 108, 128, 131, 145
P. morganii	23, 44, 88, 114
P. rettgeri	90
P. inconstans	130, 184, 185, 186, 187
Yersinia enterocolitica	69, 70, 71, 179, 180, 181,
	182, 183
Flavobacterium sp.	118, 135, 154
Moraxella sp.	167
Acinetobacter calcoaceticus	12, 28, 101, 103, 137, 138,
	114, 163, ATCC 11171, BD4
Non-classifiable enterobacteria	22, 89

satisfactory. A correlation of the fatty acid profile in repeated experiments on the same strains of Gram-negative rods has been good, as reflected by a similarity index of 0.98 or higher (Jantzen *et al.*, 1974b).

A complete medium which renders optimal growth conditions for these bacteria was used: blood agar [proteose peptone (Oxoid)] 1.5 g, sodium chloride 0.5 g, liver extract (Oxoid) 0.25 g, citrated horse blood 6%. The microbes were grown in a saturated humid atmosphere at 33°C for 20 h (exactly), harvested by saline, freeze dried, and stored in tightly stoppered tube filled with nitrogen at −20°C to stabilize the FAs during storage.

C. Derivatization

Minimum preparation of bacterial growth (cells) is necessary for pyrolysis GLC, but this produces complex chromatogram profiles and therefore requires comparison with a data bank of stored profiles to enable efficient employment (Gutteridge and Norris, 1979). We decided to determine the relative contribution of each FA.

The procedure used for derivatization, ethanolysis and extraction in hexan for GLC analysis has been described (Jantzen et al., 1974a). Seven to 12 mg of dried bacterial cells of each strain were transferred to a teflon screw-capped tube. An aliquot of 3 ml of 2 N HCl in anhydrous methanol [3 N HCl in methanol (spectroscopic grade, p. nr. 244-7, Supelco, Bellafonte, USA) was diluted to 2 N HCl with anhydrous methanol (spectroscopic grade, p. nr. 5276502, E. Merck, Darmstadt, Western Germany)] was added. The gaseous phase of the tubes was replaced by N_2 and other tubes placed at 85°C for 20 min (to methanolyse the FAs). Centrifugation was in some instances necessary to collect cell debris. The clear supernatant was removed and reduced to half its volume by a stream of N_2 and subsequently washed with methanol to remove HCl. The methyl fatty acid esters were extracted by agitating the contents three times with 1 ml of hexan. The three hexan portions were mixed, the liquid phase evaporated with a stream of N_2 and the volume adjusted with hexan to 100 μl. The hexan extracts were stored in sealed capillary tubes at −20°C.

We used a Hewlett-Packard gas chromatograph (model 5830A) with an automatic integrator of peak area (Model 18850 A GC terminal), with a carrier gas of N_2 (99.9% pure, Norgas, Oslo, Norway) at a flow of 20 ml min^{-1}. Both injector port and detector temperatures were 300°C. A 2-m long column 2 mm in inner diameter was used with 10% UCW 98 on Gas Chrom Q as the polar phase and 10% EGA-PS on Chromosorb WAW as the non-polar column material, both 80/100 mesh (Supelco). The column temperature range with UCW-98 was 140–240°C for the enterobacteria and 110–240°C for the oxidative organisms. All bacteria were examined over the range 140–210°C on EGA-PS. The temperature increment was programmed at 2°C min^{-1}. Preliminary assays with selected control strains of each species were carried out to determine the suitable temperature ranges in respect to the FA components. Results were recorded as percentage peak area of each FA in relation to the total.

D. Identification

The FAs were identified by comparisons of the retention of commercially available standards, GCL-10 4-7038, GCL-90 4-7046, A-NHI-C 4-7010 (Supelco) of previously studied bacteria of known FA composition (19, 20, 21).

The subsequently used terminology of the fatty acids lists the number of carbon atoms before the colon and number of unsaturated bonds after, e.g. C12:1 means 12 carbons and one double bond.

E. Numerical analysis

Similarity between the FA profiles of the individual strains was efficiently analysed by numerical grouping. To accommodate the quantitative data of the fatty acid composition, a Yule correlation coefficient was employed as a similarity index, and an unweighted mean pair group cluster analysis (of Sokal-Michener) used to generate phenetic dendrograms as employed was described previously (Jantzen et al., 1974b). The data from GLC were transformed according to the formula $y = ln(x + 1)$, where x represents the relative contents of the FAs and y is the computer input.

III. Results

A. Total collection

The fatty acid composition of selected bacterial strains appears in Figs 1–11. Comparisons between the bacterial isolates do require that the FA peaks are dependably differentiated, although chemical identification is not mandatory, since the primary purpose of the study was detection of the taxonomic relationships. When components differed from commercially available standards, mass spectrometry was, consequently, not employed to finalize the peak identity. Accordingly, we have listed unidentified component peaks as an X followed by serial number as done previously (Jantzen et al., 1974b,c, 1975).

The predominant components of all species were C16:0, C16:1 and C18:1. Always present were also C12:0, C14:0, C15:0 and C17:0, C17:1.

B. Escherichia coli

The enterobacteria are relatively homogenous. A GLC profile of a strain of E. coli is shown in Fig. 1.The contribution of C15:0 varies between 2% and 20%,

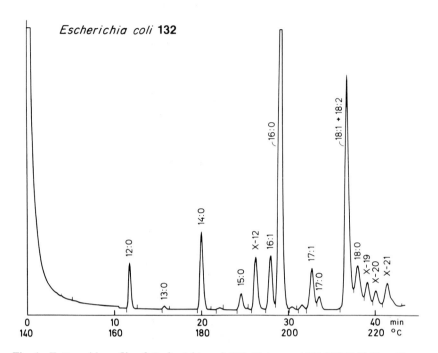

Fig. 1. Fatty acid profile of *Escherichia coli* 132. Column: 10% UCW-98/Gas Chrom Q, 200 × 0.2 cm; column temperature 140–240°C with an increment of 2°C min^{-1}; carrier gas: N$_2$ at 20 ml min^{-1}.

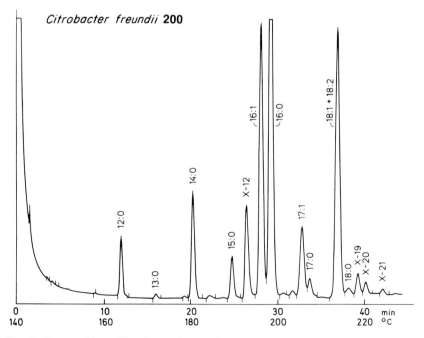

Fig. 2. Fatty acid profile of *Citrobacter freundii* 200. Column: 10% UCW-98/Gas Chrom Q, 200 × 0.2 cm; column temperature 140–240°C with an increment of 2°C min^{-1}; carrier gas: N$_2$ at 20 ml min^{-1}.

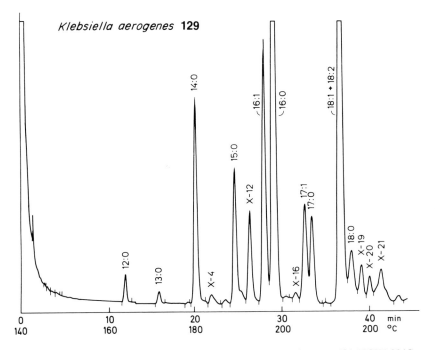

Fig. 3. Fatty acid profile of *Klebsiella aerogenes* 129. Column: 10% UCW-98/Gas Chrom Q, 200×0.2 cm; column temperature 140–240°C with an increment of 2°C min^{-1}; carrier gas: N$_2$ at 20 ml min^{-1}.

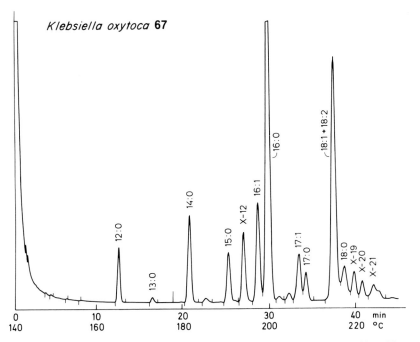

Fig. 4. Fatty acid profile of *Klebsiella oxytoca* 67. Column: 10% UCW-98/Gas Chrom Q, 200×0.2 cm; column temperature 140–240°C with an increment of 2°C min^{-1}; carrier gas: N$_2$ at 20 ml min^{-1}.

mostly 1–5%. The relative amounts of C17:0 and C17:1 vary, but they total less than C18:1 and C18:2. There is more C16:1 than X-15. Profiles similar to those observed for *E. coli* are also exhibited by other enterobacterial species.

C. *Citrobacter*

The FA profiles resemble those of *E. coli*, although less variation is seen among *Citrobacter* and the contribution of X-15 is more moderate (Fig. 2).

D. *Klebsiella*

The *Klebsiella* profiles also resemble those of *E. coli*, except that there is less C12:0 and C13:0 in *K. aerogenes* (Fig. 3). *K. oxytoca* appears to have a higher amount of C12:0 (Fig. 4), similar to the situation in *E. coli*. This is interesting considering the fact that these two species both share the ability to produce indol. *K. ozaenae* resemble *E. coli*, whereas *K. rhinoscleromatis* differs from all other klebsiellae by having little C18:2. The klebsiella species have less than 1% of the FAs with fewer (14) carbon atoms.

E. *Enterobacter*

The *Enterobacter* strains (Fig. 5) FA profiles resemble those of *E. coli*, *K. aerogenes* and *K. oxytoca*.

F. *Serratia*

The FA composition of *S. rubidae* (Fig. 6) and *S. liquefaciens* are similar, although the ratio between C16:0 and C16:1 is higher in *S. rubidae* than in *S. liquefaciens*. There is more C15:0 in *S. rubidae* than in *S. liquefaciens*.

G. *Proteus*

P. mirabilis, *P. vulgaris* (Fig. 7), *P. rettgeri* and *P. inconstans* have small contributions of components with less than C14:0. *P. morganii* (Fig. 8) resembles *K. oxytoca* in the initial 2/3 of the components.

H. *Yersinia*

Y. enterocolitica (Fig. 9) is separated into two groups on the basis of their C14:0 contents, which is either below 1.5% or above 3%.

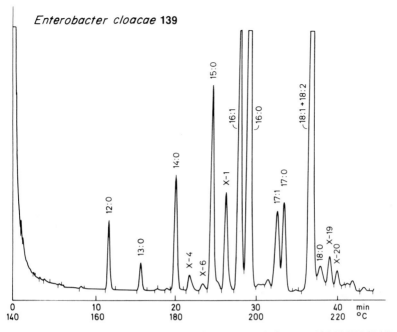

Fig. 5. Fatty acid profile of *Enterobacter cloacae* 139. Column: 10% UCW-98/Gas Chrom Q, 200×0.2 cm; column temperature 140–240°C with an increment of 2°C min^{-1}; carrier gas: N$_2$ at 20 ml min^{-1}.

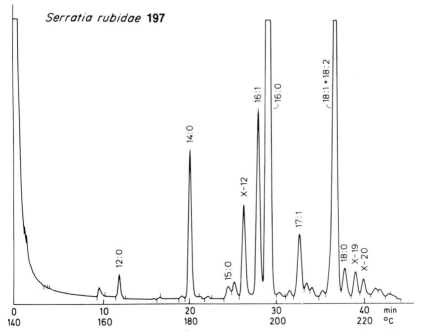

Fig. 6. Fatty acid profile of *Serratia rubidae* 197. Column: 10% UCW-98/Gas Chrom Q, 200×0.2 cm; column temperature 140–240°C with an increment of 2°C min^{-1}; carrier gas: N$_2$ at 20 ml min^{-1}.

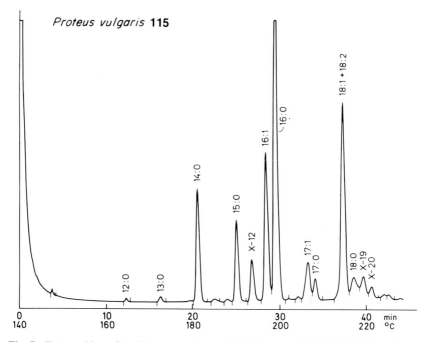

Fig. 7. Fatty acid profile of *Proteus vulgaris* 115. Column: 10% UCW-98/Gas Chrom Q, 200 × 0.2 cm; column temperature 140–240°C with an increment of 2°C min⁻¹; carrier gas: N₂ at 20 ml min⁻¹.

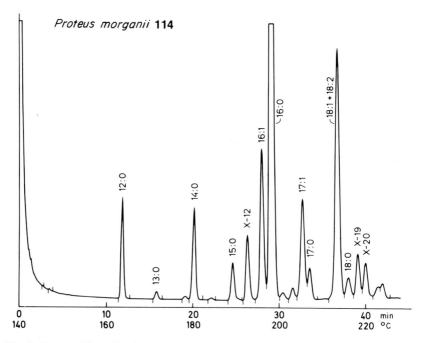

Fig. 8. Fatty acid profile of *Proteus morganii* 114. Column: 10% UCW-98/Gas Chrom Q, 200 × 0.2 cm; column temperature 140–240°C with an increment of 2°C min⁻¹; carrier gas: N₂ at 20 ml min⁻¹.

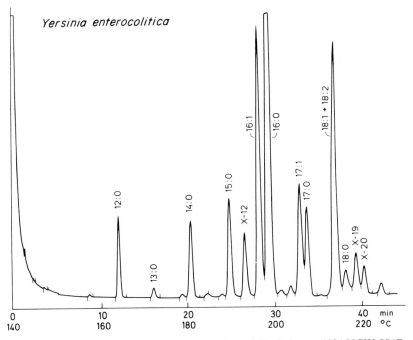

Fig. 9. Fatty acid profile of *Yersinia enterocolitica* 180. Column: 10% UCW-98/Gas Chrom Q, 200 × 0.2 cm; column temperature 140–240°C with an increment of 2°C min^{-1}; carrier gas: N$_2$ at 20 ml min^{-1}.

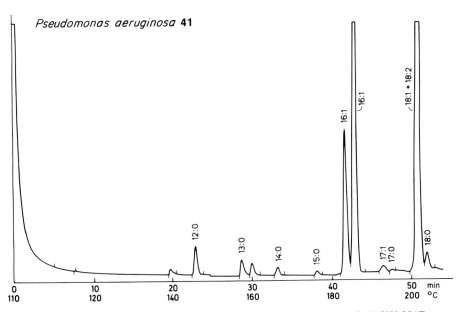

Fig. 10. Fatty acid profile of *Pseudomonas aeruginosa* 41. Column: 10% UCW-98/Gas Chrom Q, 200 × 0.2 cm; column temperature 110–240°C with an increment of 2°C min^{-1}; carrier gas: N$_2$ at 20 ml min^{-1}.

I. *Pseudomonas aeruginosa*

The four strains of *P. aeruginosa* (Fig. 10) have similar FA profiles. C12:0, C16:0, C16:1 and C18:1 dominate, whereas only modest amounts of C17:0 and C17:1 are found.

J. *Acinetobacter calcoaceticus*

The elution profiles of *A. calcoaceticus* (Fig. 11) resemble those of *P. aeruginosa*, but differ markedly from the enterobacteriae. The major components are C16:0, C16:1 and C18:0. All the pseudomonads and acinetobacters lack X-16, have a number of extra unidentified peaks (18:1) than in the enterobacteria. The only exception is strain No. 28, which although it lacks X-16 like the rest, it is the only *Acinetobacter* with X:15, and it has a C18:1 contents, which resembles the pattern of the enterobacteria.

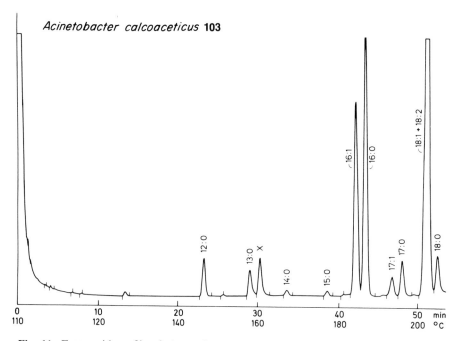

Fig. 11. Fatty acid profile of *Acinetobacter calcoaceticus* 103. Column: 10% UCW-98/Gas Chrom Q, 200 × 0.2 cm; column temperature 110–240°C with an increment of 2°C min^{-1}; carrier gas: N$_2$ at 20 ml min^{-1}.

K. *Alcaligenes faecalis* and *Moraxella*

In the single strains of *A. faecalis* and *Moraxella*, the profiles are in essence indistinguishable from the range among the enterobacteria. The *Alcaligenes* has a high ratio between C16:1 and C16:0. The contents of C17:0 and C18:1 are low.

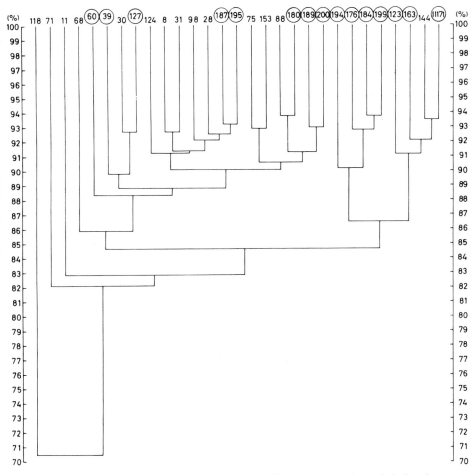

Fig. 12. Overview dendrogram of 167 strains of Gram-negative bacteria belonging to the Enterobacteriaceae and oxidative species. Dendrogram generated by the Yule correlation coefficient as similarity index and the unweighted pair group mean clustering procedure. Numbers refer to strains designations. The encircled numbers represent subgroups of bacteria (highest strain number in cluster indicated) of which dendrogram details are shown in subsequent figures.

L. *Flavobacterium*

The chromatograms of the three strains of flavobacteria were completely different. The profile of *Flavobacterium* strain No. 154 resembled that rendered by occasional enterobacteria. The profile of *Flavobacterium* strain No. 135 was not unlike that of an acinetobacter. In *Flavobacterium* No. 118, C18:1 contributed around 75%, making the profile different from all other strains examined.

M. *Numerical analysis*

For practical reasons, the dendrogram pattern of the entire collection of strains is presented for phenons above a similarity level of 94% in Fig. 12 and the further structure above 94% presented in the dendrograms of Figs 13–17 to show the finer distinctions.

The enterobacteria have a high degree of similarity with respect to FA profiles. Consequently, the high level clusters all tend to contain strains of several genera, generally without species distinction (Figs 13 and 14). There are two major exceptions among the enterobacteria. The *Y. enterocolitica* strains were gathered in one cluster (Fig. 15). The group designated 187 (signifying the circumstance that strain No. 187 is the one with the highest serial number in that cluster) is constituted mainly of *Proteus* isolates, but proteae are also clustered in phenons containing other species of enterobacteria (Fig. 16).

The strains of *Acinetobacter* and *P. aeruginosa* constitute distinct clusters (Fig. 17) separable from the enterobacteria, and there is a relatively high level of similarity between these two oxidative genera. The strain *A. calcoaceticus* No. 28 differs from the other strains of that species and is shown separately in Fig. 12. The variability among the flavobacteria is reflected by the fact that one of these strains was clustered together with strains of *Acinetobacter*.

IV. Discussion

The bacteria within the family Enterobacteriaceae are homogeneous in respect to fatty acid composition. Enterobacteria produce more FAs than the oxidative Gram-negative rods. This parallels the higher metabolic activity of enterobacteria. The similarity in FA composition may be related to the extended ability of transfer of extrachromosomal genetic material between different genera within enterobacteria. The strains of *E. coli* were distributed among clusters containing strains of all other enterobacteria species except *Y. enterocolitica*. There was no FA distinction between lactose-positive and

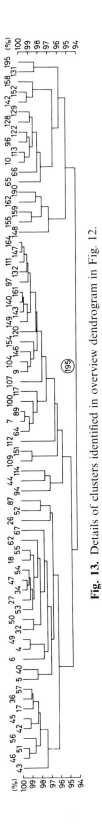

Fig. 13. Details of clusters identified in overview dendrogram in Fig. 12.

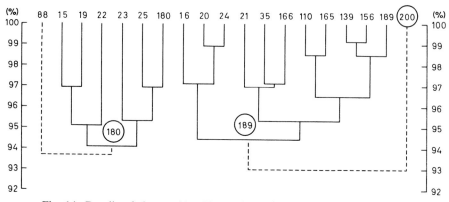

Fig. 14. Details of clusters identified in overview dendrogram in Fig. 12.

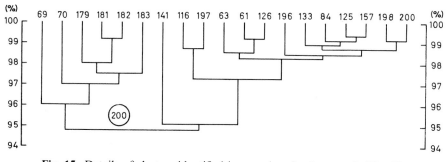

Fig. 15. Details of clusters identified in overview dendrogram in Fig. 12.

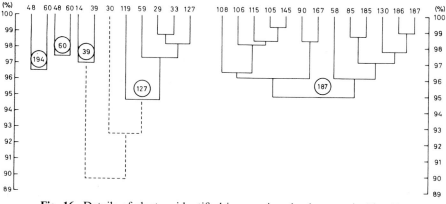

Fig. 16. Details of clusters identified in overview dendrogram in Fig. 12.

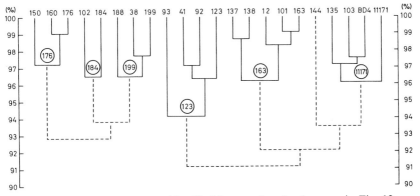

Fig. 17. Details of clusters identified in overview dendrogram in Fig. 12.

negative strains. The isolates which would formerly have been classifiable as Alcalescens-Dispar were also inseparable.

Only few exceptions of genus-specific FA contents were observed within enterobacteria. Thus, *Y. enterocolitica* and in part the *Proteus* strains were sufficiently homogeneous to produce species-specific clusters. Differentiation among the *Proteus* species has been presented previously (Vasynreko *et al.*, 1975).

The degree to which clusters were formed by strains of several enterobacterial species compare well with the results obtained from numerical taxonomy of biochemical–cultural properties of the species as demonstrated in extensive studies (Colwell *et al.*, 1974; Johnson *et al.*, 1975). Species separation of enterobacteria by traditional cultural differential methods is often not straightforward (Veron and Le Minor, 1975). DNA hybridization has demonstrated some degree of overlapping between the enterobacterial species (Colwell *et al.*, 1974). The relative FA profile similarity of most Enterobacteriacea species would, consequently, appear to reflect the comparatively close interspecies relationship recognized within this family.

The FA contents of *Proteus* strains correlate with previous findings (Mitruka, 1975), as do the findings on *Acinetobacter* (Jantzen *et al.*, 1975). The proteae tended to form their own cluster, although some of the strains merge with the body of the Enterobacteriacea. The acinetobacter FA profiles were separable from those of *Moraxella* (Jantzen *et al.*, 1975). Because few strains of the oxidative species were included in this study, the patterns of the ones studied are not understood as representing the variety of typical patterns of the respective taxons to the same extent as do the strains among enterobacteria. The results with oxidative bacteria, though, are compatible with previous findings (Jantzen *et al.*, 1974b, 1975; Mitruka, 1975) making the distinction between the oxidative and fermentative species well supported.

The acid 2-OH-12:0 found previously in *Acinetobacter* (Jantzen *et al.*, 1975) was not observed in this study. This component is distinctly separable from C12:0 on columns of UCW-98, whereas the two compounds occur together on columns of EGA-PS. The *Moraxella* profile resembled those presented previously. The pseudomonads were allocated to their own phenon and were comparatively similar to the acinetobacters: most of the latter appeared in a separate cluster.

In conclusion, GLC of cellular FA patterns is in general not suitable as a diagnostic tool for species distinction among enterobacteria. The method can be employed to distinguish between fermenters and oxidative genera, but in routine diagnostic bacteriology simpler methods are available and preferable for this purpose (Schindler *et al.*, 1980). Within the oxidative organisms, GLC of cellular FAs is more useful as a taxonomic aid (Jantzen *et al.*, 1974c, 1975).

References

Colwell, R. R., Johnson, R., Wan, L., Lovelace, T. E. and Brenner, D. J. (1974). *Int. J. Syst. Bacteriol.* **24**, 422–433.
Finegold, S. M. and Martin, W. J. (1982). "Diagnostic Microbiology", 6th edn. Mosby, St Louis, Missouri.
Gutteridge, C. S. and Norris, J. R. (1979). *J. Appl. Bacteriol.* **47**, 5–43.
Jantzen, E., Bryn, K. and Bøvre, K. (1974a). *Acta Pathol. Microbiol. Scand. Sect. B* **82**, 753–766.
Jantzen, E., Bryn, K., Bergan, T. and Bøvre, K. (1974b). *Acta Pathol. Microbiol. Scand. Sect. B* **82**, 767–779.
Jantzen, E., Bergan, T. and Bøvre, K. (1974c). *Acta Pathol. Microbiol. Scand. Sect. B* **82**, 785–798.
Jantzen, E., Bryn, K., Bergan, T. and Bøvre, K. (1975). *Acta Pathol. Microbiol. Scand. Sect. B* **83**, 569–580.
Johnson, R., Colwell, R., Sakazaki, R. and Tamura, K. (1975). *Int. J. Syst. Bacteriol.* **25**, 12–37.
Lassen, J. (1975). *Acta Pathol. Microbiol. Scand. Sect. B* **83**, 525–533.
Mitruka, B. M. (1975). "Gas Chromatographic Applications in Microbiology and Medicine". Wiley, New York.
Schindler, J., Duben, J., Hausner, O., Zikmundova, V., Lapackova, J. and Pauckova, V. (1980). *J. Appl. Bacteriol.* **49**, 331–337.
Shaw, N. (1974). *Adv. Appl. Microbiol.* **17**, 63–108.
Sutter, V. L., Citron, D. M. and Finegold, S. M. (1980). "Wadsworth Anaerobic Bacteriology Manual", 3rd edn. Mosby, St Louis, Missouri.
Uchida, K. and Iogi, K. (1972). *J. Gen. Appl. Microbiol.* **18**, 109–129.
Vasynreko, Z. P., Sinyak, K. M. and Lukach, I. G. (1975). *Proc. Acad. Sci. USSR Biol. Ser.* **6**, 827–836.
Veron, M. and Le Minor, L. (1975). *Ann. Microbiol. (Paris)* **126B**, 125–147.

Index

Contents of published volumes

370

Volume 3A

Volume 3B

Volume 4